MW00463048

SOLAR ENERGY CONVERSION SYSTEMS

System performance depends critically on how well the parts fit and work together, not merely on how well each performs when considered independently. Furthermore, a system's performance depends on how it relates to its environment–the larger system of which it is a part–and to the other systems in that environment.

Russell L. Ackoff

The scientist rigorously defends his right to be ignorant of almost everything except his specialty.

Marshall Mcluhan

Scientific disciplines are no longer thought of as dealing with different aspects of nature... The disciplines are increasingly thought of as points of view... For example, no discipline is irrelevant in efforts to solve ecological problems. Therefore, the environmental sciences include all the sciences.

Russell L. Ackoff

SOLAR ENERGY CONVERSION SYSTEMS

JEFFREY R.S. BROWNSON

ACADEMIC PRESS

Academic Press is an imprint of Elsevier
The Boulevard, Langford Lane, Kidlington, Oxford, OX5 1GB, UK
225 Wyman Street, Waltham, MA 02451, USA

First edition 2014

British Library Cataloguing in Publication Data
A catalogue record for this book is available from the British Library

Library of Congress Cataloging-in-Publication Data
A catalog record for this book is available from the Library of Congress

ISBN–13: 978-0-12-397021-3

For information on all Academic Press publications
visit our website at books.elsevier.com

Printed and bound in
06 07 08 09 10   10 9 8 7 6 5 4 3 2 1

CONTENTS

SUPPLEMENTS TO THE TEXT

For purchasers of this book:

To access solutions and software codes, please visit http://booksite.elsevier.com/9780123970213.

For instructors using this book as text for their course:

Please visit http://www.textbook.elsevier.com to register for access to the solutions and software codes.

An author-hosted site is available at http://nanomech.ems.psu.edu/textbook.php.

ACKNOWLEDGEMENTS

My family and my network of friends have been a wonderful source of support and positive encouragement during the transformation from a budding concept into an active document for teaching and learning. Thank you for helping me to achieve a new work that I hope will serve the community well. I extend special thanks to my parents for encouraging creativity, positive thinking, and perseverance, and deep thanks to my wife Ronica, and daughters, Rowan and Anya for their patience over the years as I disappeared for weekends, evenings, and even vacations to complete this text. Thanks also to my graduate and undergraduate students at the Pennsylvania State University, without whom I would not have developed the first few course notes and then full chapters to supplement the classroom experience. Your eyes played an important role correcting and clarifying the content to greatly strengthen this text for future generations. The book was also reviewed by solar and sustainability experts with decades of collective experience to ensure an accurate representation of the field. I am thankful for your time and expertise in keeping the content robust. Finally, thanks to Tiffany, Kattie, and the editorial team at Elsevier, for granting me this opportunity to develop a new creative body of work to enable the education of the next generation of solar design teams.

INTRODUCTION

01

WE ARE CALLED UPON to imagine a great new field of solar energy, one with an expansive view that includes sustainability, ethics, systems thinking, policy, and markets in addition to device and energy conversion systems engineering. As an educator and a researcher, I contribute to a field that has seen cycles of emergence and adoption, followed by dispersion and knowledge loss (largely coupled to the accessibility of geofuels). The global solar industry is exploding in growth. The solar electric industry in the USA alone has doubled in scale seven times in the last decade, and solar energy technologies have the benefit of widespread social acceptance. Solar energy has a bright future, and we must be ready for solar to emerge along new frontiers. The new wave of solar research and entrepreneurship will lead to amazing changes and diversification in applications.

Solar energy is dead. Long live solar energy!

Geofuels:

- coal [*Sun-derived],
- petroleum [*],
- natural gas [*],
- tar sands and oil shales [*],
- gas hydrates [*],
- fissile material for nuclear power [not from sunshine].

With expansion of interest and adoption, the next generation of solar research will explore systems knowledge, incorporating a great diversity of approaches, analogous to the broad fields of ecology or geology. In turn, a newly emergent field solar energy discovery within the context of the environment, society, and technology could be termed *solar ecology*, leading to the applied design and engineering of *solar energy conversion systems*.

Solar Ecology: interactive systems study of solar energy within the context of the environment, society, and technology.

Now we are a new generation of energy explorers. There are no firm "rules" to solar engagement and entrepreneurship other than those specified by the coincident emergence of policy and law. Imagine what we can do with the *patterns* of solar energy conversion systems and sustainable design, beyond photovoltaics and solar hot water panels—solar chimneys, greenhouses for power and food, solar gardens (shared solar arrays) for community power or district heating, solar water treatment, cooking with solar energy, even simple lighting design to improve indoor air quality and reduce electricity costs. There are amazing opportunities already in society and new patterns to explore for our future. I encourage *you,* the young generation of professionals in renewable energy, to explore the undiscovered potential of solar energy in society.

> "The use of clean energy technologies—that is, energy efficient and renewable energy technologies (RETs)—has increased greatly over the past several decades. Technologies once considered quaint or exotic are now commercial realities, providing cost-effective alternatives to conventional, fossil fuel-based systems and their associated problems of greenhouse gas emissions, high operating costs, and local pollution.
>
> In order to benefit from these technologies, potential users, decision and policy makers, planners, project financiers, and equipment vendors must be able to quickly and easily assess whether a proposed clean energy technology project makes sense."[1]

[1] Canada Natural Resources, editor. *Clean Energy Project Analysis: RETScreen Engineering & Cases.* Minister of Natural Resources, 2005.

DESIGN CONCEPT

This text has been designed to open up the language of solar energy conversion to a broader audience, to permit discussion of strategies for assessing the solar resource (energy exploration techniques) and for designing solar energy conversion systems (integrative design). As such, the text applies a *whole systems* approach that explores *relationships, transformation, and feedback* in solar energy conversion systems, the solar resource, society, and our supporting environment. Here, *design* is used with the systems language of *patterns*, as *pattern with a purpose*. In turn, *patterns* are used to convey relationships in systems. Underpinning the language of solar energy conversion is the central goal of Solar Design (as solar energy engineer, economist, or architect): within the scope of sustainability ethics and maintaining or increasing ecosystems

Design is pattern with a purpose.

Whole systems design: integrating systems of systems to solve for complex challenges, a transdisciplinary approach to arrive at solutions that increase client well-being and **ecosystems services** in a given locale.

services, one seeks to *maximize the solar utility of the resource for a client or stakeholders in a given locale.* This will become a well-used mantra through the entire text, so we should get accustomed to it. Each participant in the integrative teams for solar design is in turn agents of change for a sustainable future. As will be noted later in the text, solar design strategies should both increase client well-being and strengthen ecosystems services within the given locale. Throughout the text learners are encouraged to be active participants transitioning to a new energy approach with solar design strategies, and to engage in lifelong learning of solar science and technology.

Locale: what can it mean?

- Address
- Place
- Placement
- Climate regime
- Frequency
- Time horizon

There is a continuity among all solar energy conversion systems, but it was not always perceived as such. In the past, researchers in solar energy came largely from three fields: one from a mechanical engineering background (solar thermal), one from an electrical engineering background (photovoltaics), and a third from architecture (passive solar home design). The photovoltaic field tended to remain focused on device physics and component optimization independent of the outside environmental constraints or the measured solar resource constraints of a locale. To a small extent, this still continues—a tradition that we hope to shift. Those from the mechanical background and architecture tended to create a then useful distinction between *active* and *passive* solar conversion systems. From a mechanical standpoint, an active system required pumps and motors and made extensive use of fluid and heat transfer, while a passive system functioned on fields and pressure differentials that were intrinsic to the whole collector.

Participants in solar design have **agency** to effect change that will increase client well-being and strengthen ecosystems services.

Active Solar and **Passive Solar** systems are from a framework language used in building science to identify systems approaches with pumps and fans (active) from systems that operated with field and pressure gradients derived from solar gains (passive).

Today, the language of the active/passive distinction begins to wear thin, and the majority of the *active* solar field is consolidated into what we now term *Concentrating Solar Power* (CSP), for utility-scale thermal cycle electrical power conversion and industrial solar processing. At the residential and light-commercial level of solar technology, we now have commercially available solar hot water systems that have no active pumps (called bubble or geyser pumps, based on Albert Einstein and Leó Szilárd's work in a novel refrigeration cycle from 1928)[2] and we continue to have photovoltaic systems with no moving parts, driven by intrinsic electric and effective fields. Meanwhile, even properly

[2] Susan J. White. Bubble pump design and performance. Master's thesis, Georgia Institute of Technology, 2001.

designed solar homes are equipped with active circulation and control systems to force air about, and depend on the essential balance of the Sun to keep the surrounding air intake warm (at one time designed to be solely *passive solar systems*).

In this text, we explore a pattern language, or a systems logic for solar energy conversion devices that includes an *aperture*, *receiver*, *distribution mechanism*, potentially a storage mechanism, and a *control mechanism*. We will generalize them as *tubular*, *flat plate*, and *cavity* systems, and then explain the detail of their functioning as a specific *optoelectronic* (solar to electric) or *optocaloric* (solar to heat) technology.

Solar Energy Conversion Systems (SECS):

- **Aperture**
- **Receiver**
- **Distribution mechanism**
- **(Storage)**
- **Control mechanism**

The approach in the text is to inform the design process, and to reveal that *systems solutions* in solar design require a transdisciplinary and integrative design team that engages a client or stakeholders into the design and planning process. The course for which this text was developed focuses on solar energy conversion and the economics of deploying solar energy conversion systems for energy engineers and economists, but should also complement a more general degree program addressing renewable energy, systems design, and sustainability. We will examine the principles of solar energy conversion systems (SECS) and build a foundation for explaining the basic concepts and implementation of conversion processes.

Systems Solutions:

- Team effort
- Many players
- Transdisciplinary
- Engage clients early and often
- Integrative design
- Agents of change

We shall begin by describing the historical context of solar energy conversion systems and the integrative design process, followed by the properties, availability, and utility of solar radiation and geometric relationship of Sun/collector. We will include steps to evaluate the economic or sustainability criteria for decision making in energy systems deployment. As such there are strong emphases on the solar resource and solar economics, as well as applying the lens of sustainability and ecosystems services into systems design. Physical materials (absorbers, reflectors, covers, fluids) are crucial for conversion of radiant forms of energy, and special treatment is included for materials throughout the text. The latter portion of the text covers the role of specific device/collector technologies as a part of a larger systems logic for integration. These chapters will describe the principles of photovoltaic conversion (optoelectronic and optocaloric processes) and design, as well as describe the procedures for solar thermal design in residential and industrial processes, and explore solar concentration for heat and power conversions.

As was stated then in 1985 by solar research expert Ari Rabl,[3] the full extent of solar energy conversion can no longer be contained in a single textbook. Exploring developments in the solar collector alone is a great field of research. In this text, we emphasize the *systems approach* to thinking about solar energy conversion as a useful application for society. The following chapters serve to create a new framework to evaluate design and implementation of solar energy conversion linked with non-solar energy conversion technologies. The systems approach emerges from the strong foundations of solar energy conversion research, design, and practice established in the decades of the 1950s through the early 1980s, which established the core material for all solar energy conversion technologies.

[3] Ari Rabl. *Active Solar Collectors and Their Applications*. Oxford University Press, 1985.

A key realization for the newcomer to solar energy conversion is that *the technology and the economics have been changing rapidly*, and we would like to cultivate a generation of solar practitioners with *general systems skills* that will be relevant to developing technologies in 5 years, as well as in 50 years. In assimilating the content of this text, we have identified much of the systems connectivity that will prove useful to current and future scientists, engineers, economists, and policy makers, regardless of the specific solar energy conversion technology of the day.

Technological advancement is only one lever to engage adoption of a SECS by society. See the chapter on solar economics for more levers!

INSTRUCTIONAL APPROACH

For educators using this text as a complementary tool to engage students in a course on solar energy conversion systems analysis and project scoping (project design and rudimentary project finance), I have identified the educational objectives below. The target audience is for students in the third or fourth year of undergraduate degrees in engineering, physical science, and environmental/energy economics, or for graduate students in those fields. I have also sought to provide accessible content for students of policy, architectural design, and

This text is intended to be interdisciplinary.

planning, as a part of an interdisciplinary course of study. Problems have been listed at the end of each chapter, which can be modified or expanded as needed.

Where called for, I have recommended the use of software to solve multi-parameter problems and to process large data sets. As such, students are encouraged to develop critical thinking skills for addressing the problem and core material rather than how to process the arithmetic of large data. Many of the useful tools in solar energy systems design are open to the public, and I hope to demystify several of the approaches used in industry to measure and quantify the criteria for systems design. Concepts in project finance and energy analysis come together using the simulation engine SAM (Systems Advisor Model), provided freely by the US Dept. of Energy's National Renewable Energy Laboratory.[4] The simulation tool incorporates many of the desirable component-based systems simulation tools found in the powerful software TRNSYS,[5] while offering an accessible front end that works on multiple platforms to explore project estimation from beginning to advanced levels.

Software skills should be developed for design teams. Solar energy uses multi-parameter solvers in project design.

As mentioned before the text is written for an interdisciplinary course with students coming from diverse backgrounds of science, engineering, policy, business, and economics. The topics and contents have been tested over the last 5 years, teaching to classrooms of 30–50 students a session. In addition to the main body of text and several vignettes of historical or scientific significance, I have included subsections that go into significantly greater detail, explaining core concepts throughout the textbook for more advanced study such as radiant transfer physics, spherical trigonometry, meteorology, or device behavior. These detailed sections can be found in numerous other texts as well, and address fundamental concepts for solar energy science, but are often equation-heavy and completely new to an already advanced student. In my own classroom experiences I humorously coin these detailed breakdowns as "Robot Monkey" sections, given the general surprised reactions of a student cohort to an unfamiliar dense body of scientific writing that seems ordered and intended to illuminate a topic, but which simultaneously appears as though intelligent monkeys with typewriters wrote it. I would like students to be aware that these sections may be passed over the first time, but that they are encouraged to come back to these

[4] P. Gilman and A. Dobos. System Advisor Model, SAM 2011.12.2: General description. NREL Report No. TP-6A20-53437, National Renewable Energy Laboratory, Golden, CO, 2012. 18 pp; and System Advisor Model Version 2012.5.11 (SAM 2012.5.11). URL https://sam.nrel.gov/content/downloads. Accessed November 2, 2012.

[5] S. A. Klein, W. A. Beckman, J. W. Mitchell, J. A. Duffie, N. A. Duffie, T. L. Freeman, J. C. Mitchell, J. E. Braun, B. L. Evans, J. P. Kummer, R. E. Urban, A. Fiksel, J. W. Thornton, N. J. Blair, P. M. Williams, D. E. Bradley, T. P. McDowell, M. Kummert, and D. A. Arias. TRNSYS 17: A transient system simulation program, 2010. URL http://sel.me.wisc.edu/trnsys.

sections later in their lifelong learning of solar energy topics if at first the written materials appear to have been written for all intents and purposes by a well-meaning, but truly challenging Robot Monkey.

EDUCATIONAL OBJECTIVES

- Understand the historical complement of solar energy conversion with conventional heat and power technologies
- Solidify a working knowledge of solar energy conversion and systems integration given the spatial relations of moving bodies
- Explain meteorological spatio-temporal measures for estimating variance and expected values from the solar resource
- Express familiarity and competence with processes and materials for solar to thermal and electric energy conversion (optoelectronic and optocaloric transformations)
- Develop tools to assess project finance of solar goods and services to society
- Develop skills to integrate ecosystems services and sustainability in solar project design, including portfolio diversification and awareness of the solar commons in decision making
- Develop sustainable energy systems approaches to maximize solar utility for the client in a given locale
- Understand and explore the broader meanings of "solar utility," "locale," and "client/stakeholders."

SUGGESTED COURSE SCHEDULE WITH TEXT

- *Context and Philosophy of Design*
  - History of Solar Design and Ethics of Sustainability
  - Laws of Light
  - Physics of Light, Heat, Work, and Photoconversion
  - The Sun, Weather, and Uncertainty
  - Spatial Relations of Moving Bodies

- *Economics and Sustainability Criteria*
  - Measurement and Estimation of the Resource
  - Economics of Market Goods and Services
  - Project Finance
  - The Sun as Commons

- *Systems Logic of Devices and Patterns*
  - Pattern Language in SECS
    * Cavity Collectors
    * Flat Plate Collectors
    * Tubular Collectors
    * Parasoleils
  - Optocalorics
  - Optoelectronics
  - Concentration

- *Systems Integrated Solar and Project Design*
  - Integrative Design
  - Solar Utility for the Client and Locale
  - The Goal of Solar Design

COMMUNICATION OF UNITS AND A STANDARD SOLAR LANGUAGE

Language is critically important in the field of solar energy. Because of the integration of several disciplines, each with their own specialized use of scientific and technical language, the solar field calls for an effective language to communicate solar conversion and applied device design in the literature and in group discussions. Prior work exists that will help us to begin the conversation adequately. In 1978, a consortium of solar researchers came together to agree upon a set of standards for discussing solar energy conversion topics. The paper was revisited and reprinted in recent years, and it shows precedent for a detailed system of notation and language used in the solar energy world, one that has been in use for decades. The original authors have established the following observations:

Transdisciplinary: integration of common effort across several diverse disciplines, creating a holistic approach to solving systems of problems.

> "Many disciplines are contributing to the literature on solar energy with the result that variations in definitions, symbols and units are appearing for the same terms. These conflicts cause difficulties in understanding which may be reduced by a systematic approach such as is attempted in this paper…
>
> …Many disciplines are contributing to the literature on solar energy with the result that variations in definitions, symbols and units are appearing for the same terms. These conflicts cause difficulties in understanding which may be reduced by a systematic approach such as is attempted in this paper."[6]

Even in 1978 it was recognized that people would enter the solar field of science, design, and engineering from many supporting fields, and bring with them skills and language to contribute. There is a place for people from many backgrounds to expand and explore the field of SECS, and we should be aware of the common language used to communicate among ourselves, among our clients, and with the public. The authors go on to describe the vernacular for communicating values and coefficients. It was recognized that even a list of symbols and units would not be permanent or mandatory, but that a listing of terms and units could have value to the multitudes entering the transdisciplinary efforts of the solar field. We choose to use the guidelines from this paper in sections dealing with the spatial relations of moving bodies, in the physics of heat for radiative transfer, and in terms of meteorological assessment of the solar resource for quantifying variability of the resource in project design.

[6] W. A. Beckman, J. W. Bugler, P. I. Cooper, J. A. Duffie, R. V. Dunkle, P. E. Glaser, T. Horigome, E. D. Howe, T. A. Lawand, P. L. van der Mersch, J. K. Page, N. R. Sheridan, S. V. Szokolay, and G. T. Ward. Units and symbols in solar energy. *Solar Energy*, 21:65–68, 1978.

In terms of **energy**, the S.I. (*Systèm International d'Unités*) unit to be used is the joule ($J = kg\, m^2\, s^{-2}$), while the calorie and derivatives are not acceptable. There is no distinction between forms of energy, such that electrical, thermal, radiative, and mechanical energy are all measured in joules. The singular exception being the use of the watt-hour (Wh) itself an energy unit used extensively in metering of electrical energy from the grid.

Joule: S.I. unit of *energy* [J].

Watt: S.I. unit of *power* [W = J/s].

Watt-hour: commercial electricity unit of *energy*. Keep in mind that a watt-hour or a kilowatt-hour (kWh) is a unit of energy, *not* a unit of power. This is a really common mistake by students and early practitioners.

In terms of **power**, the S.I. unit is the watt ($W = J/s = (kg\, m^2)/s^3$). The watt, kilowatt (kW), megawatt (MW), gigawatt (GW), or terawatt (TW, rare) is to be applied for measures of power or the rate of energy generation/demand. Again the watt is to be used for all forms of energy and wherever one is describing instantaneous energy flow. We will not express the rate of energy production or use in other units such as J/h. In turn, the measure of the **energy flux density** shall be expressed in terms of watts per square meter (W/m^2), which we have used in terms of *irradiance* so far. Also, the measure of the **specific thermal conductance** shall be expressed in terms of $W/(m^2\, K)$. Finally, when a rate of energy flux density or the rate of energy generation/demand (power) is integrated for a specified time interval, we shall express the energy in units of J/m^2 (energy density) and joules, respectively. For example, a power generation (rate of energy production) measure of 4.1 kW would yield 14.8 MJ if maintained over the interval of 1 h. Consider that there are 3600 s/h and 0.001 MW/kW.

There is 3.6 MJ/1 kW when power is integrated over an hour of time (**1 kWh = 3.6 MJ**).

$$\begin{aligned} 1\ \text{kW} \cdot 1\ \text{h} &= [1\ \text{kW} \times 1000\ \text{W/kW} \times 1\ (\text{J/s})/\text{W}] \cdot \\ &\quad [1\ \text{h} \times 60\ \text{min/h} \times 60\ \text{s/min}] \\ &= 3,600,000\ \text{J} = 3.6\ \text{MJ}. \end{aligned} \tag{1.1}$$

A **British thermal unit (Btu)** is a common energy unit to describe the energy content in heating fuels for the USA. It is defined as the energy called for to increase the temperature of one pound of water by 1 °F. 1 Btu = 1055 J, or ~1.1 kJ. For natural gas systems, the convention is such that 1 MMBtu (one million Btu) is equivalent to 1.05 GJ.

In addition, when referring to energy for a specific interval of time one states an "hourly energy of 14.8 MJ" rather than 14.8 MJ/h, and a "daily energy of 355 MJ" rather than 355 MJ/day.

Symbols for radiance/irradiance and radiation/irradiation are split into several terms throughout this text. The energy flux density as derived from a general blackbody is given as E_b, while the specific energy flux density of

Angular Measure	Symbol	Range and Sign Convention
General		
Altitude angle	α	0° to +90°; horizontal is zero
Azimuth angle	γ	0° to +360°; clockwise from North origin
Azimuth (alternate)	γ	0° to ±180°; zero (origin) faces the equator, East is +ive, West is −ive
Earth-Sun Angles		
Latitude	ϕ	0° to ±23.45°; Northern hemisphere is +ive
Longitude	λ	0° to ±180°; Prime Meridian is zero, West is −ive
Declination	δ	0° to ±23.45°; Northern hemisphere is +ive
Hour angle	ω	0° to ±180°; solar noon is zero, afternoon is +ive, morning is −ive
Sun-Observer Angles		
Solar altitude angle (compliment)	$\alpha_s = 1 - \theta_z$	0° to +90°
Solar azimuth angle	γ_s	0° to +360°; clockwise from North origin
Zenith angle	θ_z	0° to +90°; vertical is zero
Collecter-Sun Angles		
Surface altitude angle	α	0° to +90°
Slope or tilt (of collector surface)	β	0° to ±90°; facing equator is +ive
Suface azimuth angle	γ	0° to +360°; clockwise from North origin
Angle of incidence	θ	0° to +90°
Glancing angle (compliment)	$\alpha = 1 - \theta$	0° to +90°

Table 1.1: Table of angular relations in space and time, including the symbols and units used in this text.

solar irradiance is given as G (G is for "global irradiance"). The global *solar irradiation* for the interval of a day is given as H, while the global solar irradiation for the interval of an hour is given as I.

When describing **spatial relationships** on continuous, near-spherical surfaces like the Earth and sky, Greek letters are preferred. When communicating distances, lengths, time, and Cartesian coordinates, we tend to use Roman letters. A full listing of symbols is described in Appendix B (see Table 1.1).

CONTEXT AND PHILOSOPHY OF DESIGN

02

Philosophy of science without history of science is empty; history of science without philosophy of science is blind.

Imre Lakatos

Truth is like the sun. You can shut it out for a time, but it ain't goin' away.

Elvis Presley

No root, no fruit

Bruce "Utah" Phillips

A RATIONALE AND HISTORY FOR DESIGN OF SECS

Solar energy conversion systems (SECS) call upon researchers and professionals to simultaneously assess scales of solar resource supply and use, systems design, distribution needs, predictive economic models for the fluctuating solar resource, and storage plans to address transient cycles. The spectrum of the solar resource is broad, and allows for *parallel paths* of conversion for daylighting, heating, cooling, and electrical power conversion (*solar cogeneration*). Putting it another way: SECS elements are *strongly coupled*,[1] they are not easily isolated or marginalized from the network of

[1] Etymological origin from late 13c. Old French: *cople* "married couple, lovers," from Latin *copula* "tie, connection."

energy challenges as with traditional resources for energy engineering. One does not "produce and refine" the photon (the fundamental packet of light), and due to the implicit concept of the Sun as a commons: "a cultural or natural resource accessible to the whole of society" there are no "mineral rights" to photons (although there certainly are mineral rights relative to the contingent materials that deliver the conversion technologies). The field of solar energy conversion systems is ripe for a transformative new voice in the business world (in your fields of engineering, science, economics, and design). What will be your rationale for Solar Energy Conversion Systems—how will you express the enormous potential of solar energy for conversion to work?

You are agents of change:

- **designer**
- **engineer**
- **policy maker**
- **builder**
- **client**
- **stakeholder.**

Zen rock gardens were not just for contemplation. The pale stone fields were installed on the south side of a building, functioning as diffuse reflectors for daylight to penetrate deeper inside where sutras were transcribed. They also keep the summer air cooler by reflecting the light backup to the sky.

At the core, one finds solar energy applications to be simple conversions of light to do work found useful to society. Butti and Perlin have eloquently detailed a history of solar energy use in society, demonstrating that technologies using solar energy conversion have been developing for *thousands of years*.[2] Solar energy design has been employed heavily in architecture for lighting and thermal balance in societies of the Anasazi, Greece, Romans, and in Buddhist temples of China and Japan. In fact, it is often a worthwhile experience on travels to locate the direction of the equator (which is always the direction of the Sun's zenith) and observe any collection of buildings seeming to coincide with that orientation. You may find yourself surprised by Mediterranean porticos (patios, or porches), pueblo architecture, and zen gardens that were specifically designed to take full advantage of the seasonal solar ebb and flow.[3]

For some time now, combustible products like wood, coal, and petroleum have been accepted as the fuel for our engines of technological life in society (literally, in cases). And yet fuels are not always located where we most need them, or have negative environmental and health consequences when employed on a global scale for billions of humans. We can observe general trends over history when traditional fuel combustion has been supplanted by alternative energy technologies. In particular, for this text, we shall focus on the development and design of solar energy conversion systems technologies. For periods in history when fuels have become constrained, societal innovation has turned to solar technology solutions. In the USA alone, we have historical examples of

[2] Ken Butti and John Perlin. *A Golden Thread: 2500 Years of Solar Architecture and Technology*. Cheshire Books, 1980.

[3] Baker H. Morrow and V.B. Price, editors. *Anasazi Architecture and Modern Design*. University of New Mexico Press, 1997.

There's been a huge
SOLAR ENERGY SPILL!!!
Let's go outside
& play in it!
OTHER 98%

Figure 2.1: Depiction of a massive solar energy spill causing environmental and social change. Photo by the other 98%.

solar entrepreneurship and innovation that extend into the 1800s.[4] Unfortunately, the events of solar development, particularly in the mid-Atlantic and northeast, have been largely forgotten within the USA and abroad.

We can demonstrate from history that developments cultivating the use of the Sun to ease living in society often emerge as a response to economic or fuel constraints. *Fuel constraints* can emerge for members of society in several forms, but in each case the result is a higher cost fuel supply and the pressure to seek out energy alternatives. This pressure to adapt to a new energy source is a classic economic response, but for this text we will call it an *energy constraint response*. Examples of fuel constraints include the restriction of fuel and increase in fuel costs and we observe that fuel can be restricted in several identifiable forms:

1. by nature of being *physically inaccessible* from supply chain disruptions or regional resource depletion,

[4] Frank T. Kryza. *The Power of Light: The Epic Story of Man's Quest to Harness the Sun.* McGraw-Hill, 2003.

Maryland and Pennsylvanian entrepreneurs led to the creation of the *Climax* and *Day and Night* solar water heater technologies for their respective companies in the **1890s** and **1910s**. We rest upon over one hundred years of solar hot water potential in the USA.

2. being of *limited access* due to an exceptionally high demand for fuel that outstrips supply,
3. from socially restraining policies, regulations, and laws,
4. or being accessible but only at *high risk*.

In the long-term case of **fossil fuels** there will likely be a combination of all of these factors, but we recall that fossil fuels are indeed a **non-renewable resource**, with a fixed stock of material to access.

During periods of increased fuel constraints, research into the use of solar energy is often socially advocated and the solar resource has been interpreted as *ubiquitous* and vast. Changes toward increased solar architecture use in Rome have been linked to high demand and limited access to wood in early centuries BCE, particularly influential in the development of legal rights to solar access in urban settings. The advent of the Roman *hypocausts* in the early century BCE (underfloor central heating systems, see Figure 2.2) has been linked to large-scale deforestation in the Italian peninsula, such that wood would be imported from the Caucasus region.[5]

[5] Ken Butti and John Perlin. *A Golden Thread: 2500 Years of Solar Architecture and Technology.* Cheshire Books, 1980.

Coal reserve constraints in early 19th century France were the impetus to drive Augustin Mouchot (a professor of mathematics at Lyceé de Tours) to invent solar concentration technologies to pasteurize liquids, heat water, and cook food. Limited fuel access in rural California before the 1920s and then

Figure 2.2: Example of Roman hypocaust at the Villa gallo-romaine de Vieux-la-Romaine near Caen, France. While not located in Italy, such a structure would require large quantities of wood to ensure indoor floor and bath heating in winter. Photo by User: Urban/ Wikimedia Commons/CC-BY-SA-3. January 2006.

Florida near the 1940s allowed for the development of several successful solar hot water companies. By 1941, nearly half of the homes in Miami were able to provide hot water from solar thermal systems. Devices and services developed to convert solar light to the heating of fluids and solids have been wetted for hundreds of years.

Although the term "pasteurization" was not a term coined until much later, Mouchot accomplished equivalent results with **solar concentration**.

Mouchot holds the distinction of having distilled the world's first **solar brandy** with his solar thermal concentrator system.

> "One must not believe, despite the silence of modern writings, that the idea of using solar heat for mechanical operations is recent. On the contrary, one must recognize that this idea is very ancient and its slow development across the centuries it has given birth to various curious devices."
>
> –Augustin B. Mouchot. Universal Exposition, Paris, France (1878).

In contrast, during periods in history when fuels were effectively *accessible, inexpensive, and unconstrained*, a light-induced energy transfer has been deemed *diffuse* and *insufficient* for performing technical work. *Work*, used here, includes using sunlight for sensible heat change, latent heat change, or for electricity generation (as daylight and photosynthesis were not originally incorporated into 19th century concepts of useful work). In those same regions of Florida where solar hot water was adopted, the introduction of accessible, low-cost natural gas lines into urban communities led people to abandon their solar hot water devices. There are still homes in Florida, in fact, that have remnant structures in the roofing from solar hot water installations almost a hundred years old. The California solar hot water systems faded out following the discovery of natural gas in the Los Angeles Basin[6] (see Figure 2.3).

Insolation: sunlight.

Now consider our unusual contemporary perspective that solar energy is a *diffuse* or *insufficient* technology for producing electricity or even hot water. From our earlier discussion one would see that the common view is a direct result of a surplus of accessible fossil fuels as coal, petroleum, natural gas, tar sands and oil shales, and gas hydrates. Our modern fuel constraints will emerge from limitations to access fossil fuel reserves in increasingly higher risk regions,

[6] Ken Butti and John Perlin. *A Golden Thread: 2500 Years of Solar Architecture and Technology*. Cheshire Books, 1980.

I have a friend and colleague in coal science who calls coal "concentrated sunshine."

Figure 2.3: Solar Hot Water system from Baltimore businessman Clarence Kemp—invented in 1891. Image used from the Lawrence B. Romaine Trade Catalog Collection. Mss. 107. Dept. of Special Collections, Davidson Library, UC Santa Barbara.

from limitations to access to the reserves due to environmental cautions of biome disruption (groundwater supplies, ecosystem disruptions from spills or land use) and from limitations to access the reserves due to the resulting increase in climate changing particulates and greenhouse gases with global combustion for heat and power. Those constraints will each lead to increased costs for fuel, and increased interest in substitutes like solar energy.

SYSTEMS AND PATTERNS

An important context of our modern approach to solar energy is a foundation in understanding *systems*.[7] A solar design and implementation team does more than solving singular problems, the team must work together to address *systems of problems or challenges* for solar energy deployment. When we design SECS to solve for systems of challenges, we hope to establish new patterns that work with the surrounding systems. So in essence, a solar design team will be *solving for patterns* on behalf of the clients, and within the larger constraints of the broad solar ecology embodied by the locale.[8] We will address patterns in SECS later in the text, giving this abstract concept more shape.

[7] Donella H Meadows. *Thinking in Systems: A Primer*. Chelsea Green Publishing, 2008.

Solving for pattern: solving for systems of coupled problems; a well-established phrase from sustainability author and farmer Wendell Berry.

[8] Wendell Berry. *Home Economics*. North Point Press, 1987.

By the end of your studies, you should be able to provide your own answers for the following:

1. What is a *system*?
2. What are *surroundings* relative to a system?
3. What is the relevance of a *boundary* for a system-surroundings relation?
4. What is the role of *stocks* and of *flows* in a *dynamic system*?
5. What are the main elements for *feedback* in a *dynamic system*?

A *system* is a collection of elements and other systems that are *conjugate, or coupled* in weak or strong network relations, having pattern or structure that yields an emergent set of characteristic behaviors. The systems we are concerned with are called *environmental systems*, each having general boundaries that delineate the system (inside) and the surroundings (outside). An environmental system is also an *open system*, dynamic in time and space, with flows of information, mass, energy, and entropy possible across the boundaries of the system.

In addressing a system-surroundings relationship, we implicitly define a boundary between the two. The boundary will have conditions of *permeability* to energy, mass, and entropy. When we allow permeability to energy, mass, and entropy (open systems), we observe *flows* in response to *potentials*. These patterns of flows are to be studied as *systems dynamics*, where the open systems have a continuous driving force (such as the Sun's radiation or gravity) they operate at non-equilibrium conditions.

Our nearest star, the Sun, is an amazing dynamic system a mere 93 million miles away. The boundary between the Sun system and the surrounding near vacuum of space extends well beyond the photosphere (source of most solar radiation). There is a "solar atmosphere" that is observable during a total solar eclipse, which includes the reversing layer, the chromosphere, and the corona, respectively. Beyond the solar atmosphere is the near vacuum of space, essentially the solar system *surroundings*.

Our home planet, Earth, is yet another amazing dynamic system. We associate the boundary of Earth as being between the upper atmosphere and the near vacuum of space. Looking at the relationship between Earth and Sun though, we observe

Solar Ecology: study of the paired systems of society-environment and technology.

System: a collection of elements/components and other systems that are coupled together in a network.

Surroundings: what is outside of the system boundary.

Open System: boundary is permeable to *mass* and *energy* exchange.

Potential: a driving force to flow.

Stock: stored potential.

Flow: change in mass, energy, or information with respect to time.

the coupling of gravity and radiation between the two. This reveals a larger system, with the boundary to the surroundings enclosing both celestial bodies.

Due to the gravitational influence (read: coupling) of the moon on our tidal systems, we may wish to define the boundary as inclusive of the moon for the Earth system.

SECSs, as technologies that we use in society and which integrate into the surrounding environment, are a third system. SECSs call upon researchers and practitioners to simultaneously assess scales of solar resource supply and use, systems design, distribution needs, predictive economic models for the fluctuating solar resource, and storage plans to address transient cycles. As we cannot remove the Earth or the Sun from our design process, we see that the solar panel/house/tree is a system (locale-specific technology) within a system (Earth), within a system (the Earth-Sun system).

Hence, solar energy conversion as a process calls upon designers to open their concept of the *System* to be inclusive of (1) the Sun, (2) Earth, and (3) the applied technological system in question. Putting it another way: SECS *elements* are *strongly coupled*. The field of solar energy conversion systems is ripe for a transformative new voice in the business world (in your fields of engineering, science, economics, and design).

SYSTEMS DESIGN

As an example of systems design, from long ago, we have the culture of the Chaco Canyon culture (also termed the Anasazi). You can still find the visible remains of Pueblo Bonito and others by using *Google Earth*. Architect Stephen Dent and urban planner Barbara Coleman have identified compelling evidence that the culture exhibited very effective city planning with awareness of the solar resource and meteorology. The settlement (now ruins) was established over many years in a delicate ecosystem of the American southwest, with a demonstrated awareness of the surrounding environment (in both solar and lunar cycles):

> The structures at Chaco generally show a high degree of environmental sensitivity in their response to climatic forces both at the building scale and in site planning. Generally, forms step down and open up to the south or southeast. Such configurations provide the plazas

> and most openings in the buildings with shelter from the prevailing northwesterly winter winds, shade from the hot afternoon summer sun, and warmth from the winter sun…the design works to give all parts of the structure nearly equal solar exposure during winter days…
>
> Frankly, it would surprise us if the Chacoans did *not* comprehend and mark solar geometry and its relation to natural cycles…[9]

[9] Stephen D. Dent and Barbara Coleman. *Anasazi Architecture and American Design*, chapter 5: A Planner's Primer, pages 53–61. University of New Mexico Press, 1997.

And yet there is also the evidence that the culture collapsed despite these innovative environmental planning steps. Perhaps it had to do with materials use. The buildings employed massive quantities of stone and wood that was transported more than 80 km. The Chacoans constructed 200 km of roads connecting settlements.[10] But perhaps it was a shift in the larger ecosystem to a new steady state that no longer sustained the larger culture of the Anasazi. This brings up the need to consider the sustainability of our efforts within the context of the environment and our surrounding societies.

[10] Anna Sofaer. *Anasazi Architecture and Modern Design*, chapter 8: The Primary Architecture of the Chacoan Culture–A cosmological expression, pages 88–132. University of New Mexico Press, 1997.

THE ETHICS OF SUSTAINABILITY

Sustainability must be central to new shifts in society and our supporting environment.

Sustainability has a meaning, which is developing and being shaped by researchers and industry today. As we will explore in subsequent chapters, *sustainability in energy systems* is central criterion within integrative systems design and ecosystem services. Here, we build a specific argument that sustainability must be central to new developments for large-scale energy shifts in society and our ecosystems. Consider that a "renewable energy" or an "alternative energy" source may not complete the criteria to support our societal demands as a *sustainable energy* system. Energy systems, as a field of study and application, encompass truly epic scale exchanges of mass, water, power and money. Any energy system that becomes a significant portion of our global consumption will indeed have a high likelihood of emerging as an unsustainable structure over the course of decades.

Sustainable energy: energy conversion for society applied using the framework of sustainability ethics. Entails striving for a clean and safe biome, well-being of stakeholders locally and throughout the global network, and awareness of a long-lived or closed loop time horizon of usage respecting the generations to come.

So what is *sustainability*, and why would we consider it to have a normative condition tied to ethical behavior?[11] Becker frames sustainability as "a global

[11] For an excellent extended read on this topic, please seek out C. U. Becker's text, *Sustainability Ethics and Sustainability Research* (2012) Dordrecht: Springer.

Sustainability: a global concept of ethical systems relations among society and the environment. Introducing *time horizons of assessment, dependency of locale, stakeholders* into design.

[12] Chrstian U. Becker. *Sustainability Ethics and Sustainability Research*. Dordrecht: Springer, 2012.

Sustainability relations:

1. contemporary people across the globe,
2. future generations,
3. our supporting biome.

[13] Chrstian U. Becker. *Sustainability Ethics and Sustainability Research*. Dordrecht: Springer, 2012.

concept that is used to discuss various societal fields, such as business or education, and to discuss a range of crucial environmental, societal, and global issues, such as biodiversity loss, climate change, distribution and use of nonrenewable resources, energy production and use, global equity and justice, and economic issues."[12]

Far from being a buzzword used for different goals in shifting contexts, sustainability is an important concept with enormous potential to society and our supporting ecosystems. Sustainability is also a valuable lens through which you will spend your future careers in energy systems looking. Moreover, the ethical dimension of sustainability is an intrinsic framework grounded in fundamental relationships among humans: the sustainability relations to contemporary people across the spatial expanse of the globe, to people across time into future generations, and to the biome that supports our living functions. One of the key aspects of sustainability is *agency*, where you and your design team are agents of change to enable solar energy conversion systems for the well-being of clients, stakeholders, and the supporting ecosystem in a given locale. As such, ethics of sustainability address simultaneous and systemic moral obligations to (1) contemporary global communities, (2) future generations of human society, and (3) the natural community or environment supporting life and biodiversity on Earth.[13]

The science of sustainability has thus far included several key factors, many of which you will recognize in your developing careers. First, sustainability in science and technology includes the integration across several disciplines, the integration of science and society. Transdisciplinary work is readily recognized in the solar world, as solar energy design and deployment is an integration of many disciplines in several fields of engineering (mechanical, electrical, architectural), physics and chemistry, economics and finance, policy development and systems science. Sustainability also requires reference to the *local nature of solutions*, the *coordination of solutions with time*, and the real *uncertainty* tied to incorporating solutions that require action and orientation to the system of problems. Beyond those foci sustainability must also encompass alignment and integration of science and sustainability ethics. In such a way, we ensure that our pattern solutions are tied to critical self-reflection of our diverse scientific approaches and our underlying assumptions. Consider how might you integrate sustainability ethics with your solar energy conversion systems designs and deployment.

Deploying solar technology does not necessarily equate to a sustainable energy approach. SECS can be developed as an integration of society and technology with the surrounding environment (a solar ecosystem), with an awareness and plan for the life cycle of the materials used and the environmental impact of deploying a large project such as a solar farm. SECS could also be developed in ways that are unethical, including inappropriate or even unhealthy waste disposal tied to the technology manufacturing processes, or installations of solar arrays that ignore strongly negative disruptions of the existing biomes in a sensitive ecosystem. In the science and practice of sustainability, we are both the stakeholders and the agents of change.

Solar Energy Conversion Systems can be developed either as **environmental technologies** within a supportive **solar ecosystem**, or as technologies that are ultimately destructive to the environment. Carefully consider the ethics of sustainability in project development.

If we consider the broader context of sustainability with respect to our supporting environment, Solar Energy Conversion Systems have the potential to be deployed as an *ecosystems technology* or an *environmental technology*, meaning the energy system interacts in a constructive way with the patterns of nature. In contrast, SECSs also have potential to be destructive to ecosystems and the environment—just because the solar technology is "renewable" doesn't necessitate that it will be clean, safe, and long lived among the sustainability relations. We shall see in the upcoming decades that SECS deployment will be pressed to address sustainability criteria in the same way that agricultural technology and industrial ecology are pressed to become incorporated within sustainability criteria. Wendell Berry wrote of *solving for pattern*, as "the industrial methods that have so spectacularly solved some of the problems of food production have been accompanied by "side effects" so damaging as to threaten the survival of farming."[14] From another perspective, solar energy can be engaged in solutions for society or the environment as *appropriate technologies*, where the solutions may be small-scale, energy

[14] Wendell Berry. *The Gift of Good Land: Further Essays Cultural & Agricultural*, chapter 9: Solving for Pattern. North Point Press, 1981.

- W = J/s
- kW = 10^3 W thousand
- MW = 10^6 W million
- GW = 10^9 W billion
- TW = 10^{12} W trillion.

Terawatt: a trillion watts of power—a rate of energy demand (J/s). The total human consumption of power on the planet is on the order of 16–17 TW, growing toward 30 TW in a few decades.

efficient, and locally appealing as well as increasing client well-being and resilience in their locale.

Environmental technologies and design solutions for SECS are deployed within a larger system of the "environment," within the context of a local ecosystem, whether that ecosystem be a dense urban downtown, or an open field in the plains. As agents of change, the design team introducing sustainable energy systems solutions must also be aware of the ecosystems impact of technology deployment and services. Consider that we are planning for major changes in energy exploration, changes that will occur on the terawatt scale for new solar development (or any renewable energy resource). We are ethically bound to be cognizant of our supporting biome with any such major changes in material, water, and energy use.

Ecology: the study of home and environment.

Economics: the management of home and environment.

ECOSYSTEM SERVICES

In reading this chapter and the technical chapters that follow, consider the additional dimensions of solar energy conversion as it interacts with the surrounding environment and ecosystems. We will frame our team-based integrative design process to include alignment with the manner in which a proposed SECS can deliver sustainable energy solutions in an economically sound manner, while reducing risk to the clients in the process of systems design and deployment. The relationships between society and our supporting environment, both biological and inorganic, are expressed as an *ecosystem*. The study of our ecosystems, both urban and rural, is *ecology*. In fact, the etymology of the term "ecology" demonstrates that the Greek word *oikos* (here, *eco-*) means "house, dwelling place, habitation"—in short, the study of the patterns within our environment and home and the ways in which living things interact in that environment.[15]

[15] Douglas Harper. Online etymology dictionary, November 2001. URL http://www.etymonline.com/. Accessed March 3, 2013.

The *Millenium Ecosystem Assessment* of 2005 provided summaries and guidelines for decision makers, concluding that human activity has significant and escalating impact on ecosystems biodiversity across the planet. The human impact reduces ecosystems resilience and biocapacity.[16] As the biomes of the planet are humanity's "life-support system," they provide us essential *ecosystem services* (benefits people obtain from ecosystems) that we are responsible for inheriting and potentially managing.

- **Supporting:** these are crucial and *fundamental services necessary for the production of all other ecosystem services* and include photosynthesis, soil formation, primary production of nutrients, nutrient cycling, and water cycling.

- **Provisioning:** the products obtained from ecosystems, including energy as fuel, food, fiber, genetic resources, biochemicals and pharmaceuticals, ornamental resources, and fresh water.

- **Regulating:** the benefits obtained from regulating ecosystems processes include air quality regulation, climate regulation (both micro- and macroclimate), water regulation, erosion regulation, water purification and waste treatment, disease regulation, pest regulation, pollination, and natural hazard regulation.

- **Cultural:** the non-material benefits that society gains from ecosystems as enriched spiritual lifestyles, cognitive development, reflection, recreation, and aesthetic experiences. The services include cultural diversity, spiritual and religious values, traditional and formal knowledge systems, educational values, inspiration, aesthetic values, social relations influenced by ecosystems, the sense of place, cultural heritage values, and recreation and ecotourism.

Consider, what are the complementary *ecosystem services* that will be increased or diminished in deploying the SECS for a client in a given locale? The idea of ecosystem services actually emerges as an analogous risk reduction strategy used in the field of finance called *portfolio diversification*, only the

[16] Walter V. Reid, Harold A Mooney, Angela Cropper, Doris Capistrano, Stephen R Carpenter, Kanchan Chopra, Partha Dasgupta, Thomas Dietz, Anantha Kumar Duraiappah, Rashid Hassan, Roger Kasperson, Rik Leemans, Robert M May, Tony McMichael, Prabhu Pinagali, Cristián Samper, Robert Scholes, Robert T Watson, A H Zakri, Zhao Shidong, Nevill J Ash, Elena Bennett, Pushpam Kumar, Marcus J Lee, Ciara Raudsepp-Hearne, Henk Simons, Jillian Thonell, and Monika B Zurek. Ecosystems and human well-being: Synthesis. Technical report, Millennium Ecosystem Assessment (MEA), Island Press, Washington, DC., 2005.

Ecosystem Services: the benefits that people obtain from ecosystems, or the systemic pattern of resources and processes available to provide resilient organism-environment activities.

Portfolio diversification: spreading out risk via distributing assets among options.

Ecosystem services: increased or decreased by our design choice?.

Old School: energy is a supply-side problem, a provisioning service only.

New School: energy is managing and matching demand with supply, and providing supporting, regulating, and cultural services.

ecosystem is the portfolio that we wish to avoid "crashing." By considering ecosystem services, we transform solar energy into a *environmental technology* strategy for *sustainable energy*, with deep impact on our sustainability relations. If we look through this list, and consider the "old school" role of energy supply, we can identify solar energy conversion systems fitting into *Provisioning* ecosystems services. If we are creative, then perhaps we could even assign *Supporting* ecosystem services in the case of photosynthesis—that is, if we actually integrated plant growth into our SECS project design. By thinking about portfolio diversification, or the act of reducing *risk* by investing or designing a variety of ecosystems assets into our projects, we see that there are numerous opportunities in *Regulating* and *Cultural* ecosystem services as well. Think back to the example of the Zen rock gardens, in which the rocks were providing *Regulating* and *Cultural* ecosystem services. Think of the use of green roofs or even garden roofs to cool building spaces and reduce the urban heat island effect. In this case, one might see strong porfolio diversification in terms of ecosystem services. Green roofs affect microclimate, they can lead to water purification or rain water runoff control, and they can provide a cultural backdrop for youth education and scientific cognitive development. If a green roof is garden-based then they can provide local foods, herbs, and increased social relations with their management. A photovoltaic array integrated into a green roof or into a field will benefit from the cooler microclimate of the evaporatranspiration process, producing slightly more power on a hot day, and may even add to a more fulfilling experience for the client when visiting the rooftop.

LIMITATIONS OF THE GOAL

From this review of the summarized MEA report, we begin to identify that not all values can be "maximized" for every client or group of stakeholders, and in fact the suite of problems that we are dealing with are *wicked*.[17] The role of science and society can be seen with the goal of solar energy design: to increase or maximize the solar utility for a client or group of stakeholders in a given locale. *Utility* is a term from economics referring to preferences of a client within a

[17] A **wicked problem** is likely part of a sustainable design challenge—wicked in the sense of being very challenging.

set of goods and services. Then we use *solar utility* to refer to the distinct set of goods and services that originate from the solar resource, rather than a non-solar good or service. Yes, this does open up a wide expanse of possibilities, as solar energy is a major driving force for weather, daylight, and agriculture, in addition to our constrained view of smaller technologies like photovoltaics. And yet, Becker informs us there are limits to utilitarian approaches.[18] These approaches of pure cost-benefit analyses should be familiar to those in economics and extend to the welfare economists of Pigot and Pereto (early 20th century), Bergson and Samuelson (mid 20th century), and most recently Arrow and Sen.[19] As a society we tend to focus on human welfare independent of the supporting biome, although we reason through decisions that have both market and non-market facets.

Utility (from economics) refers to the client's preferences within a set of goods and services. **Solar utility** refers to the set of goods and services that originate from the solar resource, rather than a non-solar good or service.

[18] Chrstian U. Becker. *Sustainability Ethics and Sustainability Research*. Dordrecht: Springer, 2012.

[19] Gjalt Huppes and Masanobu Ishikawa. Why eco-efficiency? *Journal of Industrial Ecology*, 9(4):2–5, 2005.

Globalization creates global competition, yet our biomes, our meteorologies, and our cultures are more localized and time dependent. The assessment of our world ecologists asserts that our supporting biome is not a viable trade in exchange for the amenities and assets of future income in an ever-expanding global society. In effect, our biomes are integral patterns of our society, and we literally cannot live without them being robust and resilient. Why is all of this thinking even required? Because policy permits technological development. Oftentimes a good design concept will be crushed because there is no code or allotment for the idea to find purchase.

Utility maximization is an old concept in economics (utilitarian). There are limits to its value in solving for pattern.

FRAMEWORK: SOLAR AS LIGHTING AID

There are numerous early cultural uses of building orientation to take advantage of solar energy gains, including the Anasazi cultures of the North American Southwest, temples in China, Korea, and Japan, and ancient Greek architecture. In all of those cultures, apertures for windows were developed to allow daylight to penetrate into the interior zones to provide both visible lighting during the day and thermal gains during cool seasons. We will look at this technology of aperture and absorbing interior space in future chapters as an example of a *cavity absorber* (also in contrast to a *flat plate system*).

If only we could focus on a technology that would permit daylight to enter, while retaining some thermal energy into the space...

Of course, the detraction for open windows in areas with winter seasons is that they allow comforting thermal heat to escape rapidly, requiring more and more fuel to maintain a warm space. This entails wood, coal, manure, or various liquid fuels to be stored for the cold season, which is a social cost as well as an economic one. Combustion in enclosed spaces tends to degrade air quality and presents a long-term health risk from exposure to particulates and volatile organics.

Vernacular: descriptive term used in building design. The culturally accepted way of building given local weather and resources developed over the generations.

It was the Roman culture near 100 CE that began using glass plates to permit light transmission while preventing major thermal losses to the interior zone. They were not the large sheets of dual or triple pane glazings that we know of today, but rather hand blown small plates for wealthy occupants, temples, or churches. In regions of Eastern Asia (China, Japan, Korea), cultures developed paper glazings to accomplish a similar goal. Where regionally available, cultures also developed the vernacular skill to combine large sheets of the layered mineral *muscovite* to make enclosed windows (see Figure 2.4). We will call all of these *glazing technologies*, and their function again serves the solar technology function to allow shortwave

Figure 2.4: Example of muscovite (mica mineral) material for window glazing. Photo in public domain from Wikimedia Commons.

irradiance transmission into the zone during the day, while trapping warm air inside during the evenings and cloudy days. The result for each culture using glazing was generally a reduction in fuel consumption, an increase in available time not tending the flame, combined with improved indoor air quality and health.

Now, we switch to a present-day emergence of solving the fuel for lighting dilemma. In many tropical areas of the developing world, inexpensive homes can be built of windowless walls covered in corrugated metal roofing. Essentially, these homes do not require windows to trap heat, but rather only require natural light to penetrate into the interior space for reading, working, and crafting. Back in 2002, mechanic Alfredo Moser from Sao Paulo, Brazil developed the idea of using a single plastic bottle filled with filtered water (with a touch of bleach to prevent algal growth) as a light pipe for interior daylighting. In fact, he developed it to light his workshop! (see Figure 2.5).

Figure 2.5: PET plastic soda bottle re-purposed to deliver light nearly equivalent to a 55W incandescent bulb on a sunny day...for very low cost. Image by Jeminah Ruth Ferrer.

See the Liter of Light website for video documentation and HowTo Instructables to create your own solar light bulb.

The non-profit team from MyShelter has since emerged in the Phillipines with *Isang Litrong Liwanag* (translated as "One Liter of Light"), initiated by a young Fillipino social and eco-entrepreneur Illac Diaz (graduate of MIT and Harvard). ILL emerged as a viral program to install one million light tubes into low-income homes by the end of 2012.[20] The Solar Bottle Bulb uses available disposable plastic soda bottles to divert daylight into the interior space, and is based on the principles of appropriate technology—technologies that are easily replicated and address basic human needs for developing communities. The costs for materials and labor are minimal, and the bottle is a waste material. The solar bottle bulb uses the optical physics described by *Snell's Law* and *total internal reflection* to create a light pipe (akin to a fiber optic). The solution captures the physics of light refraction and reflection that we shall discuss in the future chapter on device logic and pattern language.[21]

FRAMEWORK: SOLAR RIGHTS AND ACCESS

History has demonstrated that legal structures for solar energy have been in existence for many hundreds of years. *Solar rights* define access to solar energy and hold significant economic consequences. As stated by Assoc. Prof. of Law Sara Bronin (UConn School of Law),

> "Solar rights dictate whether a property owner can grow crops, illuminate her space without electricity, dry wet clothes, reap the health benefits of natural light, and, perhaps most significantly in our modern era, operate solar collectors—devices used to transform solar energy into thermal, chemical, or electrical energy."[22]

In ancient Rome and Greece, legal structures were set up in the form of easements, allocated government lands, and sometimes strict urban planning for orientation and elevation limitations on entire communities.[23] The emergence

[20] Tina Rosenberg. Innovations in light. Online Op-Ed, February 2 2012. URL http://opinionator.blogs.nytimes.com/2012/02/02/innovations-in-light/.

[21] Buzz Skyline. Solar bottle superhero. Blog, September 15 2011. URL http://physicsbuzz.physicscentral.com/2011/09/solar-bottle-superhero.html.

[22] Sara C. Bronin. Solar rights. *Boston University Law Review*, 89(4):1217, October 2009. URL http://www.bu.edu/law/central/jd/organizations/journals/bulr/documents/BRONIN.pdf.

[23] Ken Butti and John Perlin. *A Golden Thread: 2500 Years of Solar Architecture and Technology*. Cheshire Books, 1980; and Sara C. Bronin. Solar rights. *Boston University Law Review*, 89(4):1217, October 2009. URL http://www.bu.edu/law/central/jd/organizations/journals/bulr/documents/BRONIN.pdf.

of this legal structure was about enabling citizen access to the solar gains from daylighting and again, solar heat gains in the winter that would reduce consumption of fuels.

In the USA, we distinguish between *solar rights*: the option to install a specific solar energy system within residential or commercial properties otherwise subject to private restrictions, and *solar access*: ensuring that a structure or field may receive sunlight across property lines without obstruction by neighboring objects, including trees. Examples of common private restrictions are covenances or bylaws that forbid PV on roofs, or clothing drying lines anywhere, or front yard vegetable gardens, or solar powered puppies.[24] As of 2012, over 40 states have some form of solar access laws, either a Solar Easements Provision, a Solar Rights Provision, or both.[25] Exceptions are the Commonwealth of Pennsylvania, Michigan, South Carolina, and Connecticut, which are noticeably behind the times for the amount of solar systems development in their regions.

[24] You see what Suburbia is doing to our world?

[25] North Carolina State University Database of State Incentives for Renewables and Efficiency. DSIRE solar portal. URL http://www.dsireusa.org/solar/solarpolicyguide/?id=19. NREL Subcontract No. XEU-0–99515-01.

In Germany, there are laws for daylight access, guaranteeing daylighting to all working spaces within a building, such that every worker is exposed to the variation of sunlight over the course of the day. Indeed, occupational and indoor environmental health studies have confirmed that exposure to variable intensity lighting (such as from daylight) is important to our circadian systems that involve biological cycles repeating at 24-h intervals. Humans are driven by internal patterns, which are biochemically synchronized to solar cycles, making light a primary stimulus to our life cycles. Our circadian system regulates behavioral patterns of rest and activity as well as biochemical functioning at the cellular level, including production of Vitamin D from UV light.[26] Why do we irradiate food products like milk with UV light to create "Vitamin D enriched" foods? Because ultraviolet light normally converts cholesterol in our bodies into Vitamin D with exposure to daylight. As we have greatly increased our indoor activities, and in the US even removed ourselves from window spaces, we need to reintroduce Vitamin D in our diets. The UV light converts the cholesterol in milk products to Vitamin D, a technology patented at the University of Wisconsin-Madison (a home of good cheese).

[26] National Research Council. Review and assessment of the health and productivity benefits of green schools: An interim report. Technical report, National Academies Press, Washington, DC, USA, 2006. Board on Infrastructure and the Constructed Environment.

FRAMEWORK: SOLAR POWER ENTREPRENEURS

Would you believe that one of the modern world's first solar entrepreneur came from what is now Philadelphia? Frank Shuman was an eclectic inventor in the late 1800s who invented "wire glass" (chicken wire embedded between panes), "Safetee-Glass" (epoxy embedded between glass panes-variants are used in all automobile windows now), and a low-pressure steam engine.[27] Shuman later went on to form the Sun Power Company in 1910 and successfully harnessed solar power physics to generate steam pump power in Egypt in 1911.[28]

But before that, he had a small solar research facility in Tacony, PA (now incorporated as a historic neighborhood in Northeast Philadelphia). Shuman first built very large horizontal hot boxes (a type of flat plate collector technology) that consumed 1080 ft^2 (110 m^2), able to run a 4 horsepower steam engine that pumped water.[29] This was a testament to entrepreneurship and following a good scientific lead regarding the *Greenhouse Effect*, and then finding a great niche clientele. The very fact that Shuman did it in Pennsylvania, where inhabitants have been so surrounded by inexpensive fuels of coal, petroleum, and natural gas is truly wondrous. In fact, from personal experience, many individuals living in PA still truly believe that the Sun can contribute little of value due to the "cloudiness" of the state's weather.[30] This perception will change with time and with the engaged participation of emerging solar design teams. All this being said, PA is not the best locale for solar concentration (good for flat plate technologies, poor for concentrating technologies).

Now recall that Shuman was in the glass business, and so had access to high-quality Pittsburgh plate glass for covers and mirrors. Through the use of mirrors (silvered sheets of Pittsburgh plate glass) piping, and tracking cranks, the Tacony plant was able to experiment with over 10,000 ft^2 (929 m^2) of concentrating equipment, and could yield 25–50 horsepower pumping action on clear days.[31] Shuman's company eventually developed solar concentration technologies for the Nile River in Egypt, where the solar resource is stronger and there are better clear sky conditions for concentration. In Egypt, late 1800s, the coal was imported from Britain at \$15–\$40 a ton, while Frank Shuman had

[27] Frank T. Kryza. *The Power of Light: The Epic Story of Man's Quest to Harness the Sun*. McGraw-Hill, 2003.

[28] Ken Butti and John Perlin. *A Golden Thread: 2500 Years of Solar Architecture and Technology*. Cheshire Books, 1980; and Frank T. Kryza. *The Power of Light: The Epic Story of Man's Quest to Harness the Sun*. McGraw-Hill, 2003.

[29] Note that the conversion: 1 hp is equivalent to ~3/4 kW.

[30] Meanwhile, New Jersey, Maryland, and New York are going like mad to install PV power across the state. Also, the state name translates to Penn's Wood, trees being a pretty responsive solar technology. Must be different weather in these states, right? *(cough)*.

[31] Frank T. Kryza. *The Power of Light: The Epic Story of Man's Quest to Harness the Sun*. McGraw-Hill, 2003.

calculated that solar steam from his technology could compete with costs as low as $3–$4 per ton.[32]

Let's examine our historical case in terms of fuel constraints: the amount of direct solar irradiance in Egypt was very high while the cost of fuel was also very high—so avoided fuel costs by investing in solar would reap long-term benefits. In contrast, the amount of annual direct solar irradiance is lower in PA, while the cost of fuel in 1911 was also extremely low due to local coal resources in the Commonwealth. Hence, there would not be a likely market for developing concentrating solar in PA. Keep this in mind for the future of SECS design: it is often not the amount of Sun that drives the success of a technology, but rather the economic advantage of avoiding fuel costs and improving social and environmental conditions (see Figure 2.6).

We again have a modern-day analog to Frank Shuman's within India today. The SolarFire.org group was created in 2007 by Frasier Symington of Canada, with Mike Secco leading a project for concentrating oven technologies in Mexico. The group was later developed for industrial applications by Eerik and Eva Wissenz and an expanding global network of partners and investors, with an open business model to deliver low-cost solar concentration to areas of high

[32] Frank T. Kryza. *The Power of Light: The Epic Story of Man's Quest to Harness the Sun*. McGraw-Hill, 2003.

The amount of sunlight (irradiance) does not drive the adopted technology—the economic advantage of avoiding fuel costs drives SECS adoption.

Unfortunately, the events of World War I preempted any future developments from the Sun Power Company, which was largely supported by British financial backing.

Figure 2.6: The Prometheus 100 (100 sq. ft of mirror) built by Eerik and Eva Wissenz of SolarFire.org in Rajkot, India. Photo by Eerik Wissenz.

solar impact.[33] Among three modes of deployment, the concentrating technology has been developed for engaging the open source maker community with DIY (Do-It-Yourself) schematics and instructions for building, for local small-scale commercial power production, as well as for industrial-scale power production.[34]

With their Helios compound mirror concentrating technology, temperatures of >500 °C can be generated to produce high pressure dry steam for a fixed focal point of absorption. In fact, they have collaborated with Tinytech Factories in India to build concentrating mirrors from 32 to 90 m^2 in size in conjunction with 2–50 horsepower steam engines. Incidentally, the SolarFire system has the same pumping power as calculated from Frank Shuman's first concentrating plant in Tacony PA, but at a tenth of the area of collection located in Gujarat, India. The concentrating power can be used for other applications as well, such as cooking, to roast coffee, water purification, steam for industrial food and material processing, even melting aluminum.[35] Yet another case of solar energy technologies being intended as environmental technologies and appropriate technology.

[33] "Solar Fire is a modular, high temperature, fixed focal point, Solar Concentration System designed for scalability at low-cost." –SolarFire.org.

[34] Eva Wissenz. SolarFire.org. URL http://www.solarfire.org/article/history-map.

[35] Eva Wissenz. SolarFire.org. URL http://www.solarfire.org/article/history-map.

FRAMEWORK: SOLAR ECOSYSTEM SERVICES

We have reviewed the historical accounts of how solar energy across the Earth has changed lives and even lead to increased health and prosperity. Now let us redirect our attention to the surrounding environment that supports our existence, the ecosystems that are local to our living and working spaces. Any solar technology will have an impact on the ecosystem that it is deployed in, and could actually add ecosystems services to the area if designed with an awareness for landscape architecture and ecology. Solar design has not included ecosystems services as a part of the solar design concept until very recently. Presently, there are design discussions of the manner by which photovoltaic arrays can add desirable shading and foliage development in addition to power generation. Also, large-scale solar projects in the American southwest must be evaluated disturbing ecosystems support for numerous native species within the region of deployment.

Ecosystem Services:

- **Supporting:** fundamental services necessary for the production of all other services (e.g. photosynthesis, soil formation, nutrient formation and cycling, water cycling).
- **Provisioning:** the products obtained from ecosystems, including energy and fresh water, as well as food.
- **Regulating:** the benefits obtained from regulating ecosystems processes like good air quality, water quality, climate management, and erosion regulation.
- **Cultural:** the non-material benefits that society gains from ecosystems, including cognitive development, reflection, aesthetic experiences, recreation, and ecotourism.

In terms of solar energy affecting the space of the built environment or living spaces, we have shared the example of the Zen rock gardens of

Figure 2.7: The rock garden of Ryo Anji. Photo by User:cQuest/Wikimedia Commons/CC-BY-SA-3. May 2007.

Ryo-Anji from Figure 2.7. In this case, the rock garden is being used for multiple purpose manipulation of the microclimate and lighting conditions surrounding the monastery building. The rock garden is really very pragmatic, despite the marketing hyperbole offered for the calming effects of a mini sand pile and rake in your office space. If we refer to the landscape analysis of Prof. Robert Brown (University of Guelph, Canada), the rock garden is a diffuse reflective surface oriented to the south of the room for reading and copying of sutras, permitting additional light to be softly reflected into the room in the summer time. Additionally, that reflective light is energy that is *not* absorbed by the ground, keeping the air above the rock garden cooler during the day. This alone would not be sufficient, as the cool air would just sink and flow away from the building, which is why one contains the cool microclimate with steep retaining walls. Now, a reflective surface is only as good as it remains nice and white (or bright gray in this case). The region near Kyoto is hot and humid in the summer, meaning there will be vegetative growth from mosses or plants in the area, and leaf-fall onto the rock garden surface. The way to maintain such a surface is to rake the small gray stones on a regular basis, perhaps even indulging in creative patterns and allowing a few large stones to break up the features.[36] This is not to say the rock gardens of Japan are without aesthetic form, but as with many things

[36] Robert D. Brown. *Design with microclimate: the secret to comfortable outdoor spaces.* Island Press, 2010.

tied to the energy and comfort of a living space, practicality will emerge from the vernacular of hundreds of years of solar experience.

FRAMEWORK: SOLAR FOR ENERGY INDEPENDENCE

No, we will not be discussing the USA this time. Energy independence has been a big issue for all industrialized nations for many decades. In particular, we look at the country of Germany in the 1970s. What began as a small protest in 1975 regarding the potential for a new nuclear power generation plant in the southwest of Germany. The series of events in Germany lead to the development of a major solar research center. The context involved a hamlet called Wyhl, near the Kaiserstuhl wine-growing region of the Upper Rhine Plain. This area is right next to the east-west border to France and north of the Switzerland-Germany border.

There was an initial protest by the farming community of wine-growers and townspeople in 1975. Dispersal of the initial protest was reportedly handled poorly by the local law enforcement, and the bad publicity began. Subsequently, the community of Freiburg (a university town) staged a re-occupation of the Wyhl site with approximately 30,000 people—they were not disbanded. The plant was never built and the proposed area later became a nature reserve.[37] But the events opened up a much larger national debate in Germany in terms of getting more power to the expanding industrial base. Questions arose of where one stores nuclear waste, increasing the concern for the nuclear power option, along with the unfortunate events of the Three-Mile Island nuclear reactor accident in 1979 within the USA (just outside of Harrisburg, the capital of Pennsylvania). By 1980 a commission of the Bundestag proposed a major change in energy policy to opt out of nuclear power. The opinion of the German public was in affirmation, the Green Party was formed and elected to the Bundestag in 1983. The events of the Chernobyl nuclear disaster in 1986 solidified the German trend away from nuclear power (see Figure 2.8).[38]

In contrast, we note that the emerging field of solar energy engineering research effectively died in the Universities of the USA as of about 1985, coinciding

[37] Consensus: Never mix the threat of decreased water quality with the productivity of a wine region.

[38] *Spiegel Online* has a Flash-based timeline of events in the Nuclear Protest movement for Germany.

Figure 2.8: On April 24, 2010, the *KETTENreAKTION* ("Chain reAction") protest resulted in a public display against nuclear development via a 75-mile human chain (over 120,000 individuals) between two nuclear plants in Krümmel and Brunsbüttel. The image is in Uetersen, Germany on April 24, 2010. Photo by User:Huhu Uet/ Wikimedia Commons/ CC-BY-SA-3.

with the expiration of the 1978 Energy Tax Act (P.L. 95–618). Some institutions maintained small practitioner communities, but in comparison to a similarly young field of nuclear engineering the solar field was largely removed from the USA for two decades. Many solar textbooks deriving fundamentals of the solar resource are typically dated pre-1985, and our solar researchers shifted to other funded fields such as refrigeration, microelectronics, and nanotechnology (and then retired). The Energy Policy Act of 2005 (EPACT) brought solar industry back to life, amended in the Emergency Economic Stabilization Act of 2008 (P.L. 110–343), and put some vigor back into the field. Now consider, there has been a 20–25 year gap in the lineage of training solar energy scientists and engineers in the USA, and the field has since atrophied from lack of job opportunities, government funding, and a glut of fossil fuels. What do you suppose happened in Germany in the interim?

Let us consider what to do if your country was already planning to expand in electric power consumption from industrial and commercial growth, but was constrained not to use the main available fuels like coal and fissionable nuclear materials. Well, do you remember Freiburg? Germany invested in science and technology since the 1980s. In 1981, the Fraunhofer Institute for Solar Energy Systems (or Fraunhofer ISE) was founded as the first solar research institute

...and the surrounding region is wine country, not beer.

in Europe, operating independently of the nearby University of Freiburg. The southwest of Germany where Freiburg is located is also the highest in solar annual irradiation, yet this is less insolation than any part of the USA other than Alaska. The Institute has evolved into the largest center of its kind in Europe and is one of the world leaders for solar research and corporate partnerships in solar energy. The first form of energy independence can be found in strong lines of applied and basic research, much like what was once found in Bell Labs in New Jersey, patenting in the 1940s and then designing the first commercial "solar battery" (now known as *photovoltaics*) in 1954.

In addition to German research support, the *Electricity Feed Law* or *Stromeinspeisungsgesetz* was enabled in 1990, ensuring preference to electricity generated by hydropower, wind energy, solar energy, landfill gas, sewage gas, or biomass. The law mandated the purchase of renewable-sourced electricity by utility companies and provided large loans and subsidies to producers of renewable power. Next, the *Renewable Energy Act* (*Erneuerbare-Energien-Gesetz*, EEG) came into effect in 2000, and continued to invigorate the global photovoltaics industry back to life. The intent of the act was to stimulate a renewable energy economy in Germany via energy efficiency in buildings combined with a feed-in-tariff for renewable-generated electricity.

As a closing remark, the Fukushima Dai-ichi nuclear disaster in Japan led the decision by Chancellor Angela Merkel's coalition to shut down the 17 remaining nuclear power stations by 2022 (announced May 30, 2011). The researchers at the Fraunhofer ISE look forward to a bright and continued future in solar energy development in Germany. The second form of energy independence comes from government stepping in when markets cannot transition to a new technological medium.

After finishing several discussions of solar energy affecting social and ecosytems services, how does this make your creative mind expand to try new entrepreneurial arcs? Furthermore, how can we progress in studying solar technologies by designing solar energy conversion systems in such a way that add to ecosystem services while engaging social heath and economic benefits? We have obligations to ourselves, to our surrounding communities, but also to future generations and to the biome that supports us. Questioning these obligations and the long-term challenges ahead of society means that you are beginning to approach the underpinnings of the ethics of sustainability.

PROBLEMS

1. In one short statement, define the key role of the Solar Energy Engineer, Economist, or Designer. Then explain in more detail how this statement can be broadly interpreted to include or even compare daylighting, solar hot water, and solar electricity.

2. What is the relationship between fuel constraints and the perception of solar potential in a specific region of the planet, for a particular time in history (including the present). Compose a short essay providing three examples.

3. Research the energy incentives for solar hot water in China, and determine if energy is cheap or expensive for the average Chinese citizen (both heat and electric power)? Given your observation, why does China have 60% of the world's capacity of solar hot water systems deployed?

4. Compose a short essay on the value of having solar rights and solar access policies in locations with high fuel constraints. What do stakeholders gain or lose with policy development in favor of solar rights and solar access?

5. How long has solar energy been used in the USA for residential solar thermal water heating?

6. What is the advantage of a porch (patio, portico) in terms of solar building design? Why should they always be oriented toward the equator, rather than away from the equator?

7. Why would daylighting be important to the workers of Germany?

8. What are ecosystem services, and what are naturally occurring solar ecosystem services?

9. Why did solar energy for steam power return to obscurity after Frank Shuman and the Sun Power Company had success in 1911?

10. What are the broader socioeconomic, health, and ecological implications of the Solar Bottle Bulbs?

11. Find a modern example of a legal confrontation over *solar rights*. Document in a short essay.

RECOMMENDED ADDITIONAL RESOURCES

- Let it Shine: The 6000-Year Story of Solar Energy.[39]
- Anasazi Architecture and American Design (Chapter 5 and 8).[40]
- The Power of Light: The Epic Story of Man's Quest to Harness the Sun.[41]
- Solar Rights.[42]
- The Gift of Good Land: Further Essays Cultural & Agricultural.[43]
- The Story of Stuff (movie).[44]
- The Story of Stuff: How Our Obsession with Stuff is Trashing the Planet, Our Communities, and Our Health—and a Vision for Change (book).[45]
- Sustainable Energy: Choosing Among Options.[46]

[39] John Perlin. *Let it Shine: The 6000-Year Story of Solar Energy*. New World Library, 2013.

[40] Baker H. Morrow and V.B. Price, editors. *Anasazi Architecture and Modern Design*. University of New Mexico Press, 1997.

[41] Frank T. Kryza. *The Power of Light: The Epic Story of Man's Quest to Harness the Sun*. McGraw-Hill, 2003.

[42] Sara C. Bronin. Solar rights. *Boston University Law Review*, 89(4):1217, October 2009. URL http://www.bu.edu/law/central/jd/organizations/journals/bulr/documents/BRONIN.pdf.

[43] Wendell Berry. *Home Economics*. North Point Press, 1987.

[44] Annie Leonard. The story of stuff. Story of Stuff. Retrieved Oct 28, 2012, from the website: The Story of Stuff Project, 2008. URL http://www.storyofstuff.org/movies-all/story-of-stuff/.

[45] Annie Leonard. *The Story of Stuff: How Our Obsession with Stuff is Trashing the Planet, Our Communities, and Our Health–and a Vision for Change*. Simon & Schuster, 2010.

[46] Jeffreson W. Tester, Elisabeth M. Drake, Michael J. Driscoll, Michael W. Golay, and William A. Peters. *Sustainable Energy: Choosing Among Options*. MIT Press, 2005.

LAWS OF LIGHT

03

LIGHT-INDUCED energy transfer is a coupling of light and matter.[1] Another general term for the interaction and behavior of light with matter is *optics*. This chapter collects together fundamental rules of radiative transfer. Some will be trivial, while others require detailed mathematical description to convey the full import. For most readers, this will be the first time that the topics are accumulated to form a cohesive picture of solar energy as a powerful and diverse resource for energy transformations.

Chapter goal: to establish the basics or foundations of radiative transfer across the broad spectrum.

[1] With some small exceptions of extremely high concentrations of light and plasmonic devices.

In Figure 3.1, we have illustrated some simple methods for diagramming light as a directional transfer of energy. The forms will allow us to sketch out scenarios of the life cycle of light as photons travel from an emission source to an absorbing, receiving surface. As a transdisciplinary design team, being able to communicate contributions of light for multiple surfaces through diagrams will be just as important as numerical accounting of a radiative energy balance. Diagrams allow us to see where there are gaps in our current understanding of a problem, and then later allow us to assign quantitative energetic contributions to our arrows as we refine the scope of the problem.

Shortwave band: group of wavelengths emitted by the Sun as a thermal source surface (also partially reflected/scattered by the sky). [$250\,\text{nm} < \lambda < 3000\,\text{nm}$].

Longwave band: group of wavelengths emitted by thermal source surfaces of ambient terrestrial temperatures, ~180–330 K [$3000\,\text{nm} < \lambda < 50{,}000\,\text{nm}$].

In the future, when the term "light" is used, we mean electromagnetic radiation transfer. Specifically in terms of phenomena occurring via the Sun or Earth, one can confine that description to wavelengths between about 250 nm (5 eV) and 3000 nm (0.4 eV). This is the majority of received solar irradiance below the atmosphere, and is termed the *shortwave band*, while the Earth and atmosphere contribute to the lower energy *longwave band* that assists with maintaining the greenhouse effect.

The basic unit of energy for light is an **electron volt**: eV. 1 eV is 1.6×10^{-19} J. We will show later the bandgap of silicon for a photovoltaic cell is approximately 1.1 eV.

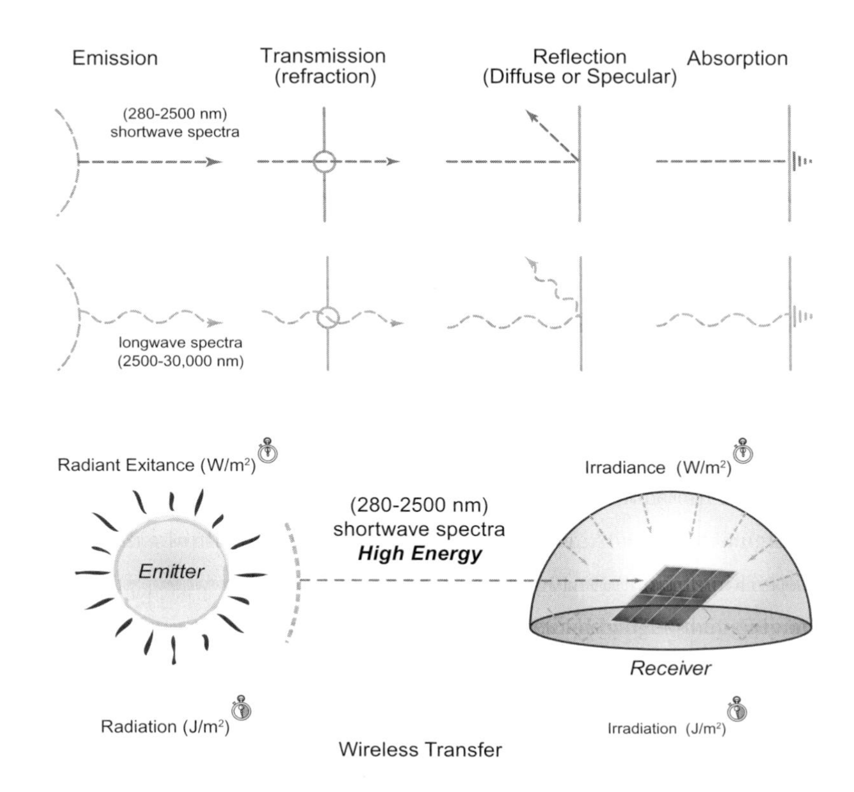

Figure 3.1: Key for future diagramming of sources for light. The convention is central to this text, and useful for quick diagramming of radiative transfer in different SECS.

Figure 3.2: The Sun-sky-collector system is an *open thermodynamic system* emitting radiant energy in this case to an optoelectronic and optocaloric receiver—a photovoltaic panel within the sky dome.

In Figure 3.1 the straight arrows symbolize shortwave light, while the longwave arrows symbolize longwave light. A curved arc symbolizes an emitting surface, while a straight line (intersecting with an arrow) signifies a receiving surface for transmission, reflection (which is another source of light), or absorption of light. A bent line signifies a reflecting or scattering surface (such as the sky or the ground), while a line intersecting a surface with a circle indicates transmission/refraction through a near transparent material (relative to that particular band). As we shall demonstrate shortly, a receiving or emitting surface can interact selectively with different bands of light.

In Figure 3.2, we have diagrammed a simple system for shortwave light (omitting the diagramming for longwave light) relative to a SECS: a photovoltaic (PV) module mounted on the ground in a field.[2] Notice that the Sun is diagrammed as one source of light for the PV module, but the sky and

[2] The example of scattered shortwave light from the atmosphere is simplified with a straight arrow emerging from the "sky dome." The intent is to convey that the shortwave light from the sky is derived from reflection, not thermal emission.

the ground are reflecting sources of solar light for the module as well (called diffuse reflective surfaces). Hence, even if a major cloud system were to block out the central beam of solar irradiation, the PV panel would still receive some contribution of shortwave light from the ground and the sky.

Relative to the basis of *optics* as light-matter interactions, we are reminded that *radiation is always directional*. Light is either *emitted or reflected from* a surface or it is *incident upon* on a receiving surface. Incident light normalized as a radiant flux per area is called *irradiance* (W/m^2; a rate or power density), and can be integrated over time to form a measure of *irradiation* (J/m^2; energy density). Put another way, irradiance is the radiant flux that impacts a receiving surface from all directions. If we change our perspective relative to the Sun as an emitter, we use the term *radiance* (W/m^2 sr) for the flux of light emitted per unit solid angle of the surface of the Sun. Emitted light per area is termed *radiant exitance*, in units of W/m^2. Light spreads out in all directions from the emitting source, and as such objects that are far from the surface receive less irradiance than objects right next to the surface. This is termed the *Inverse Square Law of Radiation*. Due to the distance of the Earth from the Sun, the intensity of that light diminishes in inverse proportion to the square of the distance between emitter and receiver.

Irradiance: radiant flux (power) incident upon a surface [W/m^2].

Irradiation: energy (integrated over a time step) incident upon a surface [J/m^2].

Radiance: radiant flux emitted by a source surface per unit solid angle [W/m^2 sr].

Radiant exitance: radiant flux (power) emitted by a source surface [W/m^2].

Radiation: energy (integrated over a time step) emitted by a source surface per unit solid angle [J/m^2 sr]. In this text radiation is exchangeable with time integrated radiant exitance.

LIGHT IS A PUMPING SYSTEM

Let's create an analogy for working with light. Energy in materials can be conceived as lots of levels (or states) from low to high. This is a quantum description of energy, the fundamental connection among both electronic and vibrational (thermal) states in matter. We likely know about photons and electrons, but what about *phonons*? Phonons are the quantum states of vibrations and oscillations existing in all materials, which emerge collectively as thermal energy (informally called heat). The quantum nature among *photons*, *electrons*, and *phonons* is linked to the macroscopic via large ensembles of particles, described collectively as phenomena of *light*, *electricity*, and *heat* (to use the colloquial term for thermal energy). Hence, all quanta of energy

All **states of matter** have **quanta of energy** associated with them. We tend to ignore this for large incoherent systems, but the statistical ensemble of quantum states allows us to link **photons** to matter for changing energy states for vibrations (**phonons**) and **electrons**.

Suggestion: relearn **thermal energy** as an ensemble of **phonons** and **electrons**, rather than as "heat." True **heat** refers to the transfer of energy.

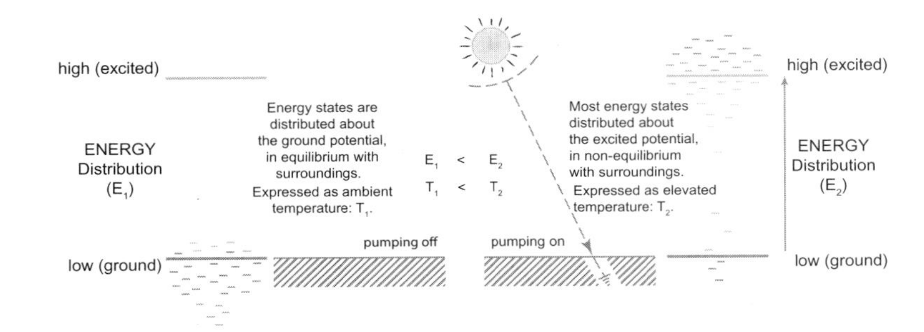

Figure 3.3: Schematic of a light current pumping a distribution of low energy states (ground states at ambient temperature) to high energy states (excited states at elevated temperature).

exist in a population or distribution of levels (filled from low to high) in steady state with their surroundings; until such time as they are perturbed by the addition of a higher quality energy current, and thereby raised up to higher levels. As we observe in Figure 3.3, light is a high quality energy current, and once absorbed by a materials, the photon energy is transformed into the electric or thermal energy form by moving low energy levels up to higher energy levels.

The room temperature sandwich that you picked up for lunch today is still vibrating with a local equilibrium of thermal energy described by an ensemble of energy states (energy levels in steady state with the surroundings). When you put that sandwich into the microwave for 60 s, you push that sandwich out of thermal equilibrium with the surrounding air. Microwave band photons are absorbed near-instantaneously by the sandwich (more precisely the water and fat molecules in the sandwich), perturbing the phonon energy levels in the sandwich to much higher states. This occurs even though the surrounding air has not been heated—wireless heating![3] So how did we accomplish remote-control heating of a sandwich using only light, and what is a useful analogy for this process across radiant energy conversion systems?

[3] Somebody call Nicola Tesla: TESLA PATENT #685,957 Apparatus for the Utilization of Radiant Energy.

From the perspective of this textbook, "light" (i.e., electromagnetic radiation, or an ensemble of photons) can be best thought of as the current performing work for a *pump*. Now you may realize that a traditional pump transforms useful currents or flows of energy like electricity or mechanical energy (called *work*), through a system to move mass or energy from an ensemble out of a *well* of low level states to new higher level states. You should also recognize that when a pump is turned off, or the work current is no longer present with the pump system, the energy transfer stops and everything falls back down the well to those low levels.

From a fluid/water perspective, we are familiar with applying a current of electricity to a pump system to move water up and out of a well in the ground. With the help of a dam for energy storage, we might then use that higher potential water to do work later like running hydroelectric power (look into *pumped storage reservoirs* for examples). From a thermal perspective (as in a heat pump) we apply a current of electricity to pump low energy ensembles (low temperature) out of a steady state "well" into high energy ensembles (high temperature; called "moving" heat). Heat pumps are used in your refrigerators as well as our home air conditioners and home energy management systems. From an electronic perspective for lasers, we apply a work current of electricity into a semiconductor to pump low energy electron ensembles out of a steady state well to high energy levels, eventually leading to stimulated emission of coherent light.

Loosely speaking, we will describe **light as a pump**, with the understanding among readers that photons from the Sun are actually a flow of high quality energy to perform *work* in tandem with the materials contained in a solar energy conversion system.

Now is the time to begin thinking of light, interacting with an optical material, as "wireless" energy transfer. Light from the Sun is a work current to pump low energy ensembles (be they thermal or electronic) out of their respective steady state wells and up to high energy levels. Photons can be collected by an absorber material to pump-up the levels of energy out of the well within a system. Those levels might me electrons (as in a photovoltaic material), or they might be thermal energy states (as in a warm rooftop in the sunshine), but when the photon is gone, the higher energy states will eventually fall back down the well to lower energy states. So long as there is a steady stream of photons (called a steady state condition), the system continues to pump up the energy of the absorber, and useful work can be collected. A *pump* requires *work* to counteract the natural tendency for high-to-low flow for energy and increasing entropy, organizing low energy states to higher energy states. In Figure 3.4 we show a simple heat pump. Absorbed photons can serve as a source of work from our surroundings, where the SECS absorber material functions as an energy conversion device (ECD) to permit low energy states inside the system to be excited by the work current

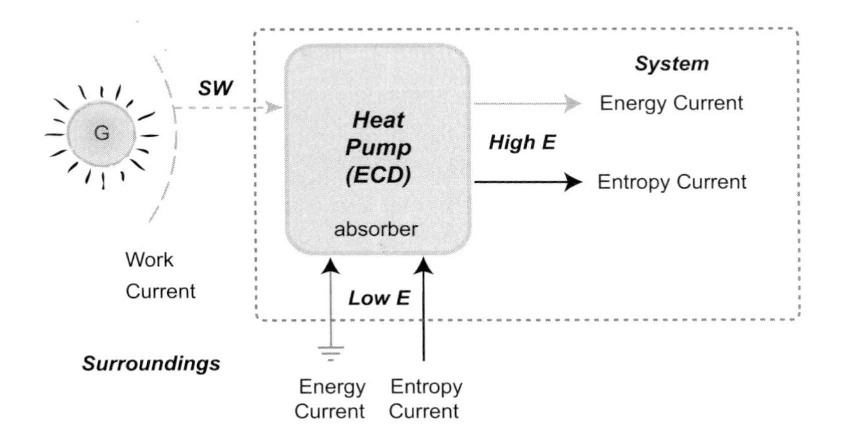

Figure 3.4: Schematic of a crude heat pump, where work is applied as a current of photons to pump "heat" from a low energy state to a high energy state. This path is organizing energy from the reservoir/surroundings into the system. All arrows indicate currents of energy or entropy.

of light. These excited states could be electrons in a semiconductor, thermal vibrations that warm a material (also called phonons), or a change in the state of a molecule (like for photosynthesis). Once we have "pumped up" the SECS, then we can collect that pumped energy falling back down, using a coupled portion of the SECS (a heat engine within the SECS) to do new work for society and the environment.

Now, given the newly developed concept of a steady state *pumping* of matter, using a light source can induce three major optical responses:

1. **Electronic response** (semiconductors: optoelectronic) [electric response in metals].
2. **Caloric response** (thermal vibrations: optocaloric).
3. **Electrochemical response** (photosynthesis and vision: optochemical or photochemical).

In Figure 3.5, we have demonstrated radiative transfer scenarios for both visual systems of solar energy conversion (using photometry terms such as illuminance) and for general solar energy conversion sytems (SECS; using radiometry terms such as irradiance).

Each of these responses can be utilized as an environmental energy conversion technology for society, *or* as a support in the larger portfolio of ecosystems services. These responses can work in complement with each other, as in photosynthesis where caloric and electrochemical responses work favorably for

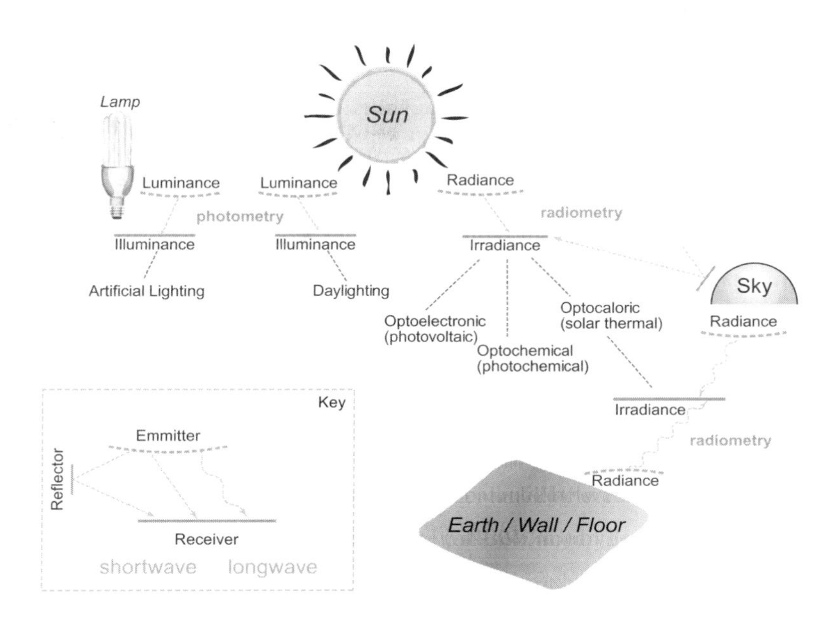

Figure 3.5: A diagram of useful technologies and general measurement terminologies associated with radiant energy conversion, and common sources of light. Shortwave reflections from Earth surfaces have not been detailed here, however diffuse ground reflection is clearly a source of light for SECSs.

life support. Or the responses can oppose each other, as in photovoltaics where electronic and caloric responses effectively work against each other in delivering total electronic current to a device.

It is up to the design team to consider using a light source effectively to achieve a sustainable goal of systems support and/or societal benefit. The more one is aware of using light as a "pump" for environmental modification, combining a response with energy storage, or conversion and transmission, the more creative the team can become in maximizing the utility of the solar resource for their client in a given region.

Consider light as a work current or **pump** for environmental modification.

For example, lighting for the use and delight of human sight, both artificial and daylighting, is a case of design seeking to have *utility* for a given *client* or group of stakeholders. And while you might not be aware of it in your home or workplace, lighting is designed for a specific *location* as well, given the orientation of the room, concepts of direct and diffuse projection of light, and the color balance of a source of light with the receiving surfaces of the surrounding materials. In the simple case of artificial lighting, the LED light source (or lamp) is an emitter, the parabolic white or metallic shell about the light is a reflector, while your rods and cones inside the eyes are the antennae receiving the photons, and converting them into useful signals, to avoid lions and tigers and bears in the forest, or to enjoy a beautiful piece of artwork in a museum.

SECS AND LIGHT

Each solar energy conversion system has several key technical components that we can identify, several simple qualities surrounding it. Those technical components relay information about the way light is directed through an aperture (opening) into a receiving surface, there stored or redistributed as converted heat/light/electricity/fuel, and the flow of energy is subject to one or more control mechanisms.

Again, **phonons** are essentially waves of vibrating atoms; they are part of the quantum level of thermal energy, as **photons** are the quantum level of light.

Recall that SECSs transform a flow of high energy photons into a flow of either electrons, phonons, or more emitted photons from low to high energy. Our high-quality energy flow like light serves as a working current to move energy/heat from low to high states, the device enabling this heat movement is termed a *heat pump*. Then the high energy materials (electrons or hot fluids) are collected from the excited collector system into the surroundings to do work in a *heat engine*.[4]

[4] See Appendix A for a review of heat pumps and heat engines.

WHAT IS A SOLAR ENERGY CONVERSION SYSTEM COMPOSED OF ON THE EARTH SIDE (RECEIVER SURFACES)?

- Aperture (gets bigger with concentration).
- Receiver (could be the cover-absorber system).[5]
- (Storage).
- Distribution Mechanism (internal to the system).
- Control Mechanism.

[5] What is the difference between an **aperture** and a **receiver**? The *aperture* is the opening to capture the maximum photons (includes a reflector or lens for concentration), while the *receiver* is the cover and absorber system. When there is no concentration, they are essentially serving the same purpose.

We can synthesize these pieces of information about light when we change our focus from the sources of energy (the emitted light from the Sun, the reflected light from the sky) to the receiving sinks for the energy to be converted: the *Solar Energy Conversion System*. The energy inputs for SECS function are shortwave photons, and the "quality" of the solar fuel can be characterized according to intensity of the light and the wavelengths of light. For each location under consideration with SECS design, the atmospheric conditions associated with clouds and the weather will change the qualities of the available shortwave irradiance.

For general irradiance we use the symbol **E** (W/m^2). For solar irradiance we make a distinction by using the symbol **G** (W/m^2).

The **directionality of light** is a key concept of radiative transfer.

LIGHT IS DIRECTIONAL

Light, as a photon, has a life cycle that includes an initial emission from a surface, followed by scattering (reflection and refraction events), and finally absorption into a final surface. A material can be employed as an *emitter* to create photons, either by thermally heating a material until the surface glows with the appropriate distribution of photons, or by exciting the material electronically until the recombination of excited charged particles results in emission lines of equivalent energy (like a laser, or LED). Next, a material can be used to *refract* light that is transmitted through it as light is refracted through a quartz prism, or as shortwave light is refracted through water in a swimming pool. Recall that during the life of a photon, light is also likely to be *reflected*, or scattered off of the surfaces of materials. So materials can be also used to specifically reflect wavelengths of light, and this reflection may be *diffuse* or *specular* in nature (or a combination of the two). When downwelling shortwave light interacts with $N_{2(g)}$ molecules in the atmosphere, the molecular material scatters the blue wavelengths of the visible spectrum, leading to a blue sky. Finally, photons are lost via the process of *absorption* by a material. The energy absorbed can lead to the three general photoconversion processes described in this chapter, as the photon "pumps" up the energetic state of the absorbing medium.

Coefficients of **radiative transfer**

- **emittance:** ε,
- **transmittance:** τ,
- **reflectance:** ρ,
- **absorptance:** α.

Downwelling is a meteorological term, referring to light the has a directionality from above coming down to the surface of Earth. **Upwelling** is light directed in the opposite sense, either via emission (longwave) or reflection of shortwave light.

The directionality of light is a key concept, particularly for diagramming the life cycle events of light. One can always diagram the generic directionality of light from the receiving sink (absorber) back to the emitting source, or from source to sink, both in the shortwave band and the longwave band. From now on we try to start with an originating emitting surface of light, and follow the life cycle path of a photon all the way to the receiving surface to be absorbed (from emission to reflection/refraction to absorption). If a band of light is partially reflected, then there is still "life" left in the photon, correct? So we keep on diagramming a reflection leading to a final absorption event. This diagramming process will then help us to identify gaps in our knowledge of light interacting with materials such as the molecules in the sky or grass on the ground. As seen in Figure 3.6, it's not a bad idea to develop a sense for light in the same way we think of other wireless communications, with a signal being transmitted from a broadcasting emitter to a receiver antenna—often with some sort of interference intermediate to the transmission path.

A **fractional coefficient** has a value from **0 to 1**.

The light scattered from the sky and reflected off the ground are also sources of light (via reflectance).

We refer to each of these life cycle events in terms of light emission, reflection, refraction (also called transmission), and absorption. In Figure 3.7 we diagram the life cycle of a photon and the respective fractional coefficients of radiative transfer: emmittance (ε), transmittance (refraction) (τ), reflectance (scattering) (ρ), and absorptance (α). Although the primary source of shortwave light is the Sun, from the perspective of a collector surface in a SECS, other important and

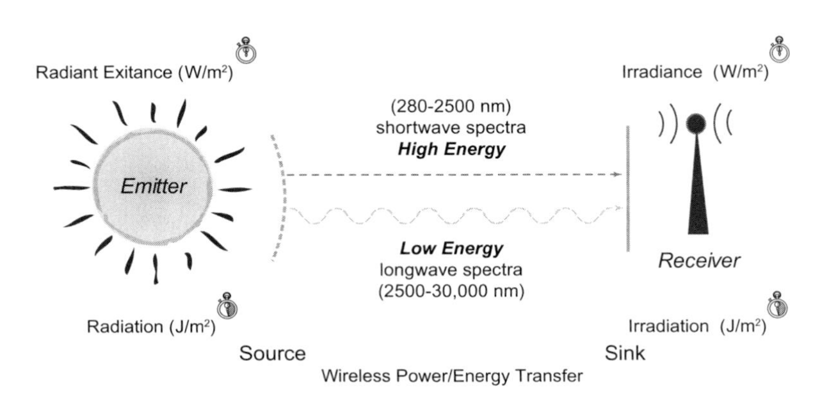

Figure 3.6: The Sun-collector system is an *open thermodynamic system* transferring radiant energy from the surface of the Sun to our collectors on Earth.

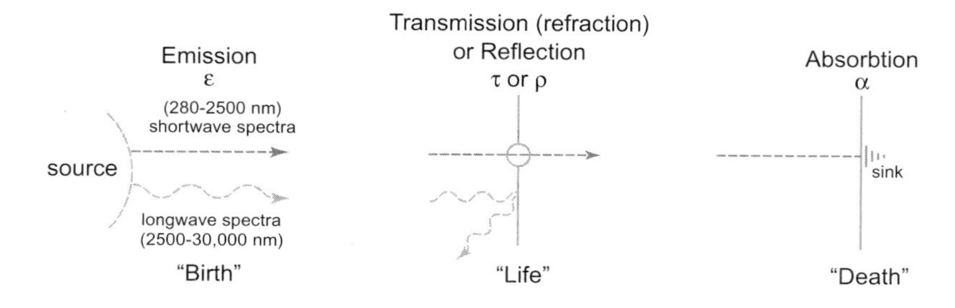

Figure 3.7: The photon has a "life cycle" that we must become aware of.

practical sources of life may be a *reflective surface*, or the light *scattered* across the dome of the sky.

The **sky** is a thermal body that emits and absorbs *longwave* light very well, as we shall confirm shortly.

Again, the useful source for *shortwave band* light initiates from the thermal *emitting surface* from the Sun ($T_{sun} \sim 5777$ K), and then the solar photons interact with the *reflecting/scattering surfaces* from the sky and ground, finally being received and absorbed by our appropriately designed SECS (a light sink). As an alternative, the *longwave band* that we encounter being absorbed by our SECS receiver is derived from thermal surfaces such as the atmosphere, the ground, the trees, walls-all objects that have surface temperatures $\sim$300 K.

Think about this: how is a laser different in light properties from the Sun?

Along with the directionality of light, there are a few other properties of light that must be mentioned in a text on solar energy. We would like to know what separates the intensity of photons in a man-made laser from that found by concentrating the photons from a broad spectrum of solar light into a central receiving tower in concentrating solar power systems. First, laser light is coherent and collimated—meaning the photons are in phase with one another (coherency) and propagating in the same direction (in parallel)—while sunlight is incoherent and spreading out in all directions from the Sun's surface and out of phase. Also, laser light is essentially monochromatic, while solar light is polychromatic, having a distribution of spectral wavelenths from a blackbody surface.[6] Because of these three properties, laser light can contain significantly more fluence (energy, J) than traditional diffuse (incoherent) light from lamps and the Sun.

[6] NASA has an educational page on lasers for the general audience.

PHYSICAL DESCRIPTORS OF LIGHT

Coherent light that is propagating in the same direction, and for which the waves are in the same phase, often associated with laser technology

Incoherent light that is propagating in a multitude of directions, out of phase, and with multiple measured frequencies or wavelengths. Incandescent and fluorescence light sources are incoherent

Monochromatic light that is confined to one wavelength

Polychromatic light that is distributed across many wavelengths

Collimated light that is propagating in the same direction, with photons traveling parallel to one another. Light from the Sun will be *partially* collimated due to great distances, but not to the same degree as collimated laser emission.

Operational Definition: defined by the technology in use, not by a scientific principle.

When we wish to convert **wavelength (nm)** to **energy (eV)**, we only need a simple equation:

$$E(\text{eV}) = \frac{1239.8(\text{nm}\cdot\text{eV})}{\lambda(\text{nm})}.$$

You could also easily use the **"count to five"** approximation of $E(\text{eV}) = 1234.5/\lambda(\text{nm})$ for our purposes in the field. There error is negligible for most of solar design.

LIGHT IS SPECTRAL

Light is a broad spectrum of energies, and those energies are characterized in terms of the periodic wavelengths. Readers are likely to be familiar with the electromagnetic spectrum, but may not be familiar with the common language of energy *bands* used by the solar community. Wavelengths of light with similar light-matter interaction properties are discussed in terms of simple groups, or *bands* within the spectrum. Bands are *operationally defined* according to properties of their emitting surface or receiving surface. For example, the "visible band" (380 –780 nm) receives the title from the limits of detection for the human eye.

High energy light (within which the band of visible light is contained) is grouped into a band and termed *shortwave spectra*. This is a bundle of wavelengths of light that are similar in that they are emitted from the Sun and create the majority of total energy density incident upon a collector on Earth's surface, given the decrease in the light power density with the average distance to Earth (1 AU, or 93 million miles, or 150 million km). Lower energy light emitted by terrestrial or atmospheric objects is termed *longwave spectra*.

This second bundle of wavelengths of light are emitted by surfaces significantly cooler than the Sun.[7] Such surfaces include the skin of our human bodies, much of the surfaces on Earth, and the effective surfaces of Earth's atmosphere (were we to simplify the atmosphere as two covering surfaces upon a thin slice of gas and particles).

[7] Although the surface of the Sun also emits **longwave spectra**, by the inverse square law those photons are effectively gone (too diffuse to be a significant contributor) at the Earth's surface.

Shortwave band: At Earth's surface, the shortwave band has been lumped together from approximately 280 nm to 2500 nm (see Table 3.1), because of the Inverse Square Law for light and the nature of our atmosphere to act as an absorber and reflector of longwave irradiance. Also, the standard materials used in our radiometric devices are limited to measuring this range, from the selective transparency of low-Fe glass covers.

Longwave band: The long wavelengths are found in the range of 2500 nm to >50,000 nm, and again our measurement devices are limited to that range.

Light, as a photon has *no mass*. Not even a little.

We can observe the spectral irradiance from the Sun in Figure 3.8. Note that there are two spectra represented, for "Air Mass 0" and "Air Mass 1.5." This nomenclature is a shorthand for irradiance outside of the atmosphere

Band Title	Spectral Range	Energy (eV)
Shortwave (AM 1.5)	250–2500 nm	–
Longwave (AM 1.5)	2500–50,000 nm	–
Radio Waves	300 mm–100 km	1.2×10^{-11}
Microwaves	0.3–300 mm	4×10^{-6}
IR (Far)	15–1000 μm	1.2×10^{-3}
IR (Mid/Long IR)	3–15 μm	8×10^{-2}
IR (Near/Short IR)	780 nm–3 μm	1.6–0.4
Visible	380–780 nm	3.3–1.6
UV	30–380 nm	3.3–40
X-rays	0.1–30 nm	1.2×10^{4}
γ-rays	1 pm–0.1 nm	1.2×10^{6}

Table 3.1: Detailed table of measured spectral bands. The energy in electron volts (eV) is representative of the first value in the range.

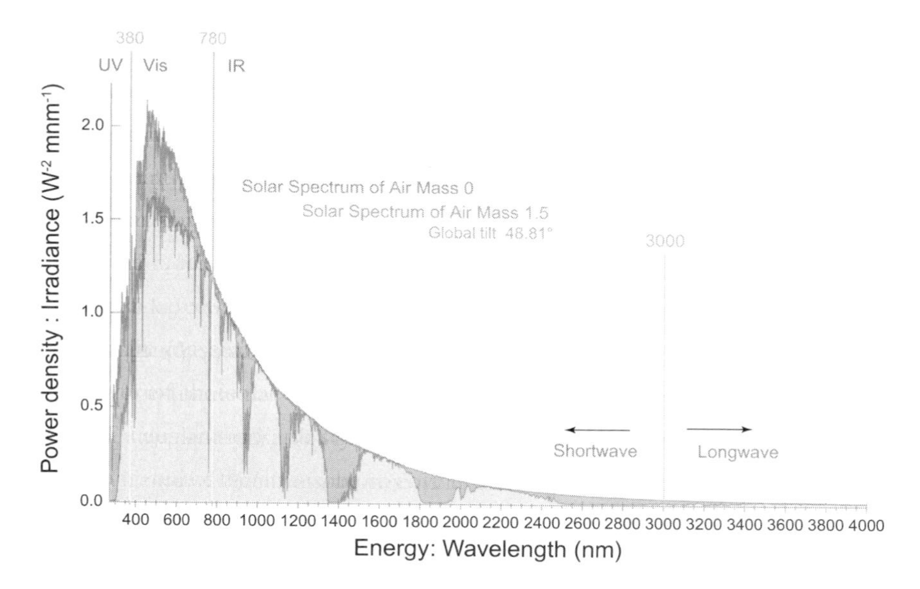

Figure 3.8: These are solar spectra calculated using SMARTS for Air Mass 0 and Air Mass 1.5.

(extraterrestrial, AM0), and a standardized representation of the irradiance filtered through a certain "thickness" for an artificially uniform, clear sky atmosphere (AM1.5; no clouds). This is an engineering representation of an air mass that temporarily ignores the curvature of Earth, and applies a simple parallel plate geometry.

This AM1.5 is used to replicate a representative irradiance condition for device testing and not for actual systems design. The sky is actually a dynamic filter for light, affected by clouds, sky chemistry, site elevation, and the altitude of the Sun positioned in the sky through the day. For our purposes, it offers an estimate of how the sky affects the spectrum of irradiance from the Sun. Of note in Figure 3.8 is that the shortwave band is reduced beneath the atmosphere, from ~3000 nm to ~2500 nm. We also observe smaller chunks filtered out from the Visible and IR sub-bands in the shortwave band. This is from the presence of gases such as water vapor, oxygen, ozone, CO, and CO_2.

AM1.5: Air mass coefficient used for testing solar technologies in the laboratory. 1000 W/m^2 of simulated solar light, characteristic of mid-latitude USA ideal clear sky conditions.

So, we have addressed that light has a source and a sink (directionality), light decreases with distance (inverse square law), and light has an electromagnetic spectrum of wavelengths, each with characteristic energy. Now, if we consider the "source" side of the photon life cycle, how does one characterize or estimate the effective number of photons emitted for an given range of wavelengths.

LIGHT DECREASES WITH DISTANCE

Light intensity decreases with distance from source to receiving surface (sink), and the rate of decrease is in proportion to the *square of the distance between emitter and receiver.* This is called the *Inverse Square Law*. The inverse square law for electromagnetic radiation describes that measured light intensity is inversely proportional to the distance squared (d^2) from the source of radiation. You might also think of the Inverse Square Law as being analogous to a volume knob on a speaker system, the intensity is turned down, without affecting the pitch of the music. As such, the Inverse Square Law does not affect the wavelengths of light being emitted from the source surface (like the Sun), and so the shape and position of the curve seen in Figure 3.8 does not change from left to right. Rather, a change in distance between the emitting and incident surfaces will shift the curve directly up or down (see Figure 3.9).

The reasoning for the inverse square law is geometric in nature. As light is emitted from a point (or sphere) like from the Sun and travels toward a receiving surface, the initial quantity of photons is spread out over an increasingly larger spherical area with distance. We can envision the areal spread with increasing distance like an inflating balloon surface. Additionally, the photons from a

Imagine a bubble getting larger and larger, spreading out over more and more area.

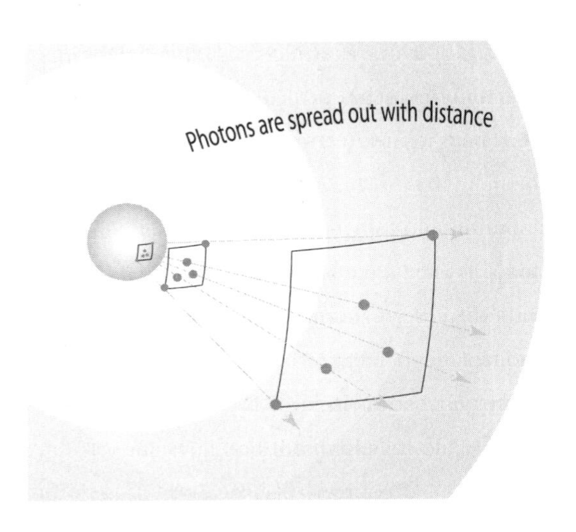
Photons are spread out with distance

Figure 3.9: Photons are distributed with respect to area, a squared unit.

spherical object like the Sun are emitted in all directions. Now, the surface area of a sphere can be determined in units of distance squared, right? So the same quantity of photons are contained within a much larger spherical surface area, effectively decreasing the irradiance on the growing surface by "diluting" the photon density.

$$G \propto \frac{1}{4\pi r^2}. \tag{3.1}$$

For our purposes the Sun's "surface" refers to the **photosphere**, which has an effective blackbody temperature of 5777 K.

The **Solar Constant** has a small variability depending on the sunspot activity in the Sun.

- $G_{sc} = 1361\ \text{W/m}^2$.

To calculate the irradiance drop from a measured value to an unknown value at a known distance, we establish that the irradiance incident from the solar source is G, and note that the surface area of a sphere is calculated by $4\pi r^2$. The average annual irradiance incident upon Earth ($d = 150$ million km) is called the *solar constant* (G_{sc}), evaluated as 1361 W/m² at the exterior of our atmosphere. The irradiance estimated at the Sun's photosphere is 6.33×10^7 W/m². The Inverse Square Law dictates that we scale the irradiance from the surface of the Sun by a factor of $\frac{1361\ \text{W/m}^2}{6.33\times 10^7\ \text{W/m}^2}$, or 2.155×10^{-5}.

The irradiance of any additional surface with respect to the Sun (like the average annual irradiance on another planet) is seen in Eq. (3.2). By using a method of ratios shown in Eq. (3.2) the proportionality terms cancel out, leaving only the ratio of squared distances (d^2).

$$\frac{G_1}{G_2} = \frac{d_2^2}{d_1^2}. \tag{3.2}$$

At the Sun's surface, $d = 0$ km and is $G = 6.33 \times 10^7$ W/m². But we cannot use $d = 0$ in this ratio relationship. Thus, if we provide that the average annual irradiance incident upon Earth ($d = 150$ million km) is 1361 W/m² at the exterior of our atmosphere, we can calculate the irradiance at the surface of Mercury ($d = 58$ million km) in Eq. (3.3). It should make sense that not only is Mercury's average irradiance larger than Earth's, but is significantly greater due to the inverse squared power relation.

$$\frac{(150 \times 10^6\ \text{km})^2}{(58 \times 10^6\ \text{km})^2} \cdot 1361\ \text{W/m}^2 = 9103\ \text{W/m}^2. \tag{3.3}$$

This same principle works for ordinary surfaces, as photographers are aware of too. But we know that there are many energies of photons that can be measured, and they play out as an electromagnetic spectrum of wavelengths.

FOUR LAWS OF LIGHT

In addition to the Inverse Square Law of non-linear decreases in light intensity with distance, we have four basic laws of light that affect our rules of energy accounting for radiative transfer.

1. **Kirchoff's Law** is one of energy balance. A surface at temperature that is at steady state will absorb light equally as well as it emits light. Light may be directional, but surfaces exchange photons in *both directions!* This is fairly important, as there is a tendency for one to only think of an absorber in one direction. Yet excellent photovoltaics will glow with a bright white light if photons are concentrated upon them (and they perform very well). This is a broadband law, meaning it applies to all energies of light (all wavelengths) relative to the emitting surface.
2. **Wien's Displacement Law** implies that the most probable wavelength emitted from a glowing surface is inversely proportional to the temperature of that emitting surface. Another way to conceptualize Wien's Law is to think of the probability density of photons being high for certain wavelengths (visible) when the Sun surface is the emitter, while being strongly displaced to the infrared band when the Earth or Atmospheric surfaces are the emitters.
3. **Planck's Law** can be generalized to mean that all objects have some internal temperature and given that temperature, they all *glow*. Planck's Law describes the statistical distribution of the emergent photons as a Bose-Einstein Distribution.
4. **Stefan-Boltzmann Law** is another energy balance proposition. The radiative energy emitted by a surface (integrated over all wavelengths) is proportional to the fourth power of the surface's absolute

Kirchoff's Law: $\alpha = \varepsilon$ @ T_{eq}.
Wien's Displacement Law:
$E_{b,max} \leftrightarrow T_{surf}$.
Planck's Law: $E_{b,\lambda}$ energy per wavelength.
Stefan-Boltzmann Law:
$E_b = \int E_{b,\lambda} d\lambda$.

temperature. This is a broadband law as well, applying to the sum of all wavelengths of light relative to an emitting surface.

The **emittance** of a theoretical blackbody is $\varepsilon = 1$. (From $\frac{e_\lambda}{e_{\lambda,b}} = \epsilon$, where e_λ is the *spectral emissive power* and $e_{\lambda,b}$ is the spectral emissive power of a theoretical blackbody).

Given that all objects *glow*, (your body, a brick of ice, even the gases in the sky), we will now demonstrate that the intensity and peak wavelengths of the spectral distribution shifts with respect to the thermal temperature of a surface. The ability of a surface to emit a photon is termed *emittance*, and the statistical distribution of photons in a space is termed a Bose-Einstein Distribution.

First we recall that the Sun's photosphere can be considered to be an opaque surface,[8] and that the effective blackbody temperature of the photosphere is 5777 K. From these points of information, we can calculate an estimate of the spectrum and most probable wavelengths emitted by the Sun's continuous radiation. In order to do so, we will take a historical step back into the 1800s, to the origins of quantum phenomena.

[8] The Sun is a miasma of incandescent plasma. Thank you, *They Might Be Giants*, for the editorial update.

KIRCHOFF'S LAW

Gustoff Kirchhoff was a German physicist responsible for coining the term "black body" in 1862, observing that:

> At thermal equilibrium with its surroundings, the emissivity (ε) of a body (or surface) equals its absorptivity (α). *KIRCHOFF'S LAW*

Prof. Kirchoff was among many people in the 1800s interested in producing light for the least amount of input energy. This should sound familiar, as we are doing the same thing with LED lighting 200 years later! As we shall see in the following chapter, the majority of thermodynamic science at the time was devoted to explaining the concept of temperature in materials such as solids, liquids, and gases. The mystery of the photon was still being revealed. The equation form of KIRCHOFF'S LAW is represented in Eq. (3.4).

$$\frac{E}{E_b} = \epsilon = \alpha. \tag{3.4}$$

Radiant energy emitted from the real surface is denoted as E (W/m^2), while that of a theoretical blackbody is denoted as E_B (W/m^2). The units cancel, and we are left with a dimensionless fractional value between zero and one. Note that the equation does not require the real surface to be a blackbody (again for E_b, $\varepsilon = 1$). Real surfaces behave more like "graybodies," meaning that their emissivity and absorptivity are some value less than one.

WIEN'S DISPLACEMENT

If the surface of each material has it's own temperature, then that particular surface temperature dictates the spread of photons across a spectrum, and for which wavelength the most probable energy state is in the spectral distribution. If the Inverse Square Law affects the displacement by intensity alone of the emitted spectrum seen in Figure 3.8 (a vertical shift with distance), then Wien's Displacement Law affects the displacement by wavelength and intensity (a non-linear diagonal shift with decreasing surface temperature).

Wilhelm Wien left us with an important empirical relation (later confirmed in theory by Planck) that we term WIEN'S DISPLACEMENT LAW. This important relation will provide us with expected values of the *most probable wavelengths* (effectively the most concentrated wavelengths) in the Bose-Einstein distribution of blackbody radiation.

The meaning of the Displacement Law is that a distribution of photons emitted from a surface at any temperature will have the same form or shape as a distribution of photons emitted from a surface at any other temperature. However all wavelengths will be displaced to lower or higher energies depending on the relative temperature of the compared surface. At the same distance of measurement, a cooler surface will emit wavelengths that are red-shifted (lower energy) relative to the spectrum emitted from a warmer surface. Also, the cooler surface spectrum will fit entirely within the spectrum of the warmer surface (if the two surfaces are compared at $d = 0$). The resulting summary equation is seen in Eq. (3.5), describing the most probable wavelength in the

When the temperature of a surface T_{eq} increases, the blackbody energy intensity increases across the distribution, and the distribution of wavelengths are displaced toward shorter wavelengths (higher energies), marked by a blue-shift for the most probable wavelength, λ_{max} to higher energy $E_{b,max}$.

distribution. Notice how we first present the equation in terms of $\lambda_{max}T$, which is important in the section on the Stefan-Boltzmann equation and estimating fractions of energy density within a band of wavelengths.

$$\begin{aligned} \lambda_{max}T &= 2.8978 \times 10^6 \text{ nm K}, \\ \lambda_{max} &= \frac{2.8978 \times 10^6 \text{ nm } K}{T(\text{K})}. \end{aligned} \tag{3.5}$$

As the Sun's thermal surface—the photosphere—is millions of kilometers from Earth, the entire shortwave spectrum will be displaced vertically (on a plot of irradiance per wavelength vs. wavelength) by a factor of five orders of magnitude (1×10^5).

Even from quick estimates for a 5777 K blackbody ($\lambda_{max} \sim 500$ nm) and a 300 K blackbody ($\lambda_{max} \sim 9660$ nm), we can internalize the scale of light (in bands) for radiation emitted from the sun and for radiation emitted from a body on Earth. Notice that the peak of the high temperature surface (e.g., the Sun) is within the shortwave band, while the peak of the low temperature surface (e.g., the atmosphere, or sky) is within the longwave band of energies.

PLANCK'S LAW

We have been thinking about shifts in the entire spectrum emitted by a thermal surface such as the Sun. Solar researchers would also like to focus in on particular wavelengths of light that are to be converted to useful energy, because the materials that we use as absorbers in SECSs are often selective to certain ranges of wavelengths. For example the useful photons in a PV cell are those with energy higher than the *band gap*, a threshold for pumping up electrons to excited states. Looking back, Kirchoff and others posed the following question, based on experimental observations:

> How does the intensity of radiation emitted by a blackbody relate to the individual wavelengths of the radiation (or the color of light) and the temperature of the observed body?

Karl Ernst Ludwig Marx Planck (born in Kiel, Germany in 1858) known publicly as Max Planck, and is considered the founder of quantum theory.

As a student in 1877, he studied under Kirchoff and was also immersed in the question of relating light intensity to frequency and temperature of a material. Prof. Planck was a theoretical physicist, and between 1899–1901 he developed a theoretical model of blackbody radiation based upon the empirical model of Wilhelm Wien. In his theoretical formulation, Planck was able to establish the dependence of the spectral emissive energy of a blackbody for all wavelengths of light ($E_{\lambda,b}$), given a known equilibrium *surface temperature* of the blackbody.

PLANCK'S LAW follows the following model (established for the surface a blackbody):

$$E_{\lambda,b} = \frac{C_1}{\lambda^5[e^{(C_2/\lambda T)} - 1]}, \tag{3.6}$$

where

$C_1 = 3.742 \times 10^8 \text{ W } \mu\text{m}^4/\text{m}^2,$

$C_2 = 1.4384 \times 10^4 \ \mu\text{m K}$

are derived from constants,

$c = 2.998 \times 10^{14} \ \mu\text{m s}^{-1},$

$h = 6.626 \times 10^{-34} \text{ J s},$

$k = 1.381 \times 10^{-23} \text{ J/K}.$

What we see in Eq. (3.6) is a way to calculate the spectral emissive power within the very small range of 1 μm or 1 nm. For a mathematical software, this equation could be used to plot the points of spectral emissive power per wavelength vs. wavelength, and integration under that curve would be a numerical method to find the blackbody emissive power.

In the 1920s, Indian physicist Satyendra Bose was able to establish a new method to "count" photons. His work was translated from English to German by personal request to Albert Einstein, who recognized the important value of Bose's work. The resulting statistical distribution of *boson* energy states (here, photons) is now called the *Bose-Einstein distribution*.

If we apply Planck's Law to the surface of the Sun, but then increase the distance from the surface to the exterior of the Earth (1 AU, ~150 million km), the Inverse Square Law dictates that we scale the irradiance by a factor of $\frac{1361 \text{ W/m}^2}{6.33\times10^7 \text{ W/m}^2}$, or 2.155×10^{-5}.

Spectral Emissive Energy: the energy per unit of area and per unit of time (i.e., the radiant exitance) in a unit wavelength interval (per nm or μm).

Wien's empirical model was only valid for blackbody radiation at high frequencies of light—or really hot things.

You can use Eq. (3.6) as the integrand when programming a numerical method of integrating the energy under a blackbody per wavelength. You can also use the equation for plotting radiant exitance per wavelength (W/m² μm) versus wavelength (μm) at the surface of a blackbody at temperature T (K).

Distributions of non-boson particles (called **fermions**) are either the familiar **Maxwell-Boltzmann distribution** for classical particles or the **Fermi-Dirac distribution** for particles obeying the Pauli exclusion principle, such as electrons.

STEPHAN-BOLTZMANN EQUATION AND FRACTIONS OF RADIATION

By integrating the equation for Planck's Law over all wavelengths, an analytical solution was also found for the total energy per unit area. Here, we see that the numerical solution which we might obtain by a math software can be condensed into a constant (σ) times the temperature of the surface raised to the fourth power.

There is an analytical solution to integrating wavelengths over the **Bose-Einstein distribution** of light energy. This solution is $E_b = \sigma T^4$.

Joseph Stefan deduced the relationship of total emissive power in a blackbody from the experimental measurements of Irish physicist John Tyndall. Stefan's student *Ludwig Boltzmann* derived the relationship using the theory of thermodynamics in 1884, and hence the calculus of the blackbody emissive power is known as the STEFAN-BOLTZMANN LAW.

$$E_b = \int_0^{\infty} E_{\lambda,b}\, d\lambda = \sigma T^4, \tag{3.7}$$

where

$$\sigma = 5.6697 \times 10^{-8}\ \text{W/m}^2\ \text{K}^4,$$

$$E_{0-\lambda,b} = \int_0^{\lambda} E_{\lambda,b}\, d\lambda. \tag{3.8}$$

Now, when we integrate Eq. (3.6) (Planck's equation) for a limited range of wavelengths 0-λ the result is Eq. (3.8). By dividing that function from Eq. (3.8) by the Stefan-Boltzmann solution for Eq. (3.7), we arrive at a fractional measure of the radiant energy relative to the total radiant energy emitted from the blackbody. The resulting integral range is from 0–λ. Also, the new integral in Eq. (3.10) has been translated, and the integral is now a function of λT rather than just a function of λ. NOTE: this is the same form that is used in Wien's Law for Eq. (3.5)!

By dividing Eq. (3.8) by Eq. (3.7), we arrive at a unitless fraction (from 0–1).

$$\frac{E_{0-\lambda,b}}{E_b} = \frac{E_{0-\lambda,bT}}{\sigma T^4} = f_{0-\lambda T}, \tag{3.9}$$

$$
\begin{aligned}
f_{0-\lambda T} &= \frac{E_{0-\lambda,bT}}{\sigma T^4}, \\
f_{0-\lambda T} &= \int_0^{\lambda T} \frac{C_1}{[\sigma(\lambda T)^5]} \frac{1}{[e^{(C_2/\lambda T)} - 1]} d(\lambda T).
\end{aligned} \tag{3.10}
$$

For evaluating a band of wavelengths, we adapt the equation for two values of λT.

Notice that we must multiply the fraction times the radiant exitance, here calculated by the **Stephan Boltzmann Law**.

$$
\begin{aligned}
E_{\lambda_1,bT-\lambda_2,bT} &= E_{\lambda,b} \cdot f_{0-\lambda T_2} - f_{0-\lambda T_1}, \\
E_{\lambda_1,bT-\lambda_2,bT} &= \int_{\lambda_1 T}^{\lambda_2 T} \frac{E_{\lambda,b}}{T} d\lambda T.
\end{aligned} \tag{3.11}
$$

Notice how we have to calculate the fraction from zero to the first wavelength of interest (λ_1), give a surface temperature T, as seen in Eq. (3.9). Then we must repeat the approach for the second wavelength of interest (λ_2) at the same temperature T. The difference between the two fractions (as a positive value) is the fraction of energy under the curve and within the band of wavelengths.

METHOD FOR CALCULATION: All of these equations seem great, but how are *you* going to use them to estimate fractions of energy within a spectral band of interest? Howell et al. have shown a numerical method to calculate any fraction for a known λT.[9] By coding the following into an algorithm for Scilab, Matlab, or your computational software of choice, you can numerically solve for sums of 10 terms (rather than "infinity").

[9] John R. Howell, Robert Siegel, M. Pinar Menguc. Thermal Radiation Heat Transfer, CRC Press, 5th ed., 2010.

$$
\begin{aligned}
\text{for known } \lambda T: x &= \frac{C_2}{\lambda T}, \\
f_{0-\lambda T} &= \frac{15}{\pi^4} \sum_{n=1}^{\infty} \left[\frac{e^{-nx}}{n} \left(x^3 + \frac{3x^2}{n} + \frac{6x}{n^2} + \frac{6}{n^3} \right) \right].
\end{aligned} \tag{3.12}
$$

As an application, one can consider the temperature of the Sun (5777 K) and the fraction of blackbody emissive power that is found between 0.38 μm and 0.78 μm.

$$
\lambda_1 T = 0.38\ \mu\text{m} \times 5777\ \text{K} = 2195\ \mu\text{m} \cdot \text{K}, \tag{3.13}
$$

VIGNETTE:
The use of light in plants and algae is unusual in that we are combining a *pump* with a *battery*. The actual protein complexes for photosystem I and photosystem II are the analogs of our light pumps, creating short-lived energy storage molecules (ATP and NADPH). In the same process of photosynthesis, the enzymes in the Calvin cycle use those short-lived energy sources to fix carbon dioxide and water and transform the solar pumped energy into *long-lived sugars complexes* (the batteries of plants). These long-lived sugars ultimately support our ecosystem and food web, and the sugars of the geologic past were the bases for the geofuels of today (75–300 million years ago). In actuality the efficiency of photosynthesis is less than 1–3%, but the act of cheaply storing that converted energy allows for great inefficiencies. Even more oddly, considering the time, temperature and pressures required, the efficiency of *making petroleum* can be estimated at approximately $10 \times 10^{-15}\%$, or "femto-percent" efficiency. Advice: Never get into an efficiency argument with a solar enthusiast.

$$\lambda_2 T = 0.78\ \mu\text{m} \times 5777\ \text{K} = 4506\ \mu\text{m} \cdot \text{K}. \tag{3.14}$$

By calculating the fractions for each using Eq. (3.12), or by looking them up in a table, we can find the total fraction between the two.

PROBLEMS

1. You are inside a room on June 21 at noon in Philadelphia, PA. The room is oriented along a north-south axis, with a south-facing window, and a lamp on in the northeast corner, you are standing near the west wall facing the center of the room. Sketch a diagram of all relevant emitter surfaces and receiver surfaces for shortwave radiance.

2. You are to evaluate the greenhouse effect on Earth. (a) In a cross-section sketch, draw the relevant emitter surfaces and receiver surfaces for longwave radiance. (b) What are the main sources (emitters) of the longwave radiance?

3. A laser pointer can emit 5 mW of monochromatic green (532 nm) light that is collimated and has a cross-sectional diameter of 5 mm (circular). (a) What is the irradiance normal to the beam? (b) Find the irradiance on a horizontal surface for the optical air mass 1.5 (AM1.5) and compare the laser irradiance with the value for a typical clear sky afternoon. (c) List three properties of unconcentrated solar irradiance that are different from laser irradiance to explain the difference.

4. Compose a script to calculate blackbody fractions by numerical approximation. You will use this script in the next question. (Scilab coding problem.)

5. The human body has a surface area of ~2 m^2, and the surface temperature of bare skin is 32 °C. (a) Given the emissivity of the human body is $\varepsilon = 0.97$, what is our radiant exitance (or radiance integrated across all angles) in *W/m^2*? (b) Now evaluate the fraction of graybody radiant exitance from humans that will pass through the "sky window" from 8–14 μm. (c) Present the exitance value in W/m^2.

6. The unit area intensity of radiation from the Sun at the photosphere is 6.33×10^7 W/m^2. (a) Check this value using the Stefan-Boltzmann Law, assuming the Sun is a blackbody emitter ($\varepsilon = 1$) with a surface temperature of 5777 K. (b) Using the Inverse Square Law and the known solar constant given for extraterrestrial Earth (1361 W/m^2), what are the solar constants calculated for the surfaces of Venus and Mars?

7. (a) Using the value for the photosphere surface temperature what fraction of extraterrestrial irradiance is found for $\lambda < 780$ nm? (b) What fraction is found for $\lambda < 3000$ nm? (c) What fraction is found between 780 nm $< \lambda < 3000$ nm (the IR sub-band of extraterrestrial shortwave irradiance)?

8. Compose a script to plot the Planck spectrum (Bose-Einstein Distributions) of the Sun as a blackbody at the photosphere, and provide the integrated irradiance (area under the curve). Compare your results to value offered in the preceding problem. (Computational software coding problem) [Your plot may have bounds from 200 nm to 50,000 nm].

9. Re-plot the spectrum of the Sun according to the Inverse Square Law, scaled to represent the extraterrestrial irradiance at the Earth's surface. (a) Integrate under the curve for the expected value of irradiance. (b) Explain why you can decrease the bounds significantly. (c) Does the most probable wavelength (found by Wien's Displacement Law) shift with the scaling? (Computational software problem) [Your plot may have bounds from 200 nm to 3000 nm].

10. Plot the spectrum of the atmosphere (the radiant exitance or glow), given an effective isotropic average temperature of 255 K. Where does the most probable wavelength seem to occur (by observation)? (a) Explain why the bounds of plotting are significantly different from those of incoming shortwave solar irradiance. (Computational software coding problem) [Your plot may have bounds from 2500 nm to 50,000 nm] Table 3.1.

RECOMMENDED ADDITIONAL READING

- Light and Color in Nature and Art.[10]
- An Introduction to Solar Radiation.[11]
- Light's Labour's Lost: Policies for Energy-efficient Lighting in support of the G8 Plan of Action.[12]
- Optics.[13]

[10] Samuel J. Williamson, Herman Z. Cummins, *Light and Color in Nature and Art,* John Wiley & Sons, 1983.

[11] Muhammad Iqbal, *An Introduction to Solar Radiation,* Academic Press, 1983.

[12] Paul Waide, Satoshi Tanishima, Light's Labour's Lost: Policies for Energy-efficient Lighting in support of the G8 Plan of Action, International Energy Agency, Paris, France, 2006 (Organization for Economic Co-Operation and Development & the International Energy Agency).

[13] Eugene Hecht, *Optics,* Addison–Wesley, 4th ed., 2001.

04 PHYSICS OF LIGHT, HEAT, WORK, AND PHOTOCONVERSION

You keep using that word. I do not think it means what you think it means.

Inigo Montoya, from The Princess Bride, Directed by Rob Reiner, Written by William Goldman (1987)

WE FIND THAT some of the most interesting obstacles are encountered on the path to *un-learning* your modern cultural knowledge of energy. For example, **Energy** and **Work** are not the same thing as **Temperature**. Also, **Heat** is not "hot" in the world of modern thermodynamics. Yet energy, temperature, and heat are frequently used together as proxies for one another, such that we assume that one requires "heat" for a "hotter" engine, for more "energy" or "power," to do more "work." This is an incorrect application of thermodynamic concepts, and tends to remove the important other transfer of energy in heat: *radiative transfer* (see Figure 4.1).

Temperature is just a characteristic measure,[1] a single value representing a great big swarm of atoms wiggling in space with *kinetic energy*. And *heat is energy transfer by conduction or radiative transfer*, not a hot body.[2] We will see that there are two other great big swarms of energy particles with kinetic and potential energy, and they also have characteristic measures that are analogous to the concept of temperature.

The kinetic energy found as the *thermal energy* of a swarm of wiggling, zipping atoms, or molecules in space can be used to apply pressure to a surface,

For the moment, ask yourself: *what is work? what is heat? what is useful energy?* Some answers can be explored in great detail at the **US DoE Energy Information Administration's** outreach site: Energy Explained.

Heat is *energy transfer* from a body of higher energy to a body of lower energy due to physical contact (meaning *conduction*, not convection) or *radiative transfer*.

Work is just energy applied to doing something we find *useful*. Call me wacky, but daylight and warm sunny days are fairly useful.

Temperature is a summary statistic of the most probable energy found in a distribution of states.

[1] A **characteristic measure** is a physical representation, or a statistical summary of the most probable energy states.

[2] Note well that heat is not convection. **Convection** is an emergent property of conduction and fluid flow.

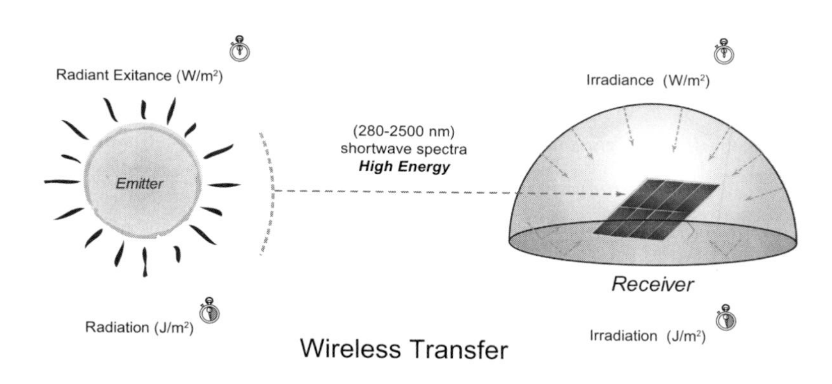

Figure 4.1: The Sun-collector system is an *open thermodynamic system* emitting radiant energy in this case to an optoelectronic and optocaloric receiver—a photovoltaic panel within the sky dome.

The swarms of energy particles are called **energy distributions**.

imparting *mechanical energy* to the fin of a turbine, and that *mechanical energy* can be used to induce a flow of electrons via a copper coil spinning inside of a magnetic field, imparting *electrical energy* inside a generator, and passed on to the next energy conversion device; all in the successive *conversions* from one *form* of energy to the next.[3]

[3] A great big tank or **stock** of any **form** of *potential energy* is called a **source** of energy. As society, we identify our major *sources* of energy as **resources**. The conversion of an initial *form* of energy to a different *form* of energy is a **transformation**. Get it?

Now, imagine that we might have multiple *forms* of **potential energy**: gravitational potential energy, electrochemical potential energy, mechanical potential energy, and nuclear potential energy. We also have multiple forms of **kinetic energy**: motion energy (big things moving, like a wind turbine blade), thermal energy (smaller things moving, like vibrating atoms and molecules), or electrical/electronic energy (even smaller things moving, like electrons and "holes").

BREAKTHROUGH: NOT LIGHT OR HEAT—BOTH

What if we had a form of energy that is both ***potential*** in form (i.e., stored) *and* ***kinetic*** in form (i.e., in motion)? We do! It is called ***radiant energy***. The electromagnetic waves of radiant energy are in constant motion (kinetic), taking all paths and traveling in transverse waves. However, at the same time, the swarm of particles of light of radiant energy (called *photons*), each embodied with a *quantum* of potential energy that is proportional to the frequency (the inverse of the wavelength) of the wavelength from the photon. Both forms in one little package. You start to see where the uncertainty begins in interpreting radiant energy for design of useful energy conversion…[4]

You're melting my mind man. What are you, some kind of Robot Monkey?

The *photon* is the fundamental particle that we deal with. A photon has no rest mass,[5] yet it does have momentum to impart *kinetic energy* that is inversely proportional to its wavelength. A photon also has *potential energy* that is inversely proportional to its wavelength. A photon is neither "hot" nor "cold," but it can exchange energy with matter to induce thermal changes, electronic changes, and other radiant energy changes among states of matter. The photon is our quantum form of energy (the basic unit of trade for energy conversion), like an atom of uranium, or a carbon-hydrogen chemical bond, or a single electron.

[4] Check out the Heisenberg Uncertainty Principle as applied to radiant energy for years of circumspection.

[5] Not even a *very little bit* of mass.

Given the concept of a steady state *pumping* of matter, using a light source can induce three responses:

1. Electronic response (semiconductors: optoelectronic) [electric response in metals].
2. Caloric response (thermal vibrations: optocaloric).
3. Electrochemical response (photosynthesis: optochemical or photochemical).

Another way of saying particles is **quanta**, as in quantum physics.

IT'S IN THE HISTORY BOOKS

How did we miss this important unique dual property of radiant energy, this key linkage between light and heat and thermodynamics? Perhaps if we go back to the historical study of energy transfer we can observe a split. The derived technologies from cultural developments where chemical fuel resources were abundant (like wood, coal, gas, and oil in western Europe) demonstrate a distinct separation of energy conceived as *heat* or *caloric*[6] and *light*.[7]

Your physics professors should be rolling their eyes at this statement and clutching their chests for the big one, because we are describing the transformation of modern science into quantum mechanics. Only about as important as the discovery of **fire** or **beer**.

Historically, *thermodynamics* developed out of need to increase the efficiency of early steam engines (a thermal to mechanical energy transformation). The European development of thermodynamics (that we find in our history books and in Wikipedia) included the study of applied energy conversion into divisions above and below a certain threshold. Energy

[6] **Heat**: originally regarded as a transfer of thermal energy from a hot body to a cooler body in the *19th century*.

[7] **Light**: originally referring to the visible, "color" portion of the electromagnetic spectrum in the *17th century*.

transformed above the threshold was considered *work* and energy below the threshold was considered *rejected heat*. In the *17th century*, the concept of work was confined to include only the ability to deliver useful *mechanical energy*. A resulting useful concept of energy transformation (conversion from one form of energy to another to yield both work and rejected energy) is that of the *heat engine*.

Work is just energy applied to doing something we find *useful*. Similarly **rejected heat** is simply energy below a certain threshold, which cannot be applied in the conceived useful application for the system at hand. Which means, in terms of design, we are mostly limited by our imaginations as to what we consider to be *useful applications* of energy.

In contrast, the science of *optics* and *photonics* gradually grew out of the life works of René Descartes, Isaac Newton, Robert Hooke, and Christiaan Huygens in the *17th century*, and Augustin-Jean Fresnel, Thomas Young, Michael Faraday, and James Clerk Maxwell in the *19th century*. But all of this work predates the quantum nature of light. Even the old energy-hungry incandescent light bulb was a remnant of *19th century* technology (invented over 130 years ago in 1879). It was only in the last century (the *20th century*) that we began to codify the field of *photonics* to include the *generation, emission, transmission, modulation, amplification, detection, and sensing of light*. Photonics included energy manipulation with semiconductor materials (early 1960s), optical fibers (developed in the 1970s), and complex thin films and surfaces that interact with different wavelengths selectively (1980s). However, the term photonics was largely applied to *information transfer* (communications) processes. Hence, light detectors were interpreted as technologies for signal capture, not power conversion.

These conceptual separations of thermal to mechanical energy conversion for work (as heat transfer) and radiant to electronic energy conversion for information transfer (as electromagnetic transfer) were based upon two historical lines of technology development (thermal-based steam engines and information-based optics and semiconductors, respectively). Over 200 years after working on steam engines, in the *21st century*, we are ready for a systems convergence of *light* and *heat*. The challenge with vintage ideas from *thermodynamics*, *heat engines*, *heat* and *light*, *optoelectronics*, and *photonics* has been to adapt the core meaning, the language of each concept to modern technologies that define new useful forms of energy: electrical power for computers and solid state lighting, or thermally warm air and water for comfort and sanitation in a home. We will see that the

Think about the following and ask if you find them **useful**: daylight, warm air, hot water, a microwave, a laser pointer, a sun tan, plants, rain, the seasons, wind. These are all products of work derived from the resource of the photon, radiant energy, and some are even from the Sun.

materials that we develop for energy conversion can be used for both work and information transfer. It all depends on the creativity of the systems designer.

LIGHT FOR A CALORIC RESPONSE

In many heat transfer textbooks *radiative transfer* is considered an essential part of the education, but at the same time *radiative transfer* is often taught as a mechanism of *energy loss*, not as a mechanism of *energy gain*. From the standpoint of a combustion engine, the radiation emitted is diffuse and insufficient component for performing work, as the source of the radiation was assumed to be of low temperature. However, from the standpoint of the Sun, the source of radiation is more than sufficient to be a source of energy gain to sustain life, generate wind and rains, and power our modern society. We have already introduced a few modern and ancient technologies demonstrating the solar heating of fluids and solids as well vetted and available for study.

The field of solar energy conversion is significantly broad (spanning many fields) that a general nomenclature is not well defined for technologies that absorb light, leading to *energy gains* as a sensible thermal increase in a substance (the measured temperature goes up in the substance) or a latent heat increase in a substance (the internal energy increases, but the measured temperature does not). Changes in temperature or latent heat in this sense can be attributed to *radiative transfer*, one of the most difficult advanced topics in heat transfer. It is our task to specify the portion of radiative transfer that can be made useful to the broad solar conversion systems audience.

Recall that light (or EM radiation) needs both an *emitter*—a source of light—and a *receiver*, or the collection device that absorbs photons to make use of the energy from the light source. Light is emitted or reflected from a surface or light is incident and transmitted or absorbed by a surface and must be diagrammed relative to an emitting/reflecting surface as well as a receiving surface. This relationship of *emitter-receiver* confirms a very important detail in solar energy conversion *EM radiation is directional.* In the case of the Sun-Earth, the emitter in the *shortwave band* is the Sun (where $T_{eq} \sim 5777$ K for the photosphere of

Yes, reflected light is a **source** of light from the perspective of the receiving surface. Once again, we need to remind ourselves that it is the directional nature that is important if we wish to make use of all sources from the sky dome and surface **albedo**.

the Sun), while the emitter in the *longwave band* for us on Earth can be the atmosphere and objects on the Earth surface (where $T_{eq} \sim 180 - 330$ K, in general).

In comparison from Wikipedia (Retrieved Jan. 27 2013): "**Optoelectronics** is the study and application of electronic devices that source, detect and control light, usually considered a sub-field of photonics. In this context, light often includes invisible forms of radiation such as gamma rays, X-rays, ultraviolet and infrared, in addition to visible light. Optoelectronic devices are electrical-to-optical or optical-to-electrical transducers, or instruments that use such devices in their operation."

Here, we present a working term broadly describing the conversion of light to thermal energy forms. As a frame of reference, we begin with the term *calorimetry*, the science of measuring the heat of physical changes (sensible heat), chemical reactions (latent heat), as well as heat capacity (specific heat). The caloric response in a material can be derived from optical methods, as we observe from the Sun heating the Earth, or a microwave heating a cup of soup. Next, the study of light and the interaction of light and matter is termed *optics*. We seek a common term to refer to the coupling of optics with caloric responses. Hence, we arrive at a logical conjugation of opto- and calor- to derive the science of *optocaloric* technologies. This term is application-based, as is optoelectronics, and should be explored for extended meaning. For example, one could use the term for solar cooling, not just for solar heating.

"Optoelectronics" implies light emission just as "radiant heating" implies light emission from the standard concept. Analogously, "optocalorics" is the study and application of devices that source, detect, and control light related to thermal behaviors. In this context, *light* includes visible and invisible forms of radiation and refers to direct thermal excitation of matter along with successive emissive radiative states that generate a caloric response.[8]

[8] In this text, we have chosen to use opto- rather than photo- due to the precedent of optoelectronics (which encompasses photovoltaics).

As an example of an optocaloric system, we will describe the Earth's atmosphere and surface at the end of this text as a *cover-absorber* system. The nature of the greenhouse effect relies on an optocaloric response of both the absorbing surface (something that we will call an opaque *Type II Graybody*) and the atmosphere which is a *Type I Selective Surface* (transparent to some wavelengths, absorbing others).

The atmosphere is "selective" in that it responds differently to various wavelengths. The atmosphere possesses a high transparency and moderately low reflectance in the shortwave band, while being relatively opaque and absorptive (high emittance, low reflectance) to longwave band wavelengths.[9] The atmosphere is also selectively transparent to a subset of wavelengths in the

[9] Admittedly, this description for the atmosphere is an oversimplification. However, it is relatively useful for our purposes in the design of SECS.

longwave band between about 8 and 13 μm (8000 and 13,000 nm). This is our *sky window*, from which we leak heat out into space, and cool things off.[10]

[10] See G. B. Smith and C.-G. S. Granqvist's superb textbook: *Green Nanotechnology: Solutions for Sustainability and Energy in the Built Environment.*

MATERIALS IN ENERGY CONVERSION SYSTEMS

A goal of integrative design and communication for new design in Solar Energy Conversion Systems is to *be adaptable to the system basis and the directionality of all electromagnetic radiation interacting with materials*. For the most part, we will typically prefer to review optics relative to the incident (absorbing, refracting, or reflecting) surface. Photons are to be accepted or transmitted to an appropriate absorbing surface—the absorber material. Optics can help us to manage photons, or even wrangle photons into confined pen-like photonic cowboys. *Whoop!*

Coupling of the work current from light to any SECS requires useful materials. In order to make use of the steady state open system that solar energy provides, we must be aware of the materials that permit technology development. All SECS technologies use physical materials to interact with electromagnetic radiation, and this is the definition of *optics*. Solar energy conversion systems can then be studied from two perspectives: (A) the materials interacting with light for electronic, electrochemical, and thermal change/storage processes, and (B) the bands of light selectively interacting with the materials (e.g., shortwave/longwave selective surfaces).

We shall see that materials have selectivity across ranges of wavelengths (bands), in terms of emittance, transmittance, reflectance, and absorptance. The selectivity in materials will allow us to create systems of materials that amplify our goals of a high utility solar energy conversion technology, and should push our design teams to think creatively about developing new technologies in the future. We shall also see that some materials *lack* the spectral selectivity that we wish to achieve in a device, as in the example of a photovoltaic device that absorbs low energy photons to warm the material (decreasing device efficiency) in addition to absorbing band gap energy photons that yield electronic states for power generation. Rather than seeing this as a downfall, we should look to larger systems-based solutions such as managing the microclimate about the integrated PV system to reduce thermal loads during the day. Again, the broader transdisciplinary integrative design team can create solutions that transcend individual materials constraints by expanding the boundary conditions for a system and ensuring a sustainable energy solution that also yields high solar utility to the client or stakeholders in the selected locale of deployment.

Why is it nobody has any sense for a microwave system as an emitter-absorber, but we use them every day? The absorbing "surfaces" are molecules of water and fat within our food (and Melmac dishes; a melamine resin). A microwave oven uses what is called a magnetron to produce collimated, concentrated microwave wavelength ($\lambda \sim 120$ mm) photons that cause absorbing molecules to wiggle and rotate. In this case, it is about the flux of photons, not the energy packed into the wavelengths.

SELECTIVE SURFACES AND GRAYBODIES: OPAQUE

Up until this point, we have been considering energy accounting for blackbodies, where the surfaces do not reflect any light in their total budget, they only absorb and emit radiant energy. We now shift to GRAYBODY ENERGY ACCOUNTING, where we allow surfaces to reflect light.

Part of collector design includes accounting for the capability of the receiving surfaces to lose energy via longwave radiation, or to gain energy from the solar shortwave band of solar light. With respect to design of incident surfaces, we consider the value of either reducing or enhancing thermal losses (depending on the SECS of interest). Again, all terrestrial surfaces will "glow" as emittance from a blackbody (or graybody) with a distribution of photon energies. The distribution will be red-shifted (as described by Wien's Displacement) to have a high probable wavelength (peaking) in the longwave band, not in the shortwave band.

For the purpose of understanding devices powered by the great fusion emitter that is the Sun, our emphasis has thus far been on the energetic *gains* from solar shortwave band light. To develop your solar energy conversion systems knowledge base, we will shift your intuitive perspective away from the source of light (the Sun will get enough emphasis later) and instead assess the materials properties of the surface receiving light (transmitting, absorbing, or reflecting).

For fields influenced by performance of combustion cycles and thermal heat transfer, the default student perspective for radiative transfer has been that of infrared light relative to the emitting surface (referred to in heat transfer texts as *thermal radiation*).[11] Hence we have a conditioned, yet inaccurate, tradition for thermal systems: "hot" materials emit longwave radiation and "cold" materials absorb shortwave radiation.[12] A similar conditioned view can emerge from studies in semiconductor systems, where optoelectronics conveys the default perspective of "cold light" relative to the emitting surface. In the SECS field we find both of these default perspectives to be insufficient and often inaccurate.

[11] Actually, the Sun is also thermal radiation coming from a much much hotter surface.

[12] Can you identify the flaw of logical omission in this statement? Reflect upon the second law of thermodynamics...

We begin by describing the fractional radiative energy balance for a general surface that has some degree of transparency. The following equation is the radiation balance for semitransparent conditions ($\tau \neq 0$):

$$\rho_\lambda + \alpha_\lambda + \tau_\lambda = 1. \tag{4.1}$$

Next, we see that Eq. (4.2) demonstrates *Kirchoff's Law of Radiation*. For any surface at thermal equilibrium, the spectral *emittance* is the same fraction as the spectral *absorptance*.

$$\alpha_\lambda = \epsilon_\lambda : \text{ For a given } T_{eq}. \tag{4.2}$$

In order to make this a slightly more interesting comparison, let us compare two graybody plates that are each emitting a majority of longwave light toward each other—assume the two surfaces have surface temperatures T_1 and $T_2 \sim 250-350\,\text{K}$. Because they are parallel the light is not lost. We can express this relation in terms of two surfaces, as diagrammed in Figure 4.2. Now, if $\epsilon_1 \neq \epsilon_2$ and/or if the temperature of surface 1 is not equal to that of surface 2 ($T_1 \neq T_2$), there will be a net energy transfer (heat). That is to say, one material will experience a net increase in energy, typically leading to a warmer body.

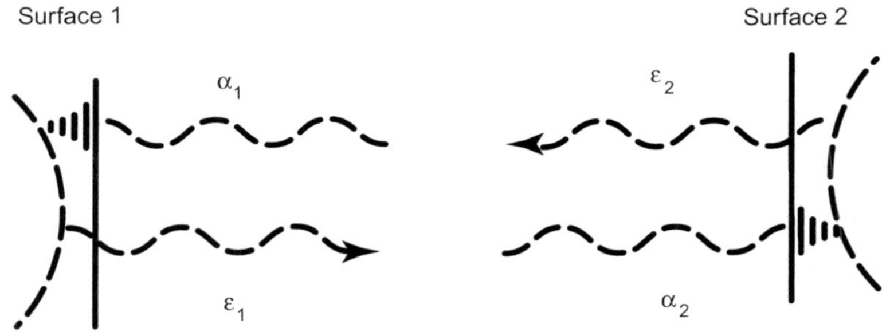

Surface 1 Surface 2
α_1 ϵ_1 ϵ_2 α_2

Figure 4.2: Diagram of longwave exchange of radiant energy between two flat plate surfaces parallel to each other. The two surfaces are at T_1 and T_2 near 300 K, while $T_1 \neq T_2$.

By definition, a blackbody material has a reflectance of zero ($\rho = 0$), as reflected light means that the light is not being absorbed. A blackbody is also opaque, and thus the transmittance would be zero as well ($\tau \rightarrow 0$).

For the following two equations, we see that each ϵ and α is tied to a particular surface, or a particular material with surface temperature T_{eq}, in accordance with Kirchoff's Law. From another perspective, if the emittance and absorptance for each surface are equivalent under thermal equilibrium, then the ratio of emittance to absorptance shows to be equal to one for both surfaces.

$$\begin{aligned} \epsilon_1 &= \alpha_1 \\ \epsilon_2 &= \alpha_2 \end{aligned} \tag{4.3}$$

or

$$\frac{\epsilon_1}{\alpha_1} = \frac{\epsilon_2}{\alpha_2} = 1. \tag{4.4}$$

In an opaque material, we notice that $\tau \longrightarrow 0$ (or nearly so), leaving an energy balance between ρ (reflectance) and *either* α (absorptance) or ϵ (emittance). Remember that each of these factors (emittance and absorptance) are unitless fractional values from 0 to 1.

We can also experience this radiative exchange using the example of the Sun ($\sim$6000 K, $\epsilon \sim 1$) and a surfer-sunbather resting on the beach of Earth ($\sim$300 K, $\epsilon \sim 0.97$). The surfer is interacting with the Sun at a distance (only 150 million km). The net excess radiant energy from the Sun (shortwave) is absorbed and transformed into thermal energy on the surface of the surfer's skin, as long as she remains outside during the sunny day. At the same time, the surfer is also a source, emitting longwave energy to her surroundings. Were she to walk inside a closed building or wait until nighttime (when the Earth surface faces away from the Sun), she would remain emitting longwave light with 97% efficiency.

For opaque materials that are of substantial thickness (like a concrete wall or aluminum sheet metal), the pragmatic approach to understanding its material optical properties is to assess the *reflectance* of the surface.[13] We know that a material can respond in terms of reflectance, absorptance (or emittance), and transmittance for each wavelength. For convenience in solar design and engineering we simplify the optical materials analyses to relevant values across shortwave (<3 μm) and longwave bands (>3 μm), thus breaking our study into incident and exitant directionality of light for solar energy conversion systems design.

[13] All practical paths to understanding opaque materials use *reflectivity*.

$$1 = \tau_\lambda + \alpha_\lambda + \rho_\lambda. \tag{4.5}$$

Combining Eqs. (4.2) and (4.5), we see that the α_λ term is equivalent to the ϵ_λ term, such that a material with a low absorptance will also be a material with low emittance, and vice versa. Also, because we have declared that this class of materials is opaque, the transmittance of the materials under consideration also goes to zero in Eq. (4.5). That simply leaves us with the ability to characterize materials based on the reflectance of the surface over the spectrum of interest, both shortwave and longwave bands. Rearranging Eq. (4.5), and setting $\tau_\lambda = 0$, we bring about the following relation:

$$\rho_\lambda = 1 - \alpha_\lambda = 1 - \epsilon_\lambda. \tag{4.6}$$

When our materials are opaque to the shortwave band and much of the longwave band, we can establish a raw convention for classifying four general optical materials. Again, in this case, the types of materials have transmittance of near zero (opaque) for both shortwave and longwave bands. From our knowledge of Kirchoff's Law Eq. (4.2), a surface at temperature that is at steady state will absorb light equally as well as it emits light. On top of that, we add the concept of an energy balance for a given surface, that is the sum of fractional optical contributions (coefficients from 0 to 1) for a specific wavelength from reflectance, absorptance (or emittance), and transmittance shall equal to one Eq. (4.5).

RADIATIVE EXCHANGE FOR ONE SURFACE

In assessing radiative transfer between surfaces, we want to use a common currency of energy units (e.g., Joules or eV), but we need to exchange among units of temperature (K) from the emitting surfaces and wavelengths (λ) from the emergent or absorbed photons. An electron volt (eV) is a common currency in optics for a unit of energy, and we will use it extensively in this text as an alternate to the Joule. The energetic quantity of an eV is defined as the amount of energy gained by moving a single free electron across an electric potential difference of one volt. A *volt* is a potential of 1 Joule per Coulomb (1 J/C) and the charge on a free electron is ($1\,e$) or 1.60×10^{-19} C, such that $1\,\text{eV} \equiv 1.60 \times 10^{-19}$ J. To convert from wavelength (λ(nm)) to energy (eV) we use Eq. (4.7) (just count to five):

$$\frac{h \cdot c}{\lambda(\text{nm})} \frac{1239.8\ \text{eV} \cdot \text{nm}}{\lambda(\text{nm})} \approx \frac{1234.5\ \text{eV} \cdot \text{nm}}{\lambda(\text{nm})}. \tag{4.7}$$

Entropy is a measure of the *dispersal of energy* (including dispersal of matter and light states) at a specific *temperature*.

All matter has a non-zero temperature, describing the distribution of heat per change in entropy of the system. As detailed in Appendix A, the change in total entropy (dS) of the system/surroundings is:

$$dS \geq \frac{1}{T} dQ, \tag{4.8}$$

where Q is the traditional symbol for heat. Temperature is the conjugate of heat (related by the *heat capacity*), and is also a measure of the average kinetic energy of particles. In this broad sense, temperature measurement is essentially establishing a summary statistic for the most probably energy states in a Maxwell-Boltzmann distribution of energetic states within a material.

Analogously, λ_{max} could be described as a summary statistic for the most probably energy states in a Bose-Einstein distribution. We have seen that both Planck's Law and the Stefan-Boltzmann Law allow us to convert thermal states described by degrees in Kelvin into summary values of energy (as Joules). In each case of wavelength conversion or temperature conversion, we are interested in accessing summary values of energy or power to simplify our accounting for net energy flux across surfaces. Heat transfer texts focus on *thermal radiation* where matter and radiation are in a state of local equilibrium.[14] The Sun is our main source of light for SECSs. That light is primarily a thermal radiation source from the photosphere, hotter than most things on Earth, right?

"Heat," as described in thermodynamics, is defined as a transfer of energy across a systems boundary due to a difference in temperature, or a gradient in temperature.[15] Thus, *heat is energy transfer*, while *thermal energy* is a form of energy due to the vibrations and movement of atoms and molecules or solid lattices. Thermal energy can be transferred across a surface or boundary, and by practical understanding we might think of that energy as being transferred

[14] There are sources of non-equilibrium radiation from synchrotron emission, stimulated emission (lasers), and Compton scattering.

There are additional non-thermal sources of photons from the chromosphere (the corona) that make the blackbody approximate diverge for the photosphere alone.

[15] If you have studied *heat transfer*, then you have heard that the very phrase is redundant because heat *means* energy transfer.

by advection, or the bulk movement of a thermal carrier fluid across space. And yet, thermodynamically, there are only two mechanisms for transfer of thermal energy: by *conduction* or by *radiation*.[16]

We make this distinction because from a solar energy perspective our primary *form* of energy that we monitor and account for is *radiant energy*, which is *transformed* into *thermal energy* via the optical absorption process. First, we can establish that there is an obvious temperature gradient between the Sun (~6000 K) and the Earth (~300 K). Due to the near-vacuum of space, the form of energy transfer (wirelessly) from Sun to Earth (and from Earth to Sun) is radiation transfer. Radiative transfer with a net energy flux in a specific direction (e.g., $-Q$ from Sun to Earth) is due to an imbalance between the radiant exitance (emitted, W/m^2) and irradiance absorbed (also W/m^2) by both surfaces. The net radiation exchange for a block of time can be conveyed as Q (in units of J) or q (in units of J/m^2). Irradiance reflects the rate of energy transfer, and we use the convention of $\dot{q}$ (units of W/m^2) to match.

We have familiarized ourselves with BLACKBODY ENERGY ACCOUNTING in the previous chapter, or accounting for the energy balance in blackbody surfaces. Now we will explore GRAYBODY ACCOUNTING, where we include more realistic surfaces that can reflect light in addition to emitting and absorbing (or transmitting). Radiative transfer with a net energy flux in a specific direction must be accounted for, because the solar designer may make use of any imbalance to perform work. For a simplified yet useful approach to exchange between opaque surfaces, we first approximate our surface as a "diffuse gray surface" with uniform thermal characteristics and uniform incident light across the surface.[17]

[16] Gregory Nellis and Sanford Klein. *Heat Transfer*. Cambridge University Press, 2009.

[17] John A. Duffie and William A. Beckman. *Solar Engineering of Thermal Processes*. John Wiley & Sons, Inc., 3rd edition, 2006; and Gregory Nellis and Sanford Klein. *Heat Transfer*. Cambridge University Press, 2009.

DIFFUSE GRAY SURFACE APPROXIMATION:

- Assume the surface emits in a diffuse manner by having emittance that is independent of direction (θ_i or γ).

- Assume the surface is "gray" if it has radiant properties independent of wavelength (λ), where $0<\epsilon<1$ (between white and black).

- Assume the surface temperature is uniform across the area assessed.
- Assume the irradiance accepted by the surface is also uniform.

From the perspective of an opaque emitting material like the Sun or our skin, the *emittance* is the glow of the surface. From the complementary perspective of the same opaque material exposed to incident light, the surface can have specular or diffuse *reflectance*. A specular reflectance is typically of a mirror (angle of incidence equal to angle of reflection), while a diffuse reflectance is typical of a white paint or chalk (reflection diffused in all angles). Accordingly reflectance can be a mixture of the two, which is how we create wall paints with flat, egg-shell, satin, or gloss finishes (from most diffuse to most specular).

The **radiosity** (J_i) is the outgoing radiant energy balance for a single surface.

Moving onto GRAYBODY ACCOUNTING for both surfaces, we need to account for both the light emission from a surface and the light reflected by a surface, called the *radiosity* of the surface. Radiosity (J_i) is the sum of emitted and reflected light for a surface i.[18]

$$J_i = \rho_i G_i + \epsilon_i E_{b,i}. \tag{4.9}$$

[18] Gregory Nellis and Sanford Klein. *Heat Transfer*. Cambridge University Press, 2009.

Recall that G refers to the global solar irradiance and E_b is a more general radiant exitance from a blackbody.

As shown earlier for opaque ($\tau \to 0$) gray surfaces the reflectance (ρ) and absorptance (α) are related as $\rho_i = 1 - (\alpha_i)$. And when we impose Kirchoff's Law according to which the absorptance and emittance must be equivalent, yielding $\rho_i = (1 - \epsilon_i)$.

Hence, Eq. (4.9) for the radiosity of a single surface can be posed in three alternate ways:

$$\begin{aligned} J_i &= \rho_i G_i + (1 - \rho_i) E_{b,i} \\ &= \text{reflected solar} + \text{emitted} \\ &= (1 - \epsilon_i) G_i + \epsilon_i E_{b,i} \\ &= (1 - \alpha_i) G_i + \alpha_i E_{b,i}. \end{aligned} \tag{4.10}$$

The net rate of radiant transfer ($\dot{Q}$) from a single surface i is therefore calculated by a balance between radiosity (J_i in W/m^2) of surface i and the irradiance (G_i in W/m^2) upon surface i (Eq. (4.11)) (see Figure 4.3).

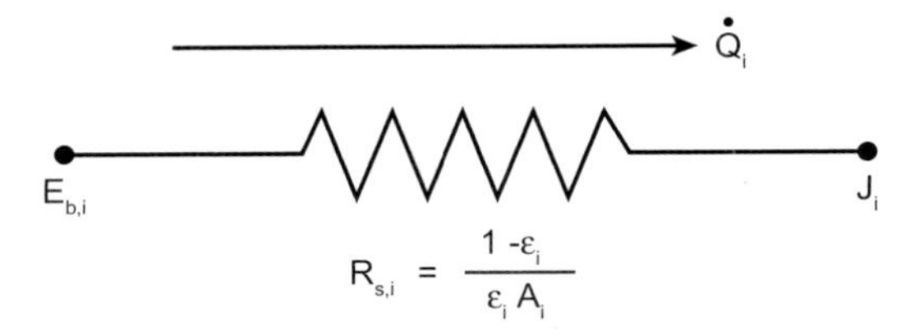

Figure 4.3: Generic thermal resistance diagram for radiative transfer. Here, $\dot{Q}_i$ refers to net rate of energy transfer between surfaces, $E_{b,i}$ is the blackbody energy emitted from surface i, $R_{s,i}$ is the surface resistance, while J_i is the final radiosity of surface i. A_i is the area, and ϵ_i is the emittance of surface i.

$$\dot{Q}_i = A_i J_i - A_i G_i. \tag{4.11}$$

It is important complementary information to recall that the irradiance for the solar photons will be largely **shortwave** in nature, while the emitted photons from Earth will all be **longwave** in nature. We may use this distinction for selective energy exchange between surfaces in SECS design.

If we temporarily assume that the surface absorbs no net energy ($\dot{Q}_i = 0$, adiabatic system), then we can rearrange Eq. (4.10) such that:

$$G_i = \frac{J_i - \epsilon_i E_{b,i}}{1 - \epsilon} \tag{4.12}$$

and insert Eq. (4.12) back into Eq. (4.10) to yield:

$$\dot{Q}_i = A_i \left(J_i - \frac{J_i - \epsilon_i E_{b,i}}{1 - \epsilon} \right) \tag{4.13}$$

$$\begin{aligned} &= A_i \left(J_i \frac{1 - \epsilon_i}{1 - \epsilon_i} - \frac{J_i - \epsilon_i E_{b,i}}{1 - \epsilon} \right) \\ &= A_i \left(\frac{-\epsilon_i J_i + \epsilon_i E_{b,i}}{1 - \epsilon_i} \right) \\ &= \left(\frac{\epsilon_i A_i}{1 - \epsilon_i} \right) (E_{b,i} - J_i) \end{aligned} \tag{4.14}$$

$$= \frac{E_{b,i} - J_i}{\left(\frac{1-\epsilon_i}{\epsilon_i A_i} \right)} = \frac{E_{b,i} - J_i}{R_{s,i}}. \tag{4.15}$$

What we have just found is that radiant transfer has a driving force, which is the blackbody emissive power minus the radiosity. The associated resistance to radiant exchange, seen in Figure 4.4, is the quantity $R_{s,i}$ (units of m^{-2}) called the surface resistance.[19] If the surface is black, then $\epsilon_i \longrightarrow 1$ and there is no surface

[19] Gregory Nellis and Sanford Klein. *Heat Transfer*. Cambridge University Press, 2009.

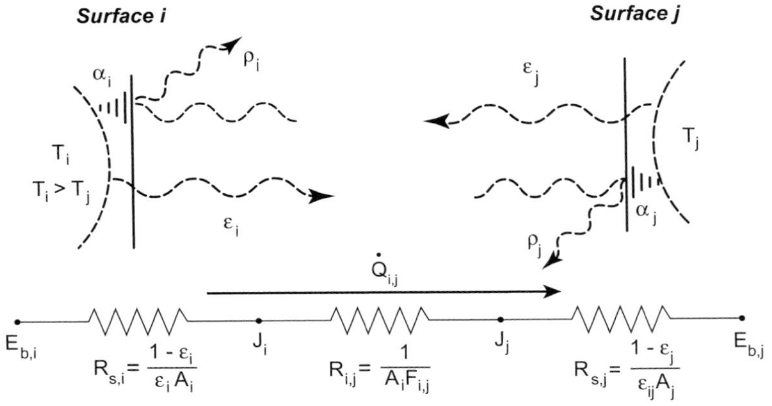

Figure 4.4: Radiative exchange between two surfaces, including thermal resistance diagram associated with the denominator of Eq. (4.18). Note how the two surfaces are of different temperatures, driving a non-zero net radiative exchange, $\dot{Q}_{i \rightarrow j}$.

Recall that true graybodies have a power output W that is proportional to the fraction of surface *emittance.*

$$W = \epsilon \cdot \sigma \cdot A \cdot T^4.$$

[20] So, shiny surfaces will not get very hot themselves from radiant transfer and can be used for SECS as light concentrating parabolic mirrors! Also, metallic lightweight space blankets will keep you warm by reflecting IR light back into your body.

resistance. If the surface is highly reflective (shiny metal), then $\epsilon_i \longrightarrow 0$ and the surface resistance approaches an infinitely large value. As such the surface will reflect all irradiance, and the surface is not in communication with its surroundings via radiative transfer.[20]

Relative to the thermodynamic equilibrium for the environmental normal conditions at Earth's surface and in Earth's atmosphere, we can describe classes of four opaque solar optics. The following table describes four very generic selective surfaces for light manipulation in the shortwave and longwave bands. The list is important in that the classes have very real opaque materials associated with them, which can be helpful in the first assessment of available materials for SECS design should a system not be available off the shelf.

Type I Graybody Low Shortwave absorptance, Low Longwave emittance: characteristic materials are pure Al, Ag, Ni, and Au (termed *specular reflectors*)

Type II Graybody High Shortwave absorptance, High Longwave emittance: characteristic materials are Parson's Black, carbon black, lampblack (essentially varieties of *soot*)

Type I Selective Surface High Shortwave absorptance, Low Longwave emittance: characteristic materials are Black chrome, Copper Black, Nickel Black (all *metal sub-oxides*), and hydrated plants

Type II Selective Surface Low Shortwave absorptance, High Longwave emittance: characteristic materials are Anodized Al, MgO, ZnO (very effective *diffuse reflectors* for cool roofs)

RADIATIVE EXCHANGE FOR TWO SURFACES

In the normal case of radiant exchange among two or more surfaces, we need some more information on how each surface is *communicating* with the other. To do this we establish the *view factor*. Consider the geometry of two diffuse surfaces (i and j), each of which emits light in all directions equally. The fraction of irradiance impinging upon Surface j given the total radiant exitance from Surface i can be posed as a geometric relation, as seen in Eq. (4.16). The fractional irradiance on Surface j derived from the total radiant exitance from Surface i is termed the *view factor* ($F_{i,j}$), and the values must exist between zero and one.

$$\begin{aligned} F_{i,j} &= \frac{\text{irradiance on Surface } j \text{ from Surface } i}{\text{total radiant exitance from Surface } i} \neq F_{j,i}, \\ F_{j,i} &= \frac{\text{irradiance on Surface } i \text{ from Surface } j}{\text{total radiant exitance from Surface } j}. \end{aligned} \tag{4.16}$$

In solar energy conversion systems, a primary surface of interest is the Sun (emitting shortwave light). But because light is directional and radiant energy is exchanged between two or more surfaces, we also point out the receiving surfaces for the Sun's shortwave light, and in return the emitting surfaces of longwave light.

The paired surfaces opposing the Sun's photosphere are going to be the Earth's terrestrial surface, our technological collection systems (like PV) on the Earth's surface, and the (often neglected) surfaces of the Earth's atmosphere. In

Although the sky is largely transparent to shortwave light, it is a significant emission source of longwave light, both up to space and down to Earth. The sky will be discussed in the next chapter.

the previous chapter on the Laws of Light, we learned of the Inverse Squared Law and how the same number of photons from the Sun get spread out over a larger and larger imaginary bubble as the distance (d) increases, leading to a loss of photon density in proportion to $1/d^2$. If we imaging that half of the Earth's surface (facing the Sun at a given moment) is intersecting with the bubble of photons spreading out into space as in Figure 4.5, we can see that many photons originating from the Sun will not be in communication with Earth (and vice versus) just due to great distances. Thus, the view factor between the Sun and Earth is tied to the Inverse Square Law of Radiation.

Additionally, the surface of Earth is curved, yet the photons incident upon the upper atmosphere will be partially collimated due to the great distance between the two surfaces. If we envision any local spot on Earth as a table top, we can imagine a vector pointing directly up to the zenith of the sky (called a *normal vector*). We can similarly imagine a beam of light coming down from the Sun to intersect with the normal vector, where the angle between the Sun's beam and the vector pointing to the sky is called an *angle or incidence*, or in this special case the

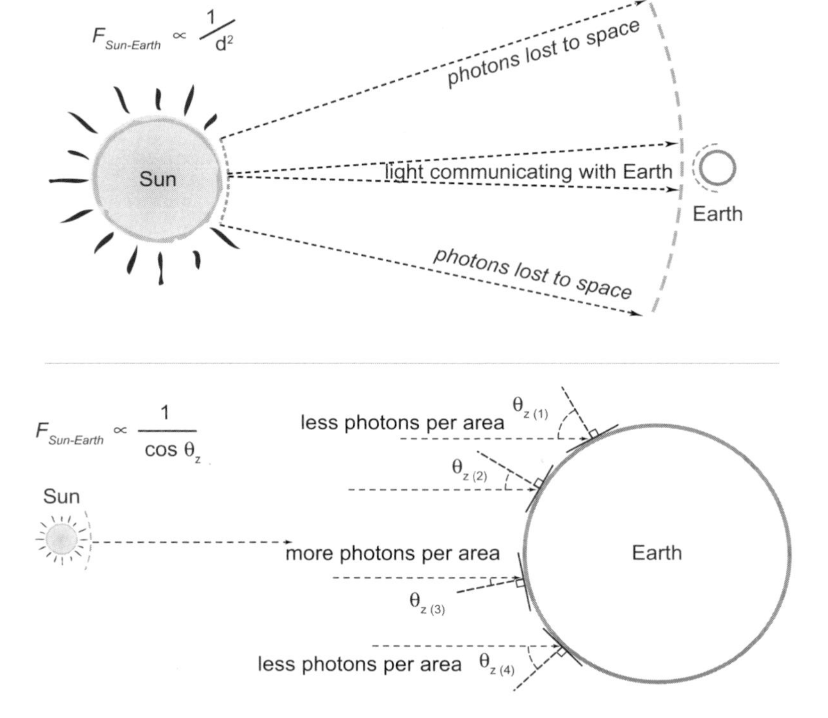

Figure 4.5: Schematic of parameters affecting view factors between the Earth and Sun.

zenith angle (θ_z). As seen in Figure 4.5 the light intensity is inversely proportional to the cosine of the angle of incidence (in terms of the density of photons per area). At higher latitudes (near the poles) the solar angle of incidence increases as the sunlight is effectively tilted away from the Earth surface. As a result, the irradiance decreases, because the same number of photons is being spread over a larger area. This is termed the *cosine projection effect*, and the view factor between the Sun and Earth is also tied to it. Both contributions combined result in a view factor $F_{(Sun-Earth)}$ that is very *very* small. In the case of Surface i being the Sun—150 million km away—the view factor has been estimated to be $F_{Sun,Earth} = 4.5 \times 10^{-10}$ (very close to zero).[21] This means that only 9 out of every 20 billion photons emitted from the Sun will eventually interact with the surface of Earth.[22] And yet we get more irradiance than we know what to do with on a daily basis!

[21] Recall that as corollary to the Inverse Square Law of Radiation most solar photons are not directed toward the Earth.

[22] Gregory Nellis and Sanford Klein. *Heat Transfer*. Cambridge University Press, 2009.

For more general cases, such as with two parallel plates, or a sphere enclosing a small convex surface, the view factor for the two internal surfaces is equal to one (both ways), while the view factor for the two external surfaces is zero. Under blackbody assumptions, $E_{b,i} = A_i\sigma T_i^4$ and $E_{b,j} = A_j\sigma T_j^4$. Under graybody conditions we now know that the radiosity of a surface must be accounted for, as A_iJ_i and A_jJ_j, each with view factors of $F_{i,j} = F_{j,i}$. Hence the radiative transfer $\dot{Q}_{i\to j}$ can be written as in Eq. (4.18).

$$\dot{Q}_{i\to j} = \frac{(J_i - J_j)}{R_{i,j}}, \tag{4.17}$$

where

$$R_{i,j} = \frac{1}{A_iF_{i,j}} = \frac{1}{A_jF_{j,i}}. \tag{4.18}$$

Given $A_iF_{i,j} = A_jF_{j,i}$ by reciprocity of view factors, $\dot{Q}_{i\to j}$ can be simplified as follows:

$$\dot{Q}_{i\to j} = -\dot{Q}_{j\to i} = A_iF_{i,j}(J_i - J_j) \tag{4.19}$$

$$= \frac{\sigma(T_i^4 - T_j^4)}{\frac{1-\epsilon_i}{\epsilon_iA_i} + \frac{1}{A_iF_{i,j}} + \frac{1-\epsilon_j}{\epsilon_jA_j}}. \tag{4.20}$$

The full expansion of radiosity for two surfaces looks pretty dense, right? And yet it does have a similar form to that of one surface seen in Eq. (4.15) emitting into space. But here we have R for two surfaces and the inverse proportionality of the area times the view factor. There is a way to draw out the most complex non-linear portions of the radiative transfer function and offer a linear relation via h_r, the radiative heat transfer coefficient.[23]

Don't panic.

[23] John A. Duffie and William A. Beckman. *Solar Engineering of Thermal Processes*. John Wiley & Sons, Inc., 3rd edition, 2006.

$$\dot{Q}_{i\to j} = A_i h_r (T_i - T_j), \tag{4.21}$$

where

$$h_r = \frac{\sigma (T_i^2 + T_j^2)(T_i + T_j)}{\frac{1-\epsilon_i}{\epsilon_i} + \frac{1}{F_{i,j}} + \frac{(1-\epsilon_j)A_i}{\epsilon_j A_j}}. \tag{4.22}$$

We are just using the simple mathematical relation for the difference of two squares here:

$$(T_i^4 - T_j^4) = (T_i^2 - T_j^2)(T_i^2 + T_j^2)$$

and

$$(T_i^2 - T_j^2) = (T_i - T_j)(T_i + T_j)$$

so

$$(T_i^4 - T_j^4) = (T_i - T_j)(T_i + T_j)(T_i^2 + T_j^2).$$

Difference of squares: $a^2 - b^2 = (a + b)(a - b)$.

Finally, there are two special cases where we can simplify our equations for radiative transfer a bit more.

SPECIAL CASE 1: What happens to Eq. (4.22) if we have two parallel plates of infinite extent facing each other, or two closely spaced layers of a sphere?[24] This scenario would mean that $A_i = A_j$ and $F_{i,j} = 1$:

[24] John A. Duffie and William A. Beckman. *Solar Engineering of Thermal Processes*. John Wiley & Sons, Inc., 3rd edition, 2006.

$$\frac{\dot{Q}_{i\to j}}{A} = \dot{q}_{i\to j} = \frac{\sigma (T_i^4 + T_j^4)}{\frac{1}{\epsilon_i} + \frac{1}{\epsilon_j} - 1}, \tag{4.23}$$

where

$$h_r = \frac{\epsilon_i \sigma (T_i + T_j)(T_i^2 + T_j^2)}{\frac{1}{\epsilon_i} + \frac{1}{\epsilon_j} - 1}. \tag{4.24}$$

SPECIAL CASE 2: What happens to Eq. (4.22) if we have a very small convex object relative to a very large enclosure? This second scenario happens for the Earth surrounded by space or for a solar collection system surrounded by the sky dome of the atmosphere. As such, $F_{i,j} = 1$, $A_i \ll A_j$, and $A_i/A_j \longrightarrow 0$.

We do remember that $\sigma = 5.6697 \times 10^{-8}\,\mathrm{W/m^2\,K^4}$, right? The **Stefan-Boltzmann constant** from the last chapter.

$$\frac{\dot{Q}_{i\to j}}{A} = \dot{q}_{i\to j} = \epsilon_i \sigma (T_i^4 - T_j^4), \tag{4.25}$$

where

$$h_r = \epsilon_i \sigma (T_i + T_j)(T_i^2 + T_j^2). \tag{4.26}$$

REFLECTIVITY AND REFLECTANCE FOR SEMI-TRANSPARENT MATERIALS

Now we move from highly opaque materials with near-zero transmittance ($\tau \to 0$) in our band of interest, to materials that permit photons to transmit through due to low reflectance and low absorptance. Reflectance (ρ) has been applied previously for opaque materials, to demonstrate the utility of simple relations among emittance and absorptance in graybody accounting. However, reflectance can also be used as a vehicle to open up our understanding of the role of transmittance in a non-opaque, or *semi-transparent* material surfaces as well.

In this section, we examine the detailed contributions of material properties in a semi-transparent material, and the coupled relationships among the fractional values of reflectance, transmittance (τ), and absorptance (α).

$$E_i = E_t - E_r. \tag{4.27}$$

Equation (4.27) is considered from the frame of reference of the incident surface. The observer is actually *in* the solar collector, and hence only measures the light that is not reflected, or $E_t - E_r$.

As seen in Eq. (4.27), when light arrives at an interface (a surface) the photons engage in a net balance of energy contributions by transmission and reflection. So where did the absorbed photons go? As the radiant energy is transmitted through a material, it has the chance of being absorbed, depending on the intrinsic optical properties of the material and the thickness of the material. Hence, the transmittance values can embody the losses of photons due to absorptance, if

That which is **not reflected** is either **transmitted** or **absorbed**. No photons are lost to the net energy balance.

Equation (4.28) is considered from the frame of reference of the observer outside of the receiving surface (up in the sky). The observer only measures the reflected light, which is the input radiant energy minus the transmitted and absorbed energy.

we pose the analysis properly. In fact, all materials have some ability to absorb photons, and so there we observe a range of behaviors in materials (over a range of thicknesses) from highly transparent to highly opaque.

$$\rho = 1 - (\tau + \alpha). \tag{4.28}$$

Reflectance is coupled with both absorptance and transmittance. The net energy relationship shown in Eq. (4.28) is a reminder of that coupled balance. When we reflect light with a material surface we also have contributions to light transmission and absorption. It's a package deal: you can't address reflectance without considering the material surface's affect on transmittance and absorptance (even if they are really small). Also, we see in Figure 4.6 that there are two additional arrows of shortwave light that emerge at the interface of Material 2. The quantity of energy or power embodied in each arrow is the difference of the total irradiance and the other two in the net energy balance. The transmittance and absorptance contributions are linked together as light propagates through the

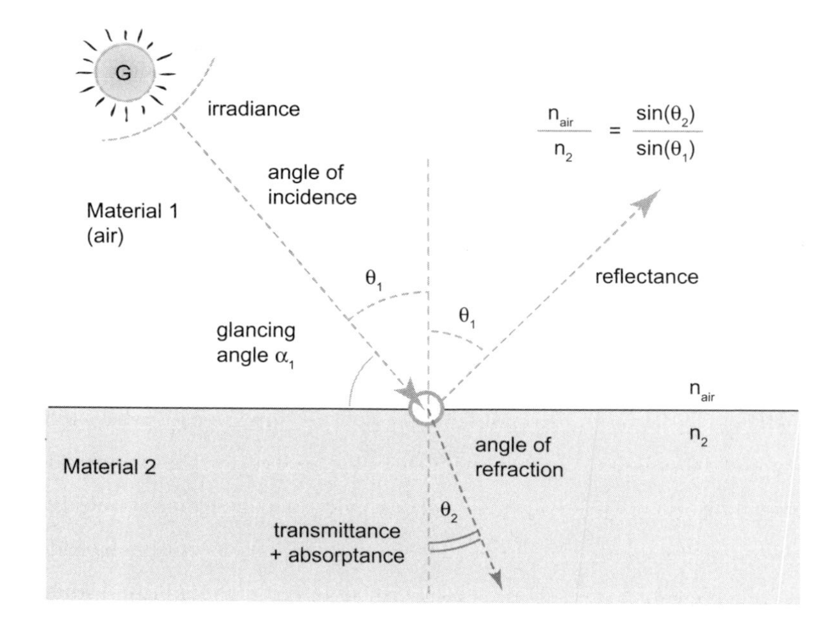

Figure 4.6: Angle of incidence from irradiance beam is the same as the angle of reflection, but different from the angle of refraction. The relation between the angle of refraction and the index of refraction in two materials is shown by Snell's Law. Note that we have also mentioned the complement of the incidence angle: the glancing angle. This will become important when discussing the *cosine projection effect* in future chapters.

material, and the difference of those contributions relative to the total irradiance will be the measure of reflectance.

Transparency is a function of material properties (indices of refraction), selective wavelengths, and sample thickness. We called materials "semi-transparent" because at some limit of thickness, within a band of interest (a range of wavelengths), *all* materials can be made to be opaque or transparent. We think of carbon as soot, being black, and having a high absorptivity in the shortwave and longwave bands. Yet if we deposit carbon as individual layers of graphene sheets (single atomic thicknesses) on glass, the surface will indeed be transparent to our visual inspection.

Yes, *all materials are semi-transparent* under various conditions of thickness and respective bands of light.

On the other side of everyday materials, consider a sheet glass in a window. Under normal circumstances (looking at the glass through the thinnest plane) the glass is visually transparent (having a high transmittance in the visual sub-band of the shortwave band). However, window glass has distinct metal impurities in it like iron, and when we view the pane of glass on edge, we often see that it is mostly opaque and green due to the thickness of the sheet when viewed on edge, and the reflecting and absorbing contributions of the metal impurities. Increased thickness can make a material opaque, while thin films of normally opaque materials (even metals) can be made to be semi-transparent. In fact, the glass industry specializes in thin films ($d < 10\,\text{nm}$) that are selectively transparent in the visible band, while being highly reflective in the infrared band (both shortwave and longwave).

The specialty glass used in solar devices is termed *low-iron glass*, because it has a higher transmittance across a broader spectrum of the shortwave band. Window glass cuts out a significant portion of the useful infrared for SECSs.

These selective surface thin films in the glass industry are called *low-E* surfaces, for low emittance (ϵ), indicating that they are correspondingly high reflectance (ρ) surfaces in the infrared range.

So what do we know about the behavior of light upon scattering inelastically off of a material surface? Well, we are probably aware that many varieties of sunglasses contain a polarizing filter in the optics. Some may even have tried to view reflected light from a lake or swimming pool through polarizing sunglasses at different orientations (turning your head to the side). Light will be more intense in specific orientations at 90° rotation, and this is a result of light polarization from reflection off the surface of a water body. Electromagnetic waves (i.e., light, photons) have relative orientations of the wave oscillations that are normal to the direction of propagation. Polarization

Air (or the sky) has an index of refraction $n=1$.

refers to the directions (vector components) in which the electric field oscillates for an electromagnetic wave, and there are two different directions of oscillation.

Parallel and perpendicular polarization vector components can also be termed *p* and *s* polarization in optics texts.

$$n_2 = n_1 \left(\frac{\sin(\theta_1)}{\sin(\theta_2)} \right). \quad (4.29)$$

Keep a mental note of θ_2 the angle of refraction. You will use it in Eq. (4.36)!

Yes, θ_1 is the same as θ, the angle of incidence in solar geometric relations.

Specular reflection implies a singular beam of light exiting the surface as in a mirror surface, as opposed to diffuse reflection where the light is scattered in many directions (as in white paint).

When a beam of photons are incident upon a material surface (a different material than the surrounding medium) and that surface has some transparency, the light will do two things at the surface: a fraction of the light will be reflected at an equal angle to the *angle of incidence* (θ_1). The equivalent angle at which light is reflected is called the *angle of reflection* for specular reflection.

A fraction of the light will be transmitted through the transparent medium at a different angle due to a change in the velocity of the photons, an *angle of refraction* (θ_2). The angle of refraction is dependent upon the index of refraction as a function of wavelength ($n(\lambda)$).[25,26] As seen in Figure 13.4 and the equation below, the two angles of incidence and refraction are related using Snell's Law. Typically, the initial medium that light is propagating through is the air from the sky, which has an index of refraction of one.

Now, in the 19th century, French engineer and son of an architect, Augustin-Jean Fresnel developed a working model to describe the wave polarization phenomena from light reflecting off of a surface.[27] The measurements of reflectivity are expressed as the ratios of radiant energy reflected (E_r) per incident radiant energy (E_i). As we shall see, *reflectivity* (r) is slightly different from *reflectance* (ρ). The expressions that Fresnel derived are for perpendicular ($r_\perp$) and parallel ($r_\parallel$) polarized components of reflected light, seen in Eqs. (4.30) and (4.31), and finally the average reflectivity for the material (r) is calculated in Eq. (4.32) (where E_i is energy incident and E_r is energy reflected).

$$r_\perp = \frac{(n_1 \cos\theta_1 - n_2 \cos\theta_2)^2}{(n_1 \cos\theta_1 + n_2 \cos\theta_2)^2} = \frac{\sin^2(\theta_2 - \theta_1)}{\sin^2(\theta_2 + \theta_1)}, \quad (4.30)$$

$$r_\parallel = \frac{(n_1 \cos\theta_2 - n_2 \cos\theta_1)^2}{(n_1 \cos\theta_2 + n_2 \cos\theta_1)^2} = \frac{\tan^2(\theta_2 - \theta_1)}{\tan^2(\theta_2 + \theta_1)}. \quad (4.31)$$

[25] Eugene Hecht. *Optics.* Addison–Wesley, 4th edition, 2001.

[26] Additionally, the likelihood of the material to extinguish (read: absorb) a photon as it traverses thicker and thicker material is related to the attenuation or extinction coefficient ($k(\lambda)$), which has relation to the absorption coefficient that we shall discuss later.

[27] English-speaking learners: *attention!* The proper way to **pronounce Monsieur Fresnel's family name** is *without the "s,"* with emphasis on the second syllable, as in frehn-*el.* There is no *frez*-nell out there in the history of solar energy, ok? There. Welcome to the insider's club, or *bienvenue.*

Careful analysis of these equations will show that the squared trigonometric functions are very similar for the two polarizations. The perpendicular polarization seen in Eq. (4.30) simplifies to a ratio of the sines squared, while the parallel polarization observed in (4.31) simplifies to a ratio of the tangents squared.

We wish to combine the two in Eq. (4.32), so the average of the two values is calculated. The result is a ratio of the radiative flux (also called intensity, in W/m^2) of reflected light (E_r) relative to the radiative flux of incident light (E_i, irradiance). For general angles of incidence, Eq. (4.29) should be used in combination with Eqs. (4.30), (4.31), and (4.32).

$$r = \frac{E_r}{E_i} = \frac{1}{2}(r_\perp + r_\parallel). \tag{4.32}$$

Again, *reflectivity* is the ratio of the radiative flux of reflected light (E_r) relative to the shortwave irradiance from the Sun (E_i). *Reflectance* is a function of both *reflectivity* and *transmittance* for a given wavelength of light ($\rho = f(r_{\lambda,\perp}, r_{\lambda,\parallel}, \tau_\lambda)$).

Reflectivity (r) is related to, but not the same as **reflectance** (ρ). In order to calculate reflectance, we must consider the contributions of energy from transmittance with respect to absorption losses.

In order to calculate **reflectivity**, one must take the average of $r_\perp$ and $r_\parallel$. For the physics crowd, the intensity of light is the square of the amplitude of the electric field in the EM wave.

There is a special case for which the reflectivity calculation is much shorter. In the singular condition when the beam irradiance occurs at normal incidence (the light beam is perpendicular to the collector surface). The angle of incidence is therefore zero. In fact, both θ_1 and θ_2 are equal to zero. As such, Eqs. (4.29) and (4.32) reduce to yield Eq. (4.33).

$$r(0°) = \frac{E_r}{E_i} = \left(\frac{n_1 - n_2}{n_1 + n_2}\right)^2. \tag{4.33}$$

Reflectance (ρ) is a function of both **reflectivity** (r) and **transmittance** (τ).

$$\rho = f(r, \tau).$$

So how do we proceed to calculate transparency, or transmittance for a given wavelength or band? We can inspect the diagram in Figure 4.6 for clues. There are two ways that we can lose energy of light along the path through that second material: first, the light can be scattered or reflected away; and second, the light can be absorbed as it passed through the second material (even in small quantities).

Let's view the engineering approximations for each, τ, α, and ρ (all losses considered) for a single cover, like a single sheet of glass or plastic. We will

That which is **not transmitted** is either **reflected** or **absorbed**. No photons are lost to the net energy balance.

cover Eqs. (4.34), (4.40), and (4.41). Over the course of examination, notice that there are equations dominated by relations to two types of transmittance estimates, those considering *energy losses due to absorption of light* (τ_α; Eq. (4.36), and those considering *energy losses due to reflection* (τ_ρ; Eq. (4.38)).

$$\tau \approx \tau_\alpha \tau_\rho. \tag{4.34}$$

τ_α: transmission considering losses due to absorption of light.

τ_ρ: transmission considering losses due to reflection of light.

θ_2: the *angle of refraction*, not the angle of incidence.

k: the extinction coefficient.

d: sample thickness.

The former mechanism, *transmission considering only losses due to absorption of light* (τ_α), can be calculated by knowing the *angle of refraction* (θ_2) and the *extinction coefficient* of the material (k), and the thickness of the material or distance that light must propagate through (d), seen in Eq. (4.36). The basis for Eq. (4.36) is described in Eq. (4.35), from the Beer-Lambert-Bouguer law of materials absorbing light. Here, the *extinction coefficient* of the material (k) is a proportionality factor. By integrating both sides, including the pathlength in the semi-transparent medium from zero to $d/\cos(\theta_2)$, we arrive at Eq. (4.36).

$$dE = -kE\,dz, \tag{4.35}$$

$$\tau_\alpha = \exp\left(-\frac{kd}{\cos(\theta_2)}\right). \tag{4.36}$$

Bear in mind that the units of cover thickness (a distance, d) must be in the same scale as those of the extinction coefficient (k).

Extinction coefficients in solar energy tend to be reported in two scales within the shortwave band (as inverse distances). In terms of inverse meters (m^{-1}) or in terms of inverse centimeters (cm^{-1}). Transparent cover materials tend to use the former, while photovoltaic materials use the latter.[28] A large extinction coefficient suggests an opaque material in the band of interest.

[28] Christiana Honsberg and Stuart Bowden. Pvcdrom, 2009. http://www.pveducation.org/pvcdrom. Site information collected on Jan. 27, 2009.

For transparent cover materials the extinction coefficient will be on the order of 4–30 m^{-1}, while photovoltaic absorbers will report extinction coefficients on the order of 10^{12}–10^{13} cm^{-1}. Photovoltaics use the *absorption coefficient*, α_a. Here, the absorption coefficient is presented as α_a to distinguish from the separate, more generic value of absorptance, α. The absorption coefficient is related to the extinction coefficient by Eq. (4.37). For λ of 500 and 100 nm, the scaling factor from k to α_a is k multiplied by 2.5×10^{-9} and 1.3×10^{-9}, yielding α_a values between 10^4 and 10^5 cm^{-1}, which are expected.

Note that both scales of extinction coefficients convey a quantifiable measure of both transmittance and absorptance in a select band of irradiance. The inverse of the extinction coefficient is the thickness of material that would be required to cause absorption losses of about 36% (decreasing by a factor of $1/e$).

$$\alpha_a = \frac{4\pi k}{\lambda}. \tag{4.37}$$

$r_{\|}$: the parallel polarized reflectivity.

$r_{\perp}$: the perpendicular polarized reflectivity.

N: the number of covers in the system.

The multiple cover scenario is important to SECSs. Windows can be double or triple panes, sometimes of both glass and polymer covers. The covers to solar thermal collectors are also multiple cover systems.

The latter mechanism, *transmission considering losses only due to reflection* (τ_ρ), can be accounted for in Eq. (13.9), for N layers or covers of semitransparent material. This equation is a clear indication that one requires the measure of reflectivity to calculate the transmittance with respect to losses from scattered light. In turn, Eqs. (4.30) and (4.31) indicate that the angles of incidence and refraction are needed as well from Eq. (4.29). There are several pieces that all loop back to the refractive indices of two materials at an interface, within a band of wavelengths of interest.

$$\tau_\rho = \frac{1}{2}\left(\frac{1-r_{\|}}{1+(2N-1)r_{\|}} + \frac{1-r_{\perp}}{1+(2N-1)r_{\perp}}\right). \tag{4.38}$$

For a single cover, $N=1$, the solution is

$$\tau_\rho = \frac{1}{2}\left(\frac{1-r_{\|}}{1+r_{\|}} + \frac{1-r_{\perp}}{1+r_{\perp}}\right). \tag{4.39}$$

In Eq. (4.40), the absorptance α is approximately equal to the total minus the transmittance that accounts for losses in absorption (the transmittance excepting absorption phenomena). Reflectance and transmittance that accounts for losses in reflectance are not included in Eq. (4.40) because the reflected energy essentially never even made it into the material to have the opportunity to be absorbed. With multiple surfaces and covers this approximation would need a more detailed model.

$$\alpha \approx 1 - \tau_\alpha. \tag{4.40}$$

In order to find τ_α, we need Eq. (4.36).

In Eq. (4.41) the reflectance ρ is a function of the energy that was lost to transmittance with respect to absorption losses and the total transmittance.

$$\rho \approx \tau_\alpha (1 - \tau_\rho) = \tau_\alpha - \tau . \tag{4.41}$$

In order to find τ_ρ, we need Eq. (4.38) or (4.39).

Now, we would like to bring all of our learning together for robust estimates of transmittance (τ), reflectance (ρ), and absorptance (α). In the equations below we provide the detailed relations derived to couple transmittance (τ), reflectance (ρ), and absorptance (α) for each component of polarization. The reasons for this become important for multiple panes of covering layers, and in complex thin films used on the surfaces of covers, reflectors, and absorbers in SECS.

For both polarizations of light of a given wavelength, λ:

$$\tau_\perp = \frac{\tau_\alpha (1 - r_\perp)^2}{1 - (r_\perp \tau_\alpha)^2}, \tag{4.42}$$

$$\tau_\parallel = \frac{\tau_\alpha (1 - r_\parallel)^2}{1 - (r_\parallel \tau_\alpha)^2}, \tag{4.43}$$

$$\rho_\perp = r_\perp (1 + \tau_\alpha \tau_\perp), \tag{4.44}$$

$$\rho_\parallel = r_\parallel (1 + \tau_\alpha \tau_\parallel), \tag{4.45}$$

$$\alpha_\perp = (1 - \tau_\alpha) \left(\frac{1 - r_\perp}{1 - r_\perp \tau_\alpha} \right), \tag{4.46}$$

$$\alpha_\parallel = (1 - \tau_\alpha) \left(\frac{1 - r_\parallel}{1 - r_\parallel \tau_\alpha} \right). \tag{4.47}$$

Transmittance is the average of the two polarizations:

$$\tau = \frac{1}{2} (\tau_\perp + \tau_\parallel) .$$

Reflectance is the average of the two polarizations:

$$\rho = \frac{1}{2} (\rho_\perp + \rho_\parallel) .$$

Absorptance is the average of the two polarizations:

$$\alpha = \frac{1}{2} (\alpha_\perp + \alpha_\parallel) .$$

In order to calculate the values for multiple covers, we must begin with knowledge of the indices of refraction (n_1 and n_2) and the angles of incidence and refraction (θ_1 and θ_2) for each cover.

Solving τ, ρ, and α for two covers:

τ_1: top cover transmittance.

τ_2: bottom cover transmittance.

$$\tau = \frac{1}{2} (\tau_\perp + \tau_\parallel) = \frac{1}{2} \left[\left(\frac{\tau_1 \tau_2}{1 - \rho_1 \rho_2} \right)_\perp + \left(\frac{\tau_1 \tau_2}{1 - \rho_1 \rho_2} \right)_\parallel \right]. \tag{4.48}$$

The equation for reflectance with two covers uses the resulting value of total system transmittance, τ, along with the transmittance calculated for the second cover, τ_2.

ρ_1: top cover reflectance.
ρ_2: bottom cover reflectance.

$$\rho = \frac{1}{2}(\rho_\perp + \rho_\parallel)\frac{1}{2}\left[\left(\rho_1 + \tau \cdot \frac{\rho_2\tau_1}{\tau_2}\right)_\perp + \left(\rho_1 + \tau \cdot \frac{\rho_2\tau_1}{\tau_2}\right)_\parallel\right]. \quad (4.49)$$

We need to find the total system transmittance, τ from Eq. (4.48) before we can solve for total system reflectance, ρ in Eq. (4.49).

The value of absorptance is finally determined by simple math from the determined values of total system transmittance, τ and total system reflectance, ρ:

$$\alpha = 1 - \tau - \rho. \quad (4.50)$$

VIGNETTE: *Light* is an unusually general term, we use it for photons in the visible spectrum but also for ultraviolet and infrared energy (shortwave and longwave spectra). Unfortunately, so is *radiation*, which can include photons, but in the most general sense for physics can also refer to high energy particles with mass that are emitted from the fission decay of a nuclear materials. For example, **gamma radiation** is composed of photons, while **beta radiation** is composed of electrons. For the purposes of this text "radiation" will mean a very specific radiant energy transfer, indicating light being emitted from a source over an interval of time (in units of J/m^2).

We work with light every day to do useful things like microwave photons cooking food and ultraviolet photons used in the dairy industry to make Vitamin D from the cholesterol already present in milk. On the odd side, we also know that the gamma radiation photons have sufficiently high energy to make silica (SiO_2, generally thought of as an "insulator") electronically conductive when encapsulating nuclear waste.

PROBLEMS

1. The glazing (glass windows) for a traditional greenhouse have $\tau = 0.90$ for the PAR band (photosynthetically active radiation: $400\,nm < \lambda < 700\,nm$). The glazing is also $\tau = 0.90$ the remainder of the IR in the shortwave band

$700\,\text{nm} < \lambda < 2500\,\text{nm}$, while $\tau = 0.60$ for the remainder in the UV band ($200\,\text{nm} < \lambda < 400\,\text{nm}$). In contrast, greenhouse glazing is almost completely opaque at longer wavelengths ($\tau < 0.02$). Assuming the Sun radiates at an effective temperature of 5777 K and the interior of the greenhouse radiates at an effective temperature of 310 K, calculate the percent (or fraction) of incident shortwave irradiance transmitted through the glass and the percent (or fraction) of longwave radiance emitted by the interior surfaces transmitted out.

2. A single solid sheet of polystyrene is proposed as a cover material. Polystyrene has an index of refraction in the visible of 1.55. The sheet thickness is $d = 2\,\text{mm}$, with an extinction coefficient of $k = 5\,\text{m}^{-1}$. Paying attention to the unit conversions, calculate the reflectance of the surface on a smooth plastic sheet for three angles of incidence: (a) 10°, (b) 45°, and (c) 60°.

3. Low-iron glass is used specifically in SECSs, as a dual pane system for solar hot water panels. For a glass sheet ($n = 1.53$) that is 3 mm thick and with an *extinction coefficient* $k = 4\,\text{m}^{-1}$ (low-iron), calculate the transmission of two covers at (a) normal incidence, and (b) at 40°. Hint: don't use the approximations. Work your way backward from Eq. (4.48) to (4.29).

4. In an evacuated tube collector system for solar thermal applications, there are no significant losses for the opaque absorber plate associated with conduction or convection. When stagnant (no mechanical pumping) the fluid inside can get quite hot.

   Assume an opaque flat plate collector (horizontal) with $2\,\text{m}^2$ of exposed surface area is absorbing $800\,\text{W/m}^2$ from the net irradiance, and the absorptance of the surface is $\alpha = 0.85$. The temperature of the surroundings is 30 °C. (a) If one were to neglect conduction and convection losses, what would the net radiation exchange be between the plate and the surroundings?) (b) What would be the equilibrium temperature of the collector?

5. [Advanced] Define temperature in the most general sense that you can find given the full body of thermodynamics and statistical mechanics.

RECOMMENDED ADDITIONAL READING

- Heat Transfer.[29]
- Design with Microclimate: The Secret to Comfortable Outdoor Spaces.[30]
- Let it Shine: The 6000-Year Story of Solar Energy.[31]

[29] Gregory Nellis and Sanford Klein. *Heat Transfer.* Cambridge University Press, 2009.

[30] Robert D. Brown. *Design with Microclimate: The Secret to Comfortable Outdoor Spaces.* Island Press, 2010.

[31] John Perlin. *Let it Shine: The 6000-Year Story of Solar Energy.* New World Library, 2013.

METEOROLOGY: THE MANY FACETS OF THE SKY

05

Look at your feet. You are standing in the sky. When we think of the sky, we tend to look up, but the sky actually begins at the Earth. We walk through it, yell into it, rake leaves, wash the dog, and drive cars in it. We breathe it deep within us. With every breath, we inhale millions of molecules of sky, heat them briefly, and then exhale them back into the world.

Diane Ackerman, A Natural History of the Senses (1991)

THE SKY is an essential part of our solar energy conversion system: it acts as a wonderful *cover* to block (by absorption or reflection) selective bands of irradiance coming in from the Sun like ultraviolet light. The sky also acts as a thermal storage medium, carrying warm or cool gases as stored fluids, and emitting that energy via longwave band radiance. The role of *clouds* within the atmosphere is the most significant contributor to decreased solar gains for a SECS, and clouds are also one of the most complicated factors to understand and predict. At the same time the sky acts also as a cover surface relative to Earth's emitting surface, absorbing longwave terrestrial radiance from the thermal body of Earth's surface, and emitting more longwave atmospheric radiance back down, to maintain a favorable temperate condition on Earth. The sky also provides the chemicals that we need to breathe (oxygen), that algae needs to make sugars (H_2O and CO_2), and the abundant moisture to generate clouds.

The sky also contains our weather and climate systems. The main energetic driver for behavior of meteorological phenomena of wind, rain, and snow is the Sun.

From the quote by Diane Ackerman at the beginning of the chapter, we can form an initial impression of the sky as voluminous, and *big*. The sky is an intermediary between the Sun and our solar energy conversion system on the ground. The sky is responsible for *depleting* solar irradiance along the path of light. It is this concept of solar energy depletion that we will study now. The sky contains molecules (gases) and particles, as well as the emergent phenomena from the aggregated suspension of water aerosols (clouds), that block some of the useful bands of irradiation for a SECS. The loss of solar energy (with respect to collection of light at the bottom of the atmosphere) when interacting with the sky is dependent upon (1) the *angle of incidence* and the *path length* of light transmitted through the sky, and (2) the *ingredients* of what is contained within the sky, in a large container called an *air mass*. We will show that the ingredients of the sky affecting solar gains in Pittsburgh, PA, were not "born" there. We will also show that over the course of the year, there are seasonal changes in the origins of the sky relative to the locale of the SECS we are to design. These seasonal changes mean the locale will be divided into multiple *regimes* based on the participating air masses.

The tilt of the Earth is reflected in the solar declination: $\delta = \pm 23.45°$, which affects our course of seasons.

In the production of geofuels such as petroleum and natural gas, we rely on geoscientists to inform the engineering and economics for decision making. Analogously, solar energy engineers and economists (and those in the wind industry) will come to rely strongly on scientists from both *meteorology* and *climatology* to inform future financial developments of large- and small-scale SECS projects. Solar energy assessment has scales that are both local and short in time, appropriate to meteorology, and those that are multiregional and which space across seasons, years, and decades, appropriate to climatology.

AIR MASSES

Remember how we described radiative exchange between two surfaces? Baby stuff. Truth be told, the real fun in radiative exchange occurs among exchanges of three surfaçes and more. We have shortwave light emitted from the Sun, then that light is partially reflected and lost by both Earth surfaces as well as the clouds and particles in the sky. We also have longwave light emitted from the atmosphere itself, as well as from the surfaces on Earth at "ground level." Our interest in solar energy conversion systems then allows us to consider additional surfaces that collect and direct light from the Sun and sky to receivers for energy conversion. There is a dynamic dance of light in the sky that makes the field of solar energy compelling and engaging! So what is this mass of air that ripples and causes the twinkling of stars at night?

First, let us separate the *engineering* concept of an Air Mass from the *meteorological* concept of the Air Mass. Recall the use of Air Mass as AM0 and AM1.5 from our description of the spectral nature of light. There we used the term to describe the effective path of light through an ideal clear sky atmosphere. This engineering description of Air Mass is used to develop a general perspective of how the sky filters out sub-bands of light and attenuates light with increasing optical depth, and to create a common simulated irradiance testing condition for solar conversion technologies. The air mass is defined in proportion to the cosine of the *zenith angle* (θ_z), which is the angle between the beam from the Sun and the vector pointing directly up to the zenith of the sky (normal vector to the horizontal).

$$\mathrm{AM} = \frac{1}{\cos(\theta_z)} \text{ [simple, parallel plate]}. \tag{5.1}$$

Gueymard has since developed an air mass model that will also apply for zenith angles >80°, as the former model tends toward infinity at $\theta_z = 90°$.[1]

$$\mathrm{AM} = \frac{1}{\cos(\theta_z) + 0.00176759 \cdot \theta_z \cdot (94.37515 - \theta_z)^{-1.21563}}. \tag{5.2}$$

But that definition of an Air Mass is not what we are looking for right now! We need to define an Air Mass in a meaningful manner for the systems science perspective, one that has been developed to reflect the changing parcels of air shifting across the surface of Earth, generating clouds, dust storms, hurricanes, and blue skies. This is the meteorological concept of the Air Mass.

[1] Matthew J. Reno, Clifford W. Hansen, and Joshua S. Stein. Global horizontal irradiance clear sky models: Implementation and analysis. Technical Report SAND2012-2389, Sandia National Laboratories, Albuquerque, New Mexico 87185 and Livermore, California 94550, March 2012. URL http://energy.sandia.gov/wp/wp-content/gallery/uploads/SAND2012-2389_ClearSky_final.pdf.

The large pancake of gases and particles that is the sky can be divided up into large volumes with common properties of chemistry, pressure, and temperature–all of which affect the degree to which solar energy is *depleted* along the particular path length through the sky. Drawing from meteorology, these large parcels are called *Air Masses*, constantly moving with turbulence due to thermal gradients. Also, while the turbulence is observed in the presence of clouds, smog, fog, and smoke, turbulence exists during clear skies as well.[2] So think of an air mass as a turbulent pancake that interacts with both shortwave and longwave light, resulting in irradiance conditions at the ground level and in orbit (for satellite remote sensing) that are constantly shifting and in need of forecasts for many SECS technologies.

[2] Théo Pirard. *Solar Energy at Urban Scale*, Chapter 1: The Odyssey of Remote Sensing from Space: Half a Century of Satellites for Earth Observations, pages 1–12. ISTE Ltd. and John Wiley & Sons, 2012.

Air masses can be grouped according to similar physical characteristics and behaviors, and their source regions. The American Meteorological Society describes an air mass as a body of air spread over a wide area of the Earth's surface, having near homogeneous properties in its horizontal extent that were established while the air was situated over a specific source region, also undergoing changes while moving away from the source region. The society also recognizes the Air Mass for its use in radiation applications, as "the ratio of the actual path length taken by the direct solar beam to the analogous path when the Sun is overhead from the top of the atmosphere to the surface. Extrapolation of surface measurements to zero air mass was the original method for estimating the value of solar irradiance at the top of the atmosphere."[3]

[3] American Meteorological Society (AMS). Glossary of meteorology. Allen Press, 2000. URL http://amsglossary.allenpress.com/glossary/. Accessed October 1, 2012.

The changes between air masses are *weather fronts*, and it should be no surprise to us that the study of the regional dynamics of air masses and accompanying fronts on a sub-annual basis is called *meteorology*.[4] It is important to solar energy conversion that we figure the time factor into our understanding, as we will find that each location under study will be under the influence of different air masses during different times of the year. In the mid-latitudes of North America, we call this periodicity *seasons*. Hence, our solar energy conversion systems are effectively in *different locales* for each period of air mass dominance[5] (see Figure 5.1).

[4] When we expand our studies of air mass behaviors to multiregional scales (larger spatial scales) and time scales in the range of decades to millennia, this is called *climatology*.

[5] Did he just imply that we have to design our systems to function for the client or stakeholders in *four* different locales instead of one?
Yes!
Boy, that's going to make the goal of solar energy engineering more complicated.
Right. So get back to work and read the section on multiple climate regimes.

Figure 5.1: Approximate synoptic weather intervals for northern mid-latitudes. Four different seasons implies four different *locales* for one geographic position.

Thanks to solar gains on the Earth surface, and the change of irradiation over seasons due to the tilt of Earth, we observe that air masses are not immobile and any fixed locale observes changes in residential air mass behavior over time. Air masses are created in *source regions*, which is where the mass acquires its characteristics of temperature and humidity, and then moves from the region of origin into new regions. Source regions have been mapped out in Figure 5.2, and they will tend to have lighter winds, allowing the air mass to accumulate the temperature and humidity conditions of the accompanying portion of the Earth's surface (the big solar energy conversion device). Hence, areas affected by jet streams will not be regions to create new air masses.

The *Bergeron classification system* is accepted and used by atmospheric science, and indicates the *genetic origin* for air masses. For the solar energy engineer and economist, this is important information to affect design constraints

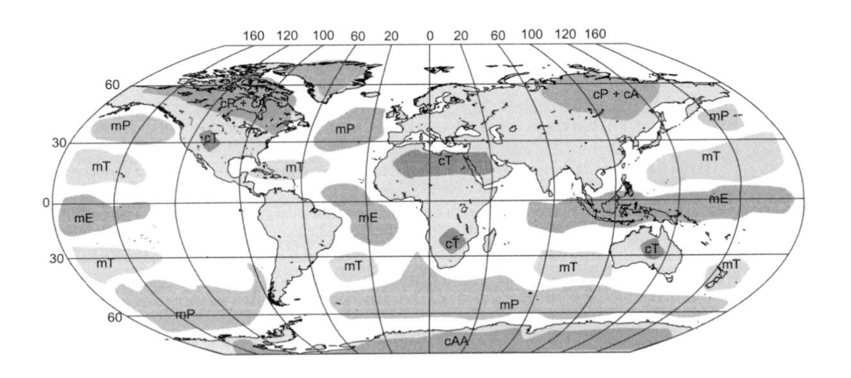

Figure 5.2: Schematic of source regions for main large air masses. Figure modified from *NASA*. User: GateKeeperX/ Wikimedia Commons/ Public Domain. November 2008.

according to the quality of the solar resource. Properties of thermal behavior and humidity are conveyed in the Bergeron classification, and the classification has a relatively simple two or three letter coding. First, air masses are labeled according to their origin above a land mass, *continental (c)* or above an ocean, *maritime (m)*. Maritime air masses will contain more humidity derived from the underlying body of water than dry continental air masses. Next, air masses are to be grouped into main *thermal* categories according to the origin: (A, P, T, E). This grouping is largely separated by jet streams and changes in latitude, which of course are each derived by the tilt of Earth relative to the projection of irradiance from the Sun. By coupling the two letters together, we can then map out the source regions for meteorological behaviors across the planet and through the seasonal shifts.

The relation of air mass to monsoon cycles is less used in continental meteorology of North America, but is highly important to solar development in Asia.

Arctic/Antarctic (A/AA): A high pressure regime that stabilizes to the far north.

Polar (P): Very common air mass origin regime for mid-latitude regions of the USA and Europe.

Tropical (T): Very common air mass origin regime for mid-latitude regions of the USA and Europe.

Equitorial (E): A high pressure regime that stabilizes near the equator.

THE SKY AS MULTIPLE CLIMATE REGIMES

Aren't you just talking about **seasons?**

Yep, we are. But we are doing it in that robot monkey sort of way that appeals to the inner engineer.

Meteorology informs us that there are common phenomena specific to spatio-temporal regions that we will call *climate regimes*. In the geographical regions in the mid-latitudes (north and south hemispheres), there tend to be *four climate regimes*. In regions affected by monsoonal swings there may be two or three climate regimes. We often notice the regime changes in terms of air temperature, but the sky regimes change as well, in terms of relative humidity, wind speeds and size of weather cells, as well as the emergent cloud behaviors. As we will show in the following sections, the basic observations in solar energy called *clearness indices* and their resulting correlations to diffuse sky models on a tilted surface are in fact emergent properties of climate regimes. The concept of climate regimes as seasons should not really be new knowledge to you, but our use of that knowledge, in terms of designing for multiple regimes is new.

Effectively, each climatic regime should be treated as a *different geographical place* throughout the year. Hence, there is not one Philadelphia in Pennsylvania, there are *four*. From the perspective of designing solar energy conversion systems, the Philadelphia skies of December 1 through February 28 is a *different regime* than the Philadelphia skies of June 1 through August 31.

If only we could remove the pesky clouds that mar our cover like bird poo on a nice clean glass window …

NO SKY MODEL

Think about an important element in our cold space estimate—the Sun. If there was no Sun, the receiver would be *even colder* than with no atmosphere. The Sun is very important in thermal balance of objects, and we shall demonstrate the extra importance of the sky. If we extrapolate the air mass to zero, we see how the sky affects solar energy conversion systems without the sky (no atmosphere). The only emitters of light in this special case would be the Sun and the surface of Earth. The surrounding space would be very cold—well below the freezing point of water.

We put photovoltaic (PV) technologies into this skyless environment on a regular basis. Satellites are regularly powered by photovoltaic systems and power storage systems. Energy is very expensive in orbit, and there are no gas stations or recharge stations.[6] In fact, the two long-term options for energy are either nuclear fission-derived power, or photovoltaic power. As we will discuss in future sections on photovoltaics, satellites are some of the most *area-constrained* systems for PV deployment, and hence make use of the most expensive high efficiency multi-junction PV cell-module systems in space.

[6] Batteries run out of reactive materials without electrolytic recharging, and fuel requires an oxidant chemical to react *and* is limited in storage.

To fully understand the role of the sky, let's first "remove" it and run the *energy accounting* for the radiative exchange between surfaces. We can

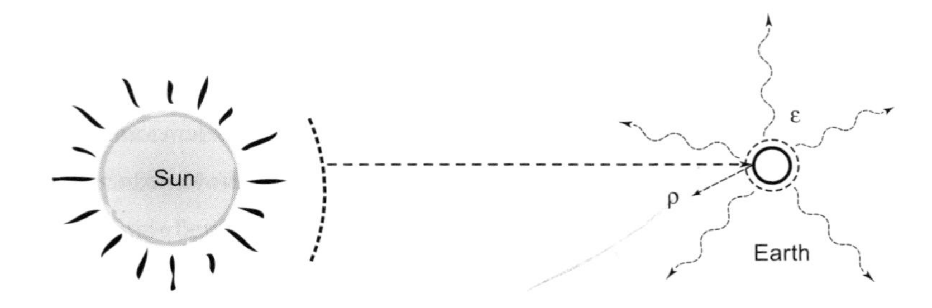

Figure 5.3: The Zero Dimensional Sun-atmosphere-Earth system energy balance. The Sun is a shortwave emitter, while the Earth is a shortwave reflector and a longwave emitter.

establish an Energy Balance Model (EBM) for Earth's climate if there were no sky. This is called a *Zero Dimensional EBM*, conveying the exchange of electromagnetic energy as it is absorbed, reflected, then transferred and stored by the Earth.[7] In this model, we will treat the entire Earth as a single point (hence, the zero dimension). The goal is to account for a steady state exchange of the energy flux that goes into the Earth and the energy flux that goes out of the Earth. If the energy flows are balanced, then the system will be said to exist in a steady state with a constant temperature. However, if the flows of energy are imbalanced we will expect the system temperature of our model Earth to change (see Figure 5.3).

[7] Excellent on-line discussions of this topic can be found at PSU and NYU.

We will check our models against the resulting temperatures that we get out of the results. So, what temperature *should* the Earth be given solar irradiation gains, and longwave light losses?

BLACKBODY ACCOUNTING (NO SKY COVER)

The Earth is exposed to an average irradiance over the course of a year. This is called the *Solar Constant* (G_{sc}) and is 1361 W/m^2.[8] Notice that the value is in irradiance units of power per area, and so represents an instantaneous flux (or a 1 s interval: 1 W = 1 J/s). If we recall the spectrum of the Sun, given the losses calculated by the Inverse Square Law, the majority of the shortwave band irradiance reaching the Earth from the Sun will be bounded by approximately 200 nm and 3000 nm. Again, the described shortwave band contains the

[8] Greg Kopp and Judith L. Lean. A new, lower value of total solar irradiance: Evidence and climate significance. *Geophys. Res. Letters*, 38(1): L01706, 2011. doi: http://dx.doi.org/10.1029/2010GL045777.

sub-bands of the UV, Visible, and Near Infrared. The Earth will also emit longwave radiation (here, termed E_e) over the period of the year, in proportion to the fourth power of the surface temperature (the Stefan-Boltzmann Law).

We will assume that there is a steady state of radiative exchange between the surfaces of the Sun and the Earth. We will also assume that they are both *blackbodies* and so do not reflect any incident light. And now we re-introduce the required energy systems balance: *gazintas* have to equal *gazoutas*![9] As with all energy conversion systems, what *goes into* the Earth's surface as energy must be equivalent to the energy that *goes out of* the Earth's surface. Or, euphemistically smashing the words together, *gazintas* = *gazoutas* at steady state.[10] By our convention, the left side of the following equations will be the energy "inputs" and the right side of the equations will refer to the energy "outputs." If the net radiative exchange is greater on the *gazintas*, then the temperature of the Earth will respond by increasing. If the net radiative exchange is greater on the *gazoutas*, the Earth temperature will decrease.

[9] Warning: this section requires a sense of humor.

Euphemisms for mass and energy transfer:
this "goes into": a *gazinta*
that "goes out of": a *gazouta*
(caveat to ESL speakers: this is slang, not proper English).

[10] Thanks to a dear friend for translating this important physical fact of nature into memorable terms.

For blackbody accounting from the perspective of Earth's surface, the case is trivial, and is proportional to surfaces communicating with each other (view factor proportionality).

$$\text{Solar Flux } (W) = G_{sc} \times (\pi r_e^2). \tag{5.3}$$

These are the inputs, where r_e is the radius of the Earth, and πr_e^2 is the averaged cross-sectional area of Earth exposed to the Sun at any instant. The Earth would appear as a disk when viewed from the Sun, having a radius r_e. Only a hemisphere is exposed to the Sun at any instant, and the Sun's light is partially collimated at great distances. The cosine projection effect distributes the same quantity of photons over a larger area to the poles, as compared to the equatorial regions, but the incident area that captures photons in two dimensions is that of a circle. Therefore, the total power absorbed by the entire surface of the Earth (the energy flowing in) at any one time can be calculated as $G_{abs} = G_{sc}$ multiplied by the area of an Earth-sized disk, and not a hemisphere.

Area of a circle: πr^2.

Average radius of Earth: r_e~6371 km.

$$\text{Earth Flux } (W) = \sigma T_e^4 \times (4\pi r_e^2). \tag{5.4}$$

Area of a sphere: $4\pi r^2$

The NASA Goddard Institute for Space Studies (GISS) in New York reported average surface temperature of Earth in 2012 was about 14.6 °C or 58.3 °F. http://www.giss.nasa.gov/research/news/20130115/. Accessed February 2, 2013.

Recall that *radiosity* (J_i) is a sum of the outgoing energy from a surface—a sum of the *gazoutas.*

These are the outputs, where $\sigma = 5.6697 \times 10^{-8}$ W/m²K⁴ as in earlier chapters, and the area radiating is the entire spherical surface of the Earth in all directions ($4\pi r_e^2$; remember this is from the perspective of the Earth surface).

At steady state *gazintas* = *gazoutas*, so:

$$G_{sc} \times (\pi r_e^2) = \sigma T_e^4 \times (4\pi r_e^2). \tag{5.5}$$

If we wish to know the effective surface temperature for Earth as a blackbody with no atmosphere, we need to pull apart the Stefan-Boltzmann equation, where surface temperature is reported to the fourth power. Solving for T_e in steps:

$$T_e^4 = \frac{G_{sc}}{4\sigma} = 6.027 \times 10^9 \text{ K}^4, \tag{5.6}$$

$$T_e^4 = \left(\frac{G_{sc}}{4\sigma}\right)^{1/4}, \tag{5.7}$$

$$T_e = 278.3 \text{ K}. \tag{5.8}$$

That's about 5 °C or 41 °F as an average global temperature—pretty cool for an average of the entire globe (compared with 2012 average of 14.6 °C or 58.3 °F). And we have yet to add in the average reflectance of Earth to shortwave irradiance. Reflected light is not absorbed, and we have learned from graybody accounting that a non-zero reflectance ($\rho > 0$) will reduce the input energy to the system, thus *reducing* the effective surface temperature of Earth below 278.3 K.

A CHANGE OF VARIABLES: 0–100 ROUND NUMBERS

Before we progress to a more detailed graybody accounting that includes *radiosity* and eventually an *atmosphere*, we will apply a change of variables, normalizing our total inputs and outputs to maxima of 100 round numbers (scaled energy values without units).[11] What we would like is an energy balance where we could just start with an assumption that 100% of the shortwave photons incident upon the Earth will serve as our *gazintas*, and 100% of the longwave photons escaping

[11] Brian Blais. Teaching energy balance using round numbers. *Physics Education*, 38(6): 519–525, 2003.

into space serve as our *gazoutas*, and then balance the rest of the values of energy transfer accordingly. In order to impart directionality—because we know that light is directional—we shall now use *gazintas* and *gazoutas* as our energetic terms of convenience. The change of variables should make it easier to read and compare the energy budgets as we increase the complexity of graybody accounting and add a simplified atmosphere to our models.

The Solar Flux is $G_{sc} \times (\pi r_e^2) \cong 1.74 \times 10^{17}$W. Doesn't 100 *gazintas* seem like a much better round number to work with?

$$G_{sc} \times (\pi r_e^2) = 100 \text{ [gazoutas]}, \tag{5.9}$$

$$100 \text{ [gazintas]} = \sigma T_e^4 \times 4(\pi r_e^2). \tag{5.10}$$

At steady state *gazintas* = *gazoutas*, so: 100 [gazintas] = 100 [gazoutas]. This equation is trivial when looking at it so directly, but we will add complexity shortly. In order to transform the value of total solar power irradiance spread over one hemisphere of Earth (in units of Watts; seen in Eq. (5.5)) to a normalized value of 100 *gazintas*, and the terrestrial radiant exitance over the sphere of Earth to a normalized value of 100 *gazoutas*, we will multiply both sides of Eq. (5.5) by the equivalent terms in Eqs. (5.9) and (5.10). Then the outgoing energy is divided by the incoming energy, seen in Eq. (5.12).

Multiplying both sides of Eq. (5.5) by the equivalent terms in Eqs. (5.9) and (5.10) is equivalent to squaring both sides of Eq. (5.5).

$b = \left(\frac{G_{sc}}{400\sigma}\right)^{1/4}$ is a normalizing constant to convert radiant power from round numbers on a scale of 0–100 to an effective surface temperature, just like we saw in Eq. (5.7).

$$G_{sc} \times (\pi r_e^2) \times 100 \text{ [gazoutas]} = \sigma T_e^4 \times (4\pi r_e^2) \times 100 \text{ [gazintas]}, \tag{5.11}$$

$$\frac{G_{sc} \times \cancel{(\pi r_e^2)}}{4 \times \sigma \times 100 \text{ [gazintas]} \cancel{(\pi r_e^2)}} \times 100 \text{ [gazoutas]} = T_e^4 \tag{5.12}$$

Rearranging the equation for temperature to the fourth power from left to right:

$$T_e^4 = 100 \text{ [gazoutas]} \times \frac{G_{sc}}{4 \times \sigma \times 100 \text{ [gazintas]}}. \tag{5.13}$$

By solving for T_e we arrive at a normalizing constant $b = 88.02\, K/\text{[gazintas]}^{1/4}$ in Eq. (5.16). In this change of variables, b acts as conversion tool for normalizing all of our values to round units, permitting an easy conversion of energy to effective surface temperature (T_e).

$$T_e^4 = 100 \text{ [gazoutas]} \times \frac{G_{sc}}{400\sigma}, \tag{5.14}$$

$$T_e = (100\ [\text{gazoutas}])^{1/4} \times b,$$

$$b = \left(\frac{G_{sc}}{400\sigma}\right)^{1/4}, \tag{5.15}$$

where $b = 88.02\ \text{K}/[\text{gazintas}]^{1/4}$. (5.16)

$$T_e = 278.3\ \text{K}. \tag{5.17}$$

Both *gazintas* and *gazoutas* are really without units, but in this unusual change of variables approach, a detailed unit analysis reveals that both are morphed into round numbers raised to the 1/4 power when solving for effective surface temperatures. Again with the brain melting! We recommend that you just suspend your disbelief for a few pages, and have some fun balancing radiative energy transfer using a normalized scale.

This confirms our initial temperature calculation developed from G_{sc} in Eq. (5.7). From here on out we can use the following equation to solve for the effective temperatures of our surfaces of interest:

$$T_e = (X\ [\text{gazoutas}])^{1/4} \times b, \tag{5.18}$$

where X is a calculated value of *gazoutas* from our successive Energy Balance Models (EBMs).

BLACKBODY + GRAYBODY ACCOUNTING (NO SKY COVER)

We will now model the Earth as a graybody. We covered the topic of *radiosity* (J_i) in the last chapter on radiative exchange, and specified that radiosity would be used to perform our graybody energy accounting, as the energy lost to a surface would be the sum of the emitted and reflected radiant energy. Reflected light from the surface of Earth is called the *albedo*. The Sun is still considered to be a blackbody, but the Earth has graybody attributes in the shortwave band (~200–3000 nm).

Albedo is a term synonymous with reflectance, but used specifically by Earth systems science and meteorology.

Recall that the *radiosity* J is the sum of outgoing energy for a surface, in particular a graybody. Here we are applying the concept of radiosity to account for the *gazoutas* with respect to the surface of Earth. The shortwave light from the Sun (G_{sc}) will be reflected in proportion to the reflectance (albedo, ρ) of the materials on the Earth's surface, while the longwave light will be emitted in proportion to the effective temperature of the Earth (which we will now term

$T_{e,2}$, for our second model of Earth with respect to the Sun) and the longwave emittance (ϵ) of the materials on the Earth's surface.

$$J_e = \rho G_{sc} + \epsilon E_e. \quad (5.19)$$

The average *albedo* of Earth (ρ_{Earth}) represents the fraction of shortwave irradiance that will be reflected back to space (as this is a zero atmosphere model too), and will have a value from 0–1. The albedo here is a measure of fractional reflectance for the materials of the Earth's surface (land, ice, water, vegetation). In our case we will use an average value of $\rho_{\text{Earth}} = 0.32$.

Next, we consider the energy flowing out of the Earth due to longwave radiant exitance, E_e. If we recall the Stefan-Boltzmann Law, we are aware that any object[12] will emit a distribution of photons at a rate given by the calculus of Stefan-Boltzmann. From our earlier reading we also know that the emitted spectrum from a lower temperature emitter like the Earth would be in the *longwave band*, bounded roughly between 3000 nm and 50,000 nm for the majority of the energy density. Hopefully by working through the problems in the previous chapter, you should now be comfortable that the incoming shortwave spectrum and outgoing longwave spectrum do not share significant overlap in the distribution of their respective photons. Note that in this version of the equation there is an additional term for the *emittance* ($\epsilon \simeq 0.97$) of the average Earth surface. The coefficient is a measure of the fractional "glow" efficiency for the Earth as a *graybody*.

[12] All objects glow.

At steady state *gazintas* = *gazoutas*, and we know that we have 100 units of inputs, with a radiosity that sums to 100 with 32 units reflected and 68 units of energy emitted. Solving for the energy of the Earth, E_e, we find that the Earth has 70 units of energy:

Radiosity is just a refined way of presenting *gazoutas*. The radiosity of the Earth is the sum of the reflected shortwave light and the emitted longwave light.

$$100 = J_e = 32 + 0.97E_e, \quad (5.20)$$

$$E_e = \frac{100 - 32}{0.97} = 70 \text{ [gazoutas]}. \quad (5.21)$$

Hence, the absorbed energy of the Earth has been reduced from 100 to 70.[13] Linked to the principle of Wien's Law, we expect that a lower power will correspond to a lower effective surface temperature. The newly modified temperature of the Earth, acting as a graybody with shortwave reflectance $\rho = 0.32$ and longwave absorbtance $\alpha \simeq 0.97$:

$$T_{e,2} = (70\ [\text{gazoutas}])^{1/4} \times b, \tag{5.22}$$

$$T_{e,2} = 255\ \text{K} = -18\ ^\circ\text{C} = 0\ ^\circ\text{F}. \tag{5.23}$$

[13] If the net radiative exchange is greater on the *gazintas*, then the temperature of the Earth will respond by increasing. If the net radiative exchange is greater on the *gazoutas*, the Earth temperature will decrease.

The normalizing constant $b = 88.02$ K is derived in Eq. (5.16).

Comparison: $T_e = 278$ K, while $T_{e2} = 255$ K: a difference of -23 K! We actually observe a surface temperature of 288 K, so our model still needs work.

In fact, -18 °C is the average temperature for **Fargo, North Dakota** in the month of **January**.

The energetic change by added reflectance to a material surface results in a drop in temperature to well below zero Celsius. This is an average temperature that would freeze water across the planet and inhibit many biochemical reactions from occurring at the surface. In such a hypothetical case, the life support on Earth would likely be focused near hydrothermal vents at the bottom of otherwise frozen oceans, supported by geothermal gradients from the Earth's mantle and core. And that's definitely *not* the environment that we currently have at Earth's surface. At least now we can see that a sky is a pretty important contribution. So let's add a cover (the sky) and see how it warms things up as a mega-solar energy conversion system.

GRAYBODY ACCOUNTING WITH A COVER

The cloudless sky is a cover that permits shortwave light to pass through, while strongly absorbing (and emitting) longwave light. The sky contains clouds that are reflective of shortwave light, which is why we see them as white.

Next, we consider an Energy Balance Model where we allow graybody accounting and then add the air mass from a sky, but we only add a simplified receiving surface where the sky is effectively free of clouds and aerosols (or where we have effectively compressed the clouds to the Earth surface). What would that air mass behave like in terms of light attenuation? In such a case, the simplest model is of a homogeneous mass of gas that hangs over the globe surface. In fact, we can temporarily think about the sky as a single *cover* (much like a sheet of glass) above the *absorber* that is Earth's surface. The clear sky must still have some properties of *absorption, reflection,*

Figure 5.4: Cumulus clouds in fair weather. The blue sky is the observation of light scattering off of oxygen and nitrogen molecules (Rayleigh scattering). The white clouds are the observations of light scattering off water droplets and dust (Mie scattering). The rocks and ground are reflecting beams of radiation up to your eyes.Photo by Michael Jastremski/ Wikimedia Commons/ CC-BY-SA-2. July 2005.

transmission, and *emission* that are predictable across the entire hemisphere of the sky (see Figure 5.4).

We bring you your first cover system, the sky: transparent in the shortwave, and absorbent in the longwave. Our graybody accounting is now expanded to include the existence of a single layer atmosphere (a cover), which has the ability to absorb and emit longwave light (a portion of the infrared). Now, any cover will have *two surfaces*, which means there is an inner and an outer surface relative to the absorber surface of the Earth. That additionally implies that there will be two surfaces emitting longwave radiation from the atmosphere, and not just one (see Figure 5.5).

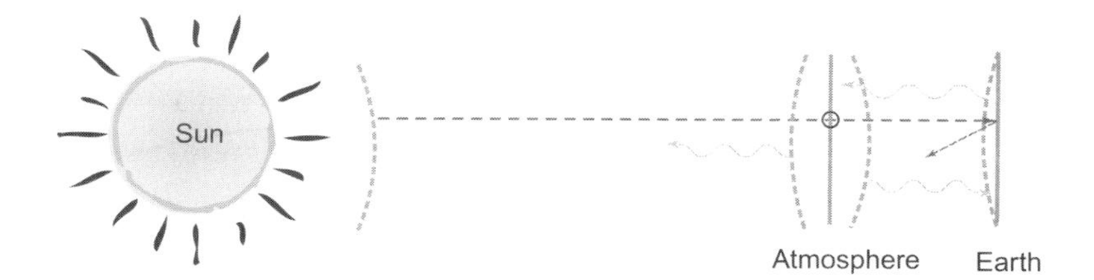

Figure 5.5: The Sun-atmosphere-Earth system energy balance. Assuming both the inner and outer surfaces of the atmosphere are longwave emitters.

There are now three bodies about which we are to balance radiant energy. The Sun, the sky, and the Earth are each thermal bodies with optical surfaces. For now we will balance the systems of energy from the perspective of a satellite out in space (just outside the atmosphere), from the perspective of a jet flying inside the atmosphere, and from the perspective of a cyclist on Earth's surface.

From satellite: Sun irradiance = SW reflectance + atmospheric emittance

From jet: Earth irradiance = atmospheric emittance

From bicycle: SW absorbance + atmospheric emittance = Earth irradiance

Each of the three systems of equations can be represented in terms of round numbers as well. The radiative exchange among three bodies:

$$\textbf{From Space}: 100 = 32 + E_a. \tag{5.24}$$

$$\textbf{From the Sky}: (0.97E_e) = 2 \times E_a. \tag{5.25}$$

$$\textbf{From Earth}: (100 - 32) + E_a = (0.97E_e). \tag{5.26}$$

$$E_a = 68\ [\text{gazoutas}],$$
$$E_e = 140\ [\text{gazoutas}],$$

and

$$T_{a,1} = (68\ [\text{gazoutas}])^{1/4} \times b = 253\ \text{K},$$
$$T_{e,3} = (140\ [\text{gazoutas}])^{1/4} \times b = 303\ \text{K}.$$

Again, $b = 88.02$ K from Eq. (5.16).

The result would then be an *average* surface temperature of $T_e = 303\ \text{K} = 30\ °\text{C}$. But this temperature is *too hot*! The average annual temperature for the long-term climate in Mumbai, India is 30 °C right now, with a much lower known global average being closer to 15 °C. So that can't be right either, and it spells a warning to all of us regarding the impact of increased CO_2 and H_2O in the atmosphere (see Figure 5.6).

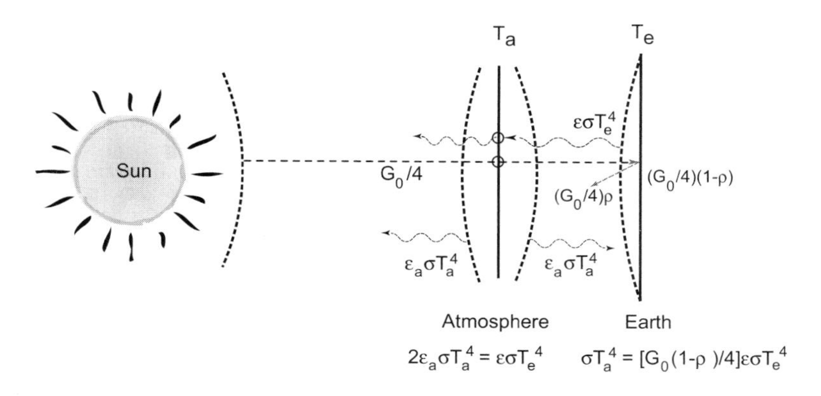

Figure 5.6: The Sun-atmosphere-Earth system energy balance with basic equations of energy balance.

GRAYBODY ACCOUNTING WITH A COVER AND A LEAKY VALVE

On a spectral level we can observe where the atmosphere is selectively transparent to a subset of wavelengths within the longwave band. The atmosphere has a selective longwave window, a band within which the longwave transmittance is quite high and from which heat leaks out into space, cooling things off. This is called the *sky window* from the problem set in the earlier chapter on light behavior. You can see the spectral window in Figure 5.7. In addition, if we inspect the figure closely, we can identify the spectral regions of transparency for the shortwave band as well.

The sky window for the longwave exists between about 8–13 μm.[14] In Figure 5.7, part (c), the absorptance of the atmosphere is represented. We see this window in the longwave band. It is interesting to note that the absorptive band to the right (13 μm) is due to CO_2, while the absorptive band to the left of the window (8 μm) is due to water vapor. So, as the CO_2 levels increase and the planet warms, more water vapor goes into the atmosphere, closing our heat escape ever-more tightly. It's not often that one gets to point out the "valve" that controls global climate temperatures. More importantly, Smith and Granqvist have proposed an innovation goal to design selective surfaces that emit radiative energy specifically within the band of the *sky window*, to specifically tune our energy emission to the frequency such that it can escape and cool things down![15]

[14] Geoffrey B. Smith and Claes-Göran S. Granqvist. *Green Nanotechnology: Energy for Tomorrow's World.* CRC Press, 2010.

[15] Claes-Göran Granqvist. Radiative heating and cooling with spectrally selective surfaces. *Applied Optics,* 20(15):2606–2615, 1981; and Geoffrey B. Smith and Claes-Göran S. Granqvist. *Green Nanotechnology: Energy for Tomorrow's World.* CRC Press, 2010.

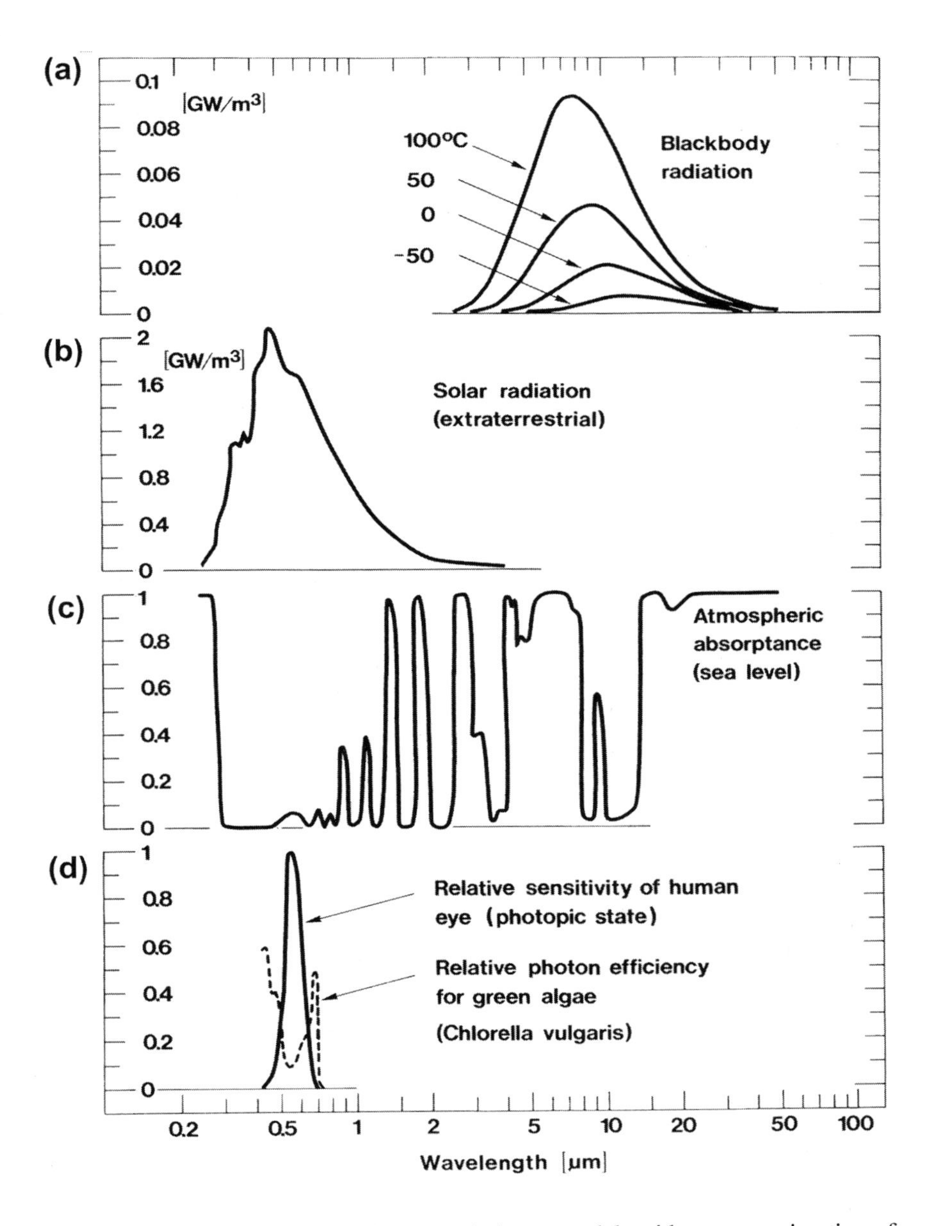

Figure 5.7: The Sun-collector system showing downwelling shortwave and upwelling longwave radiance, as well as the perception of the eye and the *sky window* from 8 to 13 μm. Image by C. G. Granqvist.

Next we repeat the three-body energy balance models with an approximation of the sky window as $\tau_\alpha \sim 0.1$ in the longwave band. The sky, as viewed from Space will permit transmission of 10% of the longwave energy from Earth. From within the sky, the inputs are only 90% of the longwave energy measured as irradiance from Earth ($0.97E_e$). From the perspective on Earth, 68 units of shortwave light are absorbed along with the units of longwave light energy emitted from one half of the sky. We still have three balances from the perspective of a satellite, a jet, and a bicycle:

$$\textbf{From Space}: 100 = 32 + E_a + (0.1 \times 0.97E_e). \quad (5.27)$$

$$\textbf{From within the Sky}: (0.9 \times 0.97E_e) = 2 \times E_a. \quad (5.28)$$

$$\textbf{From Earth}: (100 - 32) + E_a = 0.97E_e. \quad (5.29)$$

Recall that τ_α is the transmittance of a surface with respect to absorption losses.

$$\tau_\alpha = \exp\left(\frac{-kd}{\cos(\theta_2)}\right),$$

where k is the extinction coefficient, d is the cover thickness, and θ_2 is the angle of refraction.

The amazing thing is the change of −7 °C associated with radiant losses of longwave light into space.

$$E_a = 55.3 \text{ [gazoutas]}, \quad (5.30)$$

$$E_e = 127.1 \text{ [gazoutas]}, \quad (5.31)$$

and

$$T_{a,2} = (55.3 \text{ [gazoutas]})^{1/4} \times b = 240 \text{ K}, \quad (5.32)$$

$$T_{e,4} = (127.1 \text{ [gazoutas]})^{1/4} \times b = 296 \text{ K}. \quad (5.33)$$

Even so, this is still a bit cold for the atmosphere and too warm for our atmosphere. The atmosphere actually absorbs about 77% of the infrared band (including longwave) irradiance, or $\epsilon_a = 0.77$. After making this last fine-tuning adjustment, our model calculations allow us to arrive at a more comfortable result of, $T_{e,5} = T_{e,\,avg} = 288\text{ K} = +15\,°\text{C}$, which still allows Mumbai to have an average temperature of 30 °C.

Wow! That's a −7° cooling by just allowing for the atmosphere to leak radiant energy into space.

From all of this experimentation, we note two things. First, the atmosphere is a *selective surface*, in that it permits almost all shortwave irradiance to pass through as a transparent set of surfaces (there are bands of shortwave light that are absorbed by the atmosphere), while transmitting 10% and absorbing 77% of the longwave band of light from the Earth's emitting surface (and by radiative balance $1 = \tau + \alpha + \rho$: reflecting 13% of the longwave band). Second, the function of the atmosphere is both as a cover to retain thermal energy converted from radiant energy by Earth's surface, and as a valve to permit cooling of the Earth-sky system via radiative losses to space.

If I were to change the **longwave transmittance** from 0.1 to 0.08, that would have generated +1 °C of increase in the model.

Just as a caveat to the would-be modeling adventurers: we have not taken into account the full energy balance that would include convection and conduction, latent heat effects and mass flows across the atmosphere. Full Global Climate Models are massive and detailed, but these numbers do offer you an idea of the power of the solar work current delivering useful energy to pump up the thermal states on the Earth's surface as a solar energy conversion system, and the important role of the atmospheric cover.

THE ROLE OF CLOUDS

Through the chapter we have sought clever ways to avoid the discussion of the most complicated components in the atmospheric system: clouds. Clouds have the ability to scatter and even focus light, yet they are ephemeral and dynamic systems in and of themselves.

The word for associating playful shapes to the forms of clouds? **nephelococcygia**.

Clouds are amazing emergent features of the atmosphere, they can be viewed from the ground as well as from satellite imagery. Clouds have a rich array of sizes, shapes, and dynamics—all composed of microscopic droplets of buoyant water. Those water droplets scatter light from the Sun and the surrounding sky and are major contributors to reducing the solar budget from moment to moment, and across hours and days.

Clouds develop in both active and passive environments, as well as from updrafts near the Earth's surface. *Cumuliform clouds* are the familiar puffy, cauliflowered morphologies of clouds, formed in dynamically active, unstable air. The buoyancy of the air enhances their vertical growth, and their diameters are of a similar dimension to their height above the ground. As we will see in Figure 5.10, the size of cumuliform clouds extends across many orders of magnitude in distance and time. Cumuliform clouds tend to be found near cold fronts of an air mass, and include: *cumulis humilis, cumulus mediocris, cumulus congestus, and*

Clouds near or behind a cold front tend to be **cumuliform**.

Clouds near a warm front tend to be **stratiform** and extend from 100s to 1000s of km.

cumulonimbus (otherwise associated with thunderstorms). Stratocumulus clouds are somewhat different, in that they tend to form from surface updrafts and wind shear turbulence, and are not often caused by buoyancy.[16]

Clouds forming in statically stable air are termed *stratiform*. These cloud types are the sheet-like layers that cover wide ranges of the sky, and are associated with warm fronts. They are dynamically passive in that vertical growth of these clouds is suppressed by buoyant forces. Stratiform only exist while there is an external process lifting the clouds to overcome buoyancy. This process can be advection of humid air along and up the warm-front surface, can be spatially extensive (1000s of km), and they are not coupled by convection with the ground underneath them. Stratiform clouds include: *cirrus, cirrostratus, cirrocumulus, altostratus, altocumulus, stratus, and nibostratus*.[17]

[16] Roland B. Stull. *Meteorology for Scientists and Engineers.* Brooks Cole, 2nd edition, 1999.

[17] Roland B. Stull. *Meteorology for Scientists and Engineers.* Brooks Cole, 2nd edition, 1999

CLEAR SKY MODELING

We have just observed the crucial role of the sky in SECS, and discussed the importance of clouds in reducing the solar budget. Now, we would like to dial it all back from the massive global scale to a single locale that a client or stakeholders are interested in. How do we model the transparency of a hypothetical sky that is homogeneous and "clear" of emergent phenomena such as clouds? How can we apply concepts from atmospheric science to predict a clear sky projected across the sky dome?

Transparency of the atmosphere is a function of several physical materials properties such as the gas chemistry (CO_2, CO, N_2, O_2, O_3), water vapor content ($H_2O_{(g)}$), and aerosol concentration (measured as AOD, aerosol optical depth). The measured transparency is also affected by the solar altitude angle (at a given time) and the elevation of the site relative to sea level. At solar noon for a clear day, roughly 75% of the extraterrestrial irradiance can be transmitted through the atmosphere. This means that even under ideal clear sky conditions, a quarter of light can be scattered and absorbed as it passes through the atmosphere.[18] As a part of the *integrative design process*, it is highly encouraged that solar design teams incorporate *atmospheric scientists and meteorologists*, so that

Assume the **clear sky** as a baseline condition. We all have the sky around us, and then the clouds decrease the local solar budget further from the clear sky baseline.

If the **uncertainty** of a model is high, and the loss from an inaccurate model is also high: we would call this high **risk**.

[18] Matthew J. Reno, Clifford W. Hansen, and Joshua S. Stein. Global horizontal irradiance clear sky models: Implementation and analysis. Technical Report SAND2012-2389, Sandia National Laboratories, Albuquerque, New Mexico 87185 and Livermore, California 94550, March 2012. URL http://energy.sandia.gov/wp/wp-content/gallery/uploads/SAND2012-2389_ClearSky_final.pdf.

meteorological inputs are in fact accurate and updated for all climate regimes of the study in question.

Each of the following tools are models for an ideal clear sky, and each must be viewed within the balance of acceptable uncertainty. All the same, modeling the clear sky with the proper physical parameters is relatively straightforward, when provided with accurate meteorological inputs characteristic of the air mass in question.

Clouds are probably the most important factor in reducing terrestrial irradiance, however particles in the atmosphere can contribute significantly for certain SECS. In particular, the import of *aerosols* cannot be underestimated in solar energy assessments for concentrating systems. Aerosols are particles of solid or liquid suspended in air. To be suspended for an extended period of time in the sky, the particles tend to be quite small in diameter (nm to μm scale). Examples include water droplets, salt crystals, soot, and terrestrial minerals. When water aerosols aggregate sufficiently, the emergent phenomena of clouds or fog result. We note that in many portions of the developed world, aerosol loadings can be small, and the influence on non-concentrating systems like flat-panel PV can be less than 5%. However in Asia, particularly southeast China, the role of aerosols is a major seasonal contributor reducing the solar budget. Aerosol scattering strongly affects the *beam component* of solar irradiance via the Direct Normal Irradiance (DNI), and research has shown that beam estimation from empirical models is of low utility where time steps decrease below the diurnal periodicity.[19] Also, aerosols are present even on "clear sky" days when visible clouds are absent. Again and again, do not trust your eyes to estimate the solar resource—measure, measure, measure!

For the mid-latitudes of the USA, at solar noon for day of clear skies, observation suggests that approximately 25% of the downwelling shortwave irradiance is scattered and absorbed upon interacting with the Earth's atmosphere.[20] So in fact, the atmosphere is not completely transparent to the shortwave band in the real world (surprise!). If we consider the longer path length that light must propagate through during mornings and evenings, we can therefore expect the attenuation of light (optical losses) will increase in the early and late hours of the day. Light that comes in a direct path from the Sun is called direct normal irradiance (DNI, where we use the term beam normal and the irradiance symbol $G_{b,n}$).

Aerosols: particles of solid or liquid suspended in air (nm to μm scale diameter). Composition of aerosols in the atmosphere includes water droplets, salt crystals, soot, and terrestrial minerals (dirt).

Beam component: the contribution of the solar budget from the bright direct beam from the Sun.

DNI: Direct normal irradiance ($G_{b,n}$). The beam of light normal to the surface of the Sun—crucial to solar tracking systems.

[19] C. A. Gueymard. Temporal variability in direct and global irradiance at various time scales as affected by aerosols. *Solar Energy*, 86:3544–3553, 2013. doi: http://dx.doi.org/10.1016/j.solener.2012.01.013.

[20] Matthew J. Reno, Clifford W. Hansen, and Joshua S. Stein. Global horizontal irradiance clear sky models: Implementation and analysis. Technical Report SAND2012-2389, Sandia National Laboratories, Albuquerque, New Mexico 87185 and Livermore, California 94550, March 2012. URL http://energy.sandia.gov/wp/wp-content/gallery/uploads/SAND2012-2389_ClearSky_final.pdf.

However, downwelling irradiance interacts with the molecules and particles in our atmosphere as well, and can be scattered in all directions. The portion of this redirected irradiance that is cast about the sky dome and then down toward Earth is termed *diffuse irradiance*. The diffuse irradiance from the sky dome is identified as a separate *component* source of light, and explains why we observe light even when we are out of view of the *beam component* while in the shade. Solar power tends to be derived solely from the *diffuse component* of irradiance during overcast days.

Specifically for a horizontal surface (like a table or a parking lot), the sum total of shortwave downwelling irradiance is called global horizontal irradiance (GHI). This is a general case of study, as most global irradiance measurements have been made using a solar metric device called a *pyranometer* that is mounted horizontally. In complement, diffuse irradiance is measured with a pyranometer having a shading tool to eliminate beam irradiance.

Direct normal irradiance (DNI): small acceptance angle (2.5°) measure of irradiance relative to a plane perpendicular to surface of the Sun. DNI ($G_{b,n}$) tracks the Sun, and doesn't have the same meaning as the *beam component* relative to the plane of array ($G_{b,t}$).

Global horizontal irradiance (GHI): measure of global (total) irradiance relative to a horizontal surface facing up to the sky dome.

Plane of Array (POA): measure of the global irradiance relative to the general orientation of the aperture surface for the solar energy conversion system.

COMPONENTS OF LIGHT

Recall that light is directional. Even if the clear sky is refracting and reflecting sunlight down to us, we only care about that last bounce imparting direction toward our SECS. Hence, we document the sky, the snow, the pavement, each as *sources* of light for our SECS. A clear sky is a function of the solar altitude angle (α_s) (the complement being the zenith angle θ_z), site altitude, aerosol

Components of irradiance relative to **GHI** and **POA**:

- Beam horizontal (G_b)
- Sky diffuse horizontal (G_d)
- Circumsolar diffuse
- Horizon diffuse
- Ground reflectance (ρ_g, albedo)
- **Sum**: Global or total horizontal Irradiance (G)

Often we will just link the beam and circumsolar components together, and label both *beam*.

Recall that light scattered/reflected off of the ground is called the *albedo* (ρ), with a fractional range from 0–1.

concentration, water vapor, pressure, and composition of the sky at the given locale. We can break up the sky dome and the terrestrial surface into multiple pieces, or *components* of light sources. Summing all components together for shortwave light impinging upon a collector, we have the *global* or *total* irradiance.

The air chemistry in the path of that beam will scatter the energy into an additional small cone of light, called the *circumsolar* component of the sky dome. Next, the scattering events in the sky dome during the day produce a blue or white hue across the majority of the hemispherical surface. We call this source of light the *sky diffuse* component of irradiance. As the path length increases for scattering in the sky, we also may observe a subtle brightening along the lower disk of the sky dome, which we call horizon brightening, associated with the *horizon diffuse* component. Finally, the reflectance of the ground (albedo) will contribute to any collectors that are not horizontally mounted (tilted or vertical systems). As we indicated earlier, diffuse irradiance will also include the reflection of light from the ground, which depends on the surface albedo (making snow days fairly bright days). We call this the *ground reflected* component of the sky dome–affected by the *albedo* of the reflective surfaces.

In Table 5.1 we see examples of material albedos in the shortwave band, but presented as average graybody metrics. In reality, most materials have reflectances that vary selectively across the shortwave or longwave spectra.

Table 5.1: A list of material albedos averaged over the shortwave band.

Material	Albedo (ρ)
Fresh asphalt	0.04
Worn asphalt	0.12
Conifer trees	0.08–0.15
Deciduous trees	0.15–0.18
Bare soil	0.17
Green grass	0.25
Desert sand	0.40
Fresh concrete	0.55
Old snow	0.40–0.60
Fresh snow	0.80–0.90

Notice in the next Figures 5.8 and 5.9 how plants can reflect light selectively, while roofing tiles (tar-based), asphalt (also carbon), and sea water are quite poor reflectors across the shortwave spectrum.

The projection of DNI onto the specific case of a horizontal surface is a simple geometric relation, seen in Eq. (5.34). The resulting value is the beam irradiance value of G_b, derated or increased in proportion to the cosine of the zenith angle (θ_z). Again, note that the beam irradiance is specific to a horizontal surface, while DNI is a value measured from a plane perpendicular (said to be "normal") to the direction of the Sun.

$$G_b = G_{b,n} \cos(\theta_z). \tag{5.34}$$

However, many of our intended SECS are not going to be horizontal. If the surface in question is tilted with an azimuth orientation in some direction

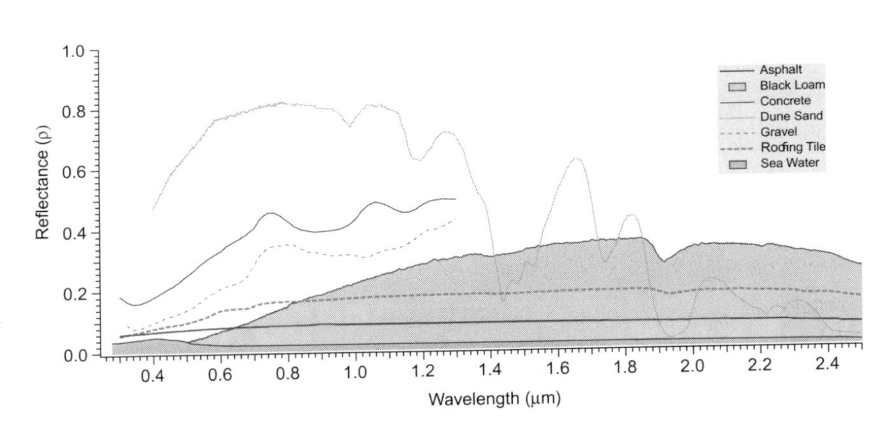

Figure 5.8: Spectrally selective reflectances from inorganic materials and sea water.

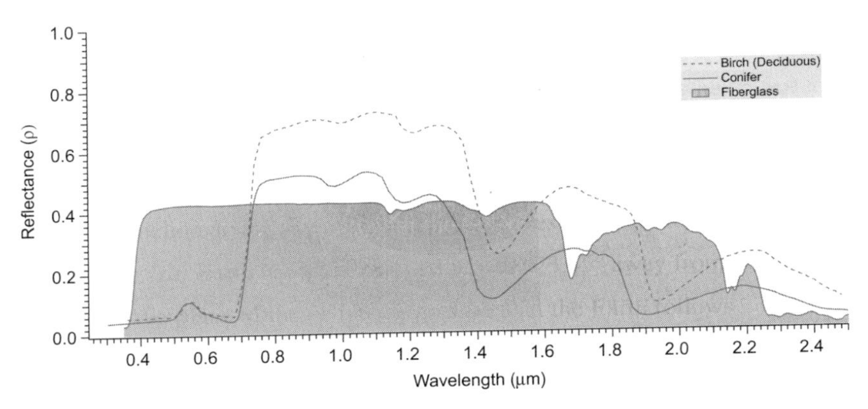

Figure 5.9: Spectrally selective reflectances from tree types and from fiberglass.

N/E/S/W, then we can sum the total contributions of global or total irradiance for a specific *Plane of Array* (POA). That is, if one knows the DNI component measurement or estimation, which will become the subject of study in Chapter 8 on measurement and estimation.

For the purposes of clear sky modeling for a horizontal surface (not tilted up or vertical), we will limit our models to include beam and diffuse components without ground contributions. It should be apparent that the global irradiance is the sum of the components. Some of the strongest models incorporate the physics of the sky chemistry interacting on a *spectral basis* with solar irradiance (each wavelength is assessed for direct and diffuse contributions), but the first that we shall discuss relates to an integration of the shortwave band energies for the beam component only.

HOTTEL CLEAR SKY BEAM MODEL (1976)

This model is for demonstration only to indicate the role of the atmosphere and the angle of refraction. The Hottel model is no longer used in practice.

Hottel introduced an empirical model of a clear sky for *direct normal irradiance* (DNI) (the beam component, on surface pointing directly at the Sun) irradiance based on calculations of atmospheric transmittance from US Standard Atmosphere data of 1962. The parameters a_0, a_1, and k were then provided for general atmosphere that were "clear" or had "urban haze" (isn't that an interesting euphemism), as a function of the altitude for the US location in question.[21] What is important about this work is the form of Eq. (5.35).

[21] William B. Stine and Michael Geyer. *Power From The Sun*. William B. Stine and Michael Geyer, 2001. Retrieved January 17, 2009, from http://www.powerfromthesun.net/book.htm.

$$G_{b,n} = G_0\left[a_0 + a_1 \exp\left(\frac{-k}{\cos(\theta_z)}\right)\right] \quad (\mathrm{W/m^2}). \tag{5.35}$$

Let's compare this equation to Eq. (5.36) which conveys the fraction of transmission losses due to absorption of light (τ_α) through a transparent cover, discussed in our earlier discussion on semi-transparent materials.

$$G_0 \cdot \tau_\alpha = G_0\left[\exp\left(\frac{-kd}{\cos(\theta_2)}\right)\right] \quad (\mathrm{W/m^2}). \tag{5.36}$$

The value of τ_α can be calculated by knowing the *angle of refraction* (θ_2) and the *extinction coefficient* of the material (k), as well as the thickness of the material or distance that light must propagate through (d). Extinction coefficients in solar energy tend to be reported within the shortwave band as inverse distances, and so by multiplying by a distance one results in a dimensionless numerator (as Hottel indicated). For transparent cover materials the extinction coefficient will be on the order of 4–30 m^{-1}. The zenith angle in Eq. (5.35) has nearly the same meaning as the angle of refraction in Eq. (5.36) (refraction is assumed to be very small here). Notice the exponential form of both equations, and compare how Hottel was able to generate a somewhat useful model of a clear sky just by using the principle of transmittance.

Having explored the earlier model for the DNI beam normal component ($G_{b,n}$) and the beam component for horizontal surfaces in the shortwave band, we can move on the diffuse horizontal irradiance (DHI) calculation. Or can we? As it turns out, the approximation of diffuse irradiance is a non-trivial pursuit. One can directly measure the total downwelling irradiance, and project the DNI data onto beam irradiance on a horizontal plane. That at least is a simple geometric matter of Eq. (5.34).

Muneer has investigated several of these diffuse sky models.[22] After review, all models appear to have a regional character associated with seasons, clouds, air mass source regions—regional characters that we known as synoptic, mesoscale, and microscale weather (or meteorology). Now, being that many of the places that humans live in experience something less than a clear day, estimating the diffuse component is challenging, and full of uncertainties. But what can we pull out of this brief discussion for useful design purposes?

[22] T. Muneer. *Solar Radiation and Daylight Models.* Elsevier Butterworth-Heinemann, Jordan Hill, Oxford, 2nd edition, 2004.

BIRD CLEAR SKY MODEL (1981)

Richard Bird and peers developed the *Bird Clear Sky Model* at the Solar Energy Research Institute (SERI), what is now known as the Dept. of Energy National Renewable Energy Laboratory (NREL, pronounced "*en*-rell").[23]

[23] R. E. Bird and R. L. Hulstrom. Simplified clear sky model for direct and diffuse insolation on horizontal surfaces. Technical Report SERI/TR-642-761, Solar Energy Research Institute, Golden, CO, USA, 1981. URL http://rredc.nrel.gov/solar/models/clearsky/.

The Bird model requires specific input data, listed below:

- solar constant (G_0, (W/m^2))
- zenith angle (θ_z)
- surface pressure (P, (mbar))
- ground albedo (ρ_g)
- precipitable water vapor ([H_2O] (cm))
- total ozone ([O_3] (cm))
- turbidity at $\lambda = 500$ nm and/or at 380 nm
- aerosol forward scattering ratio (0.84 recommended)

The model has since been incorporated into a number of openly available codes, including a standard spreadsheet format for experimentation and exploration, done by Daryl Myers at NREL. (See RReDC Clear Sky and Simple Bird Model.) The output of the model is a "clear sky" estimate for the total (global horizontal irradiance: GHI), direct normal irradiance (beam, DNI) and diffuse irradiance across a spectral range covering 122 wavelengths (irregularly spaced) from 305 nm to 4000 nm. The model calculates these conditions for a single point in solar time, given the latitude (ϕ), longitude (λ), and Time Zone. Hence, one would need to plot multiple points to gather the arc of component-based irradiance across the hours of the day. As this is a clear sky model, the effect of clouds, trees, mountain ranges, and urban shading are not yielded.

Another modern solar energy sky simulation tool is SMARTS, developed by Dr. Christian Gueymard of Solar Consulting Services.[24] However, SMARTS is largely used for research investigations, and a non-spectral tool was later developed to model the contributions of diffuse and beam components for clear

[24] Christian Gueymard. Simple model of the atmospheric radiative transfer of sunshine, version 2 (smarts2): Algorithms description and performance assessment. Report FSEC-PF-270-95, Florida Solar Energy Center, Cocoa, FL, USA, December 1995.

skies. This tool is called REST2, and similar to the Bird Clear Sky model accepts atmospheric inputs of:

- pressure,
- precipitable water (or temperature and relative humidity),
- reduced ozone and NO_2 path lengths,
- as well as the very important scattering factors of Ångström's exponents of scattering for wavelengths above and below 700 nm, and Aerosol Optical Depth.

REST2 also outputs estimates of illuminance and PAR, or Photosynthetically Active Radiation in the same file.

The REST2 model will then output estimations of diffuse, beam (DNI), and global plane of array (POA) irradiance.[25]

[25] Christian A. Gueymard. Rest2: High-performance solar radiation model for cloudless–sky irradiance, illuminance, and photosynthetically active radiation–validation with a benchmark data set. *Solar Energy*, 82:272–285, 2008.

SPATIO-TEMPORAL UNCERTAINTY

There is *uncertainty* tied to incorporating solutions that require action and orientation to the system of problems. Recall that our potential to develop SECS as ecosystems technologies is linked to the science of sustainability. Sustainability science in solar energy requires integration and alignment among multiple disciplines, integration of science and society, and specifically requires solutions that reference the *locale* of interest along with *coordination of solutions in time*.[26]

[26] Christian U. Becker. *Sustainability Ethics and Sustainability Research*. Dordrecht: Springer, 2012.

Let's discuss some really certain phenomena on our planet first. For the next several billion years, we have a high certainty that the Sun will rise in the morning, and will set in the evening everywhere on Earth because of the relative orbit of Earth and Sun. Yes, there are a few regional exceptions occurring beyond the Arctic Circle ($\phi = +66.6°$) and the Antarctic Circle ($\phi = -66.6°$), but for the vast majority of humanity living between those extremes, the Sun will rise, peak, and set every day of the year. The sunrise and sunset, along with the peak at solar noon are termed *diurnal events*. Diurnal events mean daily events, and in

The term diurnal has an etymological origin in Latin: combining *dies* for "day" with *-urnus*, a descriptive suffix that denotes time.

fact *circadian* phenomena in your biochemistries are diurnal events, because the term circadian means (*circa* + *diem*) "about the day." So diurnal events happen regardless of whether or not we have a sky on Earth—another way of framing this is that diurnal events occur independent of the *weather*.

Meteorological (weather) events are less certain events, as they involve moving masses of atmospheric chemistry interacting with the energy transfers among Earth and Sun. On the other hand, climatological trends can be more certain than most meteorology: seasons will happen annually in similar ways from decade to decade, the Earth's spin has predictable wobbles that influence perturbations in the annual energy balance on a geologic scale, and climate involves very large spatial data sets too. You will notice that we continue to refer to locations and the timing of events with respect to certainty, or the lack of it, uncertainty. Inside of the field of SECS design and deployment, we are mainly concerned with quantifying the uncertainty linked to meteorological phenomena, although climatological changes can affect our systems in minor ways over the typical lifetimes of 30–50 years for a SECS.[27]

ROBOT MONKEY DOES SPACE-TIME

Weather occurs on multiple scales of space and time. In order to discuss anything "spatio-temporal" in nature, we need to step back to an earlier period in the history of meteorology. Among his many other talents and accomplishments, Sir Geoffrey Ingram Taylor (b. 1886–d. 1975) was able to present a hypothesis to link turbulent fluid phenomena that occur over time and the same phenomena that occur within a confined space. The application of *Taylor's Hypothesis* allows me to work either in units of time or in units of distance, and still have a mechanism to draw the two together.

Taylor's hypothesis states that a series of changes in time for a fixed place, like you observing a thunderstorm passing by directly overhead, is due to the

[27] A. McMahan, C. Grover, and F. Vignola. *Solar Resource Assessment and Forecasting*, chapter "Evaluation of Resource Risk in the Financing of Project." Elsevier, 2013.

passage of an unchanging spatial pattern over that locale—that is, the clouds of the thunderstorm advecting (floating) across the sky.[28] When an observer is fixed and only looking overhead to watch the changing phenomena, we call this the *Eulerian frame of reference*. You could also imagine an observer floating along with a big cumulus cloud as it moves from the space over your house, onto your neighbor's house, and then onto the next neighborhood. When the observer moves with the meteorological phenomena *instead of remaining fixed*, it is termed the *Langrangian frame of reference*.

Taylor was part of a British delegation to the Manhattan Project to predict blast wave behavior, designed supersonic airplanes, *and* provided key theories to the field of meteorology.

[28] G. I. Taylor. The spectrum of turbulence. *Proc Roy Soc Lond*, 164:476–490, 1938.

The assumption of Taylor's hypothesis is that a change in measurement over time (like a periodic signal of *dark-bright-dark-bright-dark*) results from the lateral change in the conditions across region of the meteorological event (e.g., *cloud-sky-cloud-sky-cloud*). While not isolated to meteorology, Taylor's hypothesis permits a time series of irradiance observations over fixed locations to be converted into an equivalent translation across space (at the advective or propagation speed to the corresponding spatial pattern). Hence, *all time scales are also spatial scales* so long as the advective wind speed is much greater than the time scale of the evolving meteorological event being investigated, as is often the case.

Eulerian reference frame: the observer is *stationary*, and velocities of meteorological phenomena are related to a fixed frame. Turbulence manifests as an observed series of waves in time.

Lagrangian reference frame: the observer *moves with the fluid* element (an air parcel). Turbulence manifests as a cloud moving across the sky.

In general, there is a need for understanding meteorological scales within the solar industry and the associated energy industries and markets. *Intermittency* is an emergent property of dynamical systems consisting of aperiodic, and potentially chaotic phase alternations. Solar technologies such as photovoltaic and wind power generation are linked directly with intermittent meteorological phenomena, including irradiance, air temperature, and wind speed, designating them as a class of *non-dispatchable* power sources. At times, the renewable energy industry substitutes the more generic term of "variability" to describe the irregular power fluctuations from intermittent meteorological phenomena.

Intermittency: emergent property of dynamical systems. Consists of aperiodic, potentially chaotic phase alternations.

Non-dispatchable: power sources that cannot adjust their output on demand.

The really interesting questions in distributed generation of renewable energy revolve around how we collectively address intermittency/variability across many time scales, using current and emerging technologies and techniques.[29] There are many systematic coordination challenges in the renewable energy industry for power generation, spanning orders of magnitude in spatial and temporal scales.

[29] Alexandra von Meier. Integration of renewable generation in California: Coordination challenges in time and space. 11th International Conference on Electric Power Quality and Utilization, Lisbon, Portugal, 2011. IEEE: Industry Applications Society and Industrial Electronics Society. URL http://uc-ciee.org/electric-grid/4/557/102/nested.

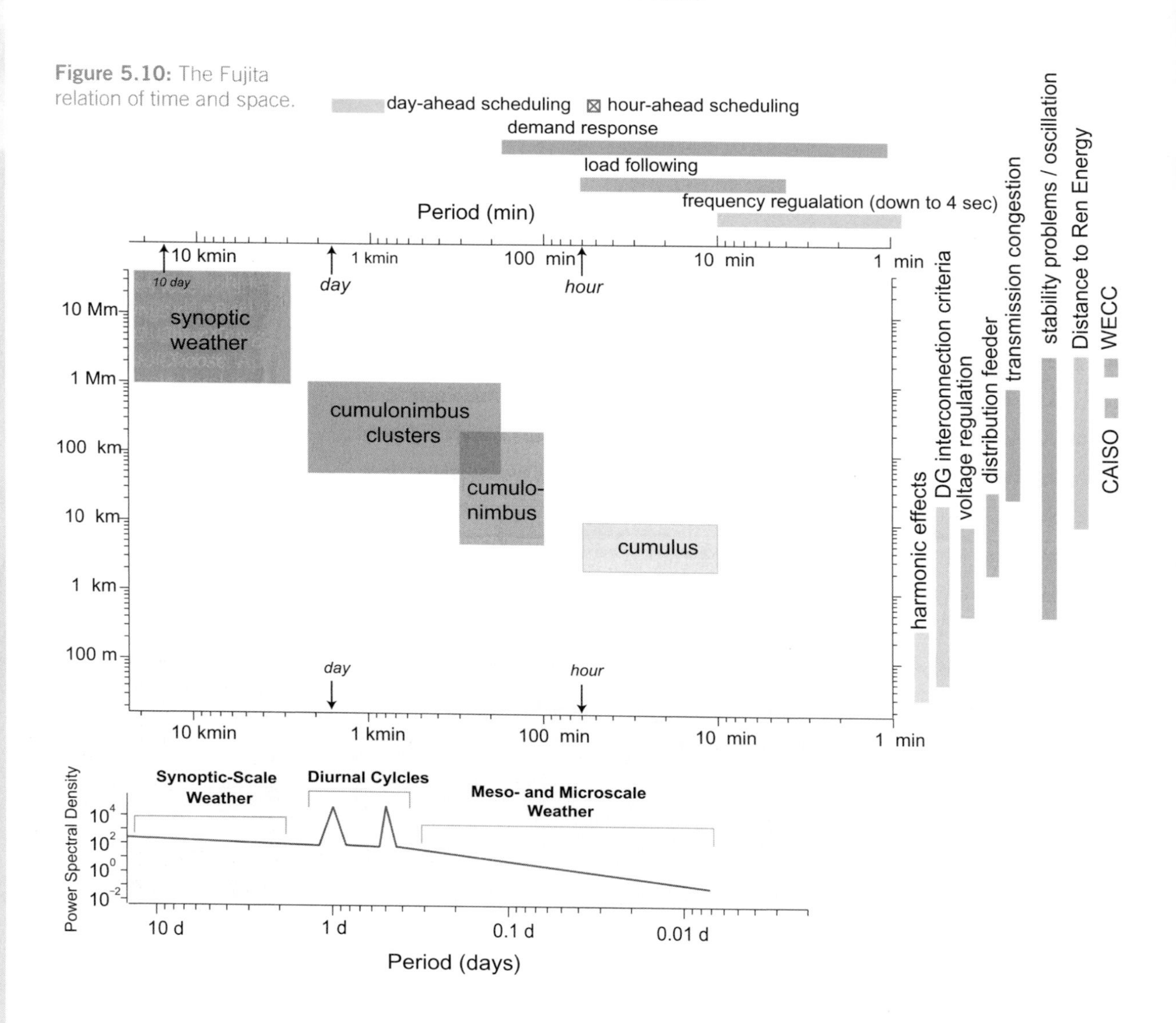

Figure 5.10: The Fujita relation of time and space.

These coordination challenges highlight the importance of successful planning and adaptation to integrate renewable energy sources in the future.[30]

We have included Figure 5.10 as a common reference scaling for time *coupled* with spatial distances, understood within meteorology. Analysis of data focuses on weather-driven phenomena occurring at longer and shorter periods than the diurnal peaks (not weather-driven). The lower left sub-figure is a simplified sketch of a Power Spectral Density (PSD) plot, and illustrates

[30] J. Rayl, G. S. Young, and J. R. S. Brownson. Irradiance co-spectrum analysis: Tools for decision support and technological planning. *Solar Energy*, 2013. doi: http://dx.doi.org/10.1016/j.solener.2013.02.029.

measures of statistical variance in irradiation values across logarithmic time scales. By integrating any interval under that curve we observe a measure of the statistical variance (σ^2) for the site in question.

The central diagram in Figure 5.10 blocks out the Fujita relations of major cloud and weather phenomena diagonally, according to both time (horizontal) and spatial (vertical) scales—again, the scales are logarithmic. At the upper range (the upper left), the field of meteorology encompasses large regions of >1000 km distances and somewhat large periods of 30–90 days. This large scale is called *synoptic scale weather*, and can be observed in seasonal shifts down to events occurring at periods larger than diurnal events. Moving from the center down toward the lower right are scales of *mesoscale* and *microscale weather*, and the cloud phenomena associated with them.[31]

Around the perimeter of Figure 5.10, we show the intersections of time and spatial scales with meteorological phenomena with respect to planning, managing the power grid, and electricity markets. Regulatory and technological parameters for power systems management and markets occur over time scales of years, to days, to less than 1 min (more than 6 orders of temporal magnitude). Power responses within the regional power grids, natural stability problems within the grid, transmission congestion, and regulatory scopes tend to occur over spatial scales of thousands of kilometers down to tens of meters (greater than 5 orders of spatial magnitude).[32] The crucial systematic and complementary scales for renewable energy resources are also those of meteorological phenomena, as demonstrated in the spatio-temporal scales of cloud features by Fujita. By applying the average translation speed of 17 m/s, we can convert the spatial scales of variability associated with transmission congestion—distances of 25–1000 km—to be relevant for meteorological phenomena within time scales of 25 s to 980 min (16 h). At an alternate distance scale, the harmonic effects that propagate within the electrical power grid for distances of 30–300 m would be relevant for meteorological phenomena within time scales of 1.8–18 s.

Again Taylor's Hypothesis "works" when the moving air that is the sky (i.e., wind speeds) are significantly faster than the evolution time of a cloud or a

[31] T. Theodore Fujita. Tornadoes and downbursts in the context of generalized planetary scales. *Journal of Atmospheric Sciences*, 38(8):1511–1534, August 1981.

The peaks in a **Power Spectral Density** (PSD) plot indicate daily or diurnal irradiance phenomena (e.g., sunrise/solar noon/sunset and harmonics of those). They are strong features, however they are actually mathematical artifacts of Fourier decomposition (fitting a non-sinusoidal diurnal cycle with a series of sinusoids).

[32] Alexandra von Meier. Integration of renewable generation in California: Coordination challenges in time and space. 11th International Conference on Electric Power Quality and Utilization, Lisbon, Portugal, 2011. IEEE: Industry Applications Society and Industrial Electronics Society. URL http://uc-ciee.org/electric-grid/4/557/102/nested.

Regulatory and technological parameters: long-term planning for transmission and generation, power contracts, day-ahead scheduling/markets, hour-ahead scheduling/markets, service restoration, and demand response.

thunderstorm of clouds. To recap, clouds are very important to solar energy resource assessment, as they make solar energy prediction uncertain. Yet clouds have time scales of evolution that are frequently much slower than the speed of the wind. So big clouds will have big periods of influence or long lives, and hence large distances of influence on a SECS. In contrast, small puffy clouds will live for much shorter periods, and hence affect smaller distances on a specific SECS.

Congratulations!
Robot Monkey grants you one glowing banana for getting through the mindstretch

PROBLEMS

1. What is the sky dome in solar energy?

2. Describe the difference between a meteorological air mass and the empirical air mass used to calibrate short-term performance of solar technologies (AM1.5).

3. How does one calculate AM0 (extraterrestrial irradiance)?

4. What is the meaning of "clear sky" in solar energy conversion? Specify what parameters are included and excluded from a clear sky calculation.

5. What is *Taylor's Hypothesis* (also called the frozen turbulence hypothesis), and how can it be applied as an important relation in solar energy resource assessment?

6. What is *variance*, and how can it be applied as an important relation in solar energy resource assessment?

7. Explain the methodology to assessed networked solar resource assessment for power systems.

RECOMMENDED ADDITIONAL READING

- Meteorology for Scientists and Engineers[33]

- A First Course in Atmospheric Radiation[34]

[33] Roland B. Stull. *Meteorology for Scientists and Engineers*. Brooks Cole, 2nd edition, 1999.

[34] Grant W. Petty. *A First Course in Atmospheric Radiation*. Sundog Publishing, 2nd edition, 2006.

SUN-EARTH GEOMETRY

06

[Harrison's "equation of time"] table enabled the clock's user to rectify the difference between solar, or "true" time (as shown on a sundial) with the artificial but more regular "mean" time (as measured by clocks that strike noon every twenty-four hours). The disparity between solar noon and mean noon widens and narrows as the seasons change, on a sliding scale. We take no note of solar time today, relying solely on Greenwich mean time as our standard, but in Harrison's era sundials still enjoyed wide use.

Dave Sobel's Longitude: The True Story of a Lone Genius Who Solved the Greatest Scientific Problem of His Time (2007).

Diunior et excellentior sit Triangulorum sphæricorum cognitio, quam fas sit eius mysteria omnibus propalare.—The nature of comprehending spherical triangles is so divine and elevated that it is not appropriate to share these mysteries with everyone.

Tycho Brahe De Nova Stella (1573)

THINK, before watches and computers were synchronizing time via satellite, in accordance with an international standard set by atomic clocks in national laboratories across the globe, our cultures had

After completing this chapter, we wish you congratulations and bid you welcome to the club of mysteries that Tycho Brahe referred to.

powerful thought and design invested in the linkage between astronomy, place, and time. We can surmise from an informal poll that our contemporaries in science and engineering (and science writing) no longer view *solar time* as relevant to "modern" society, even though solar time is equivalent to the sequencing of diurnal events that we observe. In solar design we require that our concepts transition into a solar time frame of reference. For the solar design professional, it is far easier to just find solutions first in solar time, and then correct our answers back to mean time for the public.

We have established that time and space are essential relations in solar resource assessment, and we have argued that in keeping with a sustainability ethic for design in SECS, the precepts of *sustainability science* should be in our minds to ensure that our *solving for patterns* are tied to critical self-reflection of our diverse scientific approaches and our underlying assumptions. Again, among other factors sustainability science is developed in reference to the *locale* and the local nature of solutions as well as coordinating our solutions for ecosystems services (like provisioning and regulating services) *with respect to time*. One cannot develop a strong design project for SECS without holding in mind the locale and the relative positioning of the Sun and Aperture with respect to time.

Sustainability in design and systems thinking has enormous potential to lift up society and our supporting ecosystems. You will likely spend your entire careers developing sustainability as an integral ethic to your design principles.

[1] Chrstian U. Becker. *Sustainability Ethics and Sustainability Research.* Dordrecht: Springer, 2012.

Now consider the topic of communication. As solar energy design is a part of sustainability science and environmental technology, we must communicate our work across many disciplines, and we must be able to bridge the communication among science and society.[1] So how do we "orient" ourselves in both language and coordinates such that we can communicate across disciplines and audiences regarding:

- our position on Earth relative to others,
- the orientation of our SECS relative to the Sun,
- the time of day relative to the Prime Meridian,

- the times of shadowing from the beam irradiation relative to surrounding trees and buildings, and
- the animation and tracking of our SECS with respect to the Sun?

These are all relevant and challenging, and each is within our grasp to learn and then hone to become strong career skills as professionals in the solar industry.

Given the goal of solar design, we will have three tools that we can leverage to affect the solar utility for a client in a given locale:

1. Reduce the *angle of incidence* on an aperture/receiver.
2. Reduce the *cosine projection effect* on an aperture/receiver (the extreme angles of incidence or low glancing angles).
3. Reduce losses from *shadowing* on an aperture/receiver (when desired).

However, in order to leverage these three tools, we need to know where and when the Sun will be relative to our collector system.

Goal of Solar Design:
- Maximize the solar utility.
- For the client.
- In a given locale.

EXPERIMENT WITH A LASER

In fact, this might be a good time to grab a laser pointer and experiment with a dark wall or table. Take a moment to consider what happens to a surface when you tilt a laser pointer from an angle perpendicular (also called "normal" in geometry) to your receiving surface. The perpendicular angle is called a zero angle of incidence ($\theta = 0°$). The compliment to that angle (90° minus the angle of incidence) we will call the *glancing angle* (here, α). Now tilt the pointer up or down, while holding the same focal point on the receiving surface. As you tilt the pointer away from "normal," you are *increasing* the angle of incidence and *decreasing* the glancing angle. The cosine of the angle of incidence is proportional to the density of photons on the receiving surface.

A good bright flashlight will work as well as a pointer for this experiment.

And in compliment, the sine of the **glancing angle** is proportional to the density of photons on the incident surface.

When the angle of incidence increases (the light is tilted away from the receiver), the irradiance of light *decreases* in proportion to the cosine of the glancing angle. This response is termed the *cosine projection effect*. Another mathematical representation is that the cosine projection effect represents the scalar projection of the beam intensity normal to the laser (E) onto the beam intensity normal to the receiving surface ($\|E\| \cos\theta$, found in the dot product relations of the two vectors).

For example, when the Sun is at a 30° altitude angle, the complementary zenith angle is 60°. The angle of incidence for this case is equal to the zenith angle, and the cosine of 60° is one half (0.5), meaning that half of the photons would reach the equivalent area of one square kilometer if the surface were angled 60° away from the Sun. If one were to take the inverse of incidence angle, we would attain the proportional area increase needed to collect the same number of photons: $1/\cos\theta = 2$. The cosine projection effect is why the polar regions are much colder than the equatorial regions. Also, shallower glancing angles mean that the Sun must pass through a larger Air Mass too ($\gg$AM 1.0), which absorbs and scatters some of the irradiance. What does this mean to a solar engineer and design team? MINIMIZE θ.

$60° \cdot \pi/180° = 1.047$ rad.

Although most receiving surfaces are fixed, the Sun moves throughout the day, and over the course of the year. One of our tasks as a design team for systems solutions is to find effective compromises in collector orientation that will minimize the angles of incidence during the optimal hours of solar conversion—during the middle of the day when the Sun is at its apex in the sky dome and we have lots of solar irradiation. Researchers have created entire maps to guide designers in regions of the USA and Europe.[2] An additional criterion in systems design is to develop solutions for the client or stakeholders in a specific locale that bring about a high daily glancing angle to minimize the cosine projection effects during the extremes of seasons (the two Solstices), or in the early/late hours.

[2] M. Lave and J. Kleissl. Optimum fixed orientations and benefits of tracking for capturing solar radiation in the continental united states. *Renewable Energy*, 36: 1145–1152, 2011. and T. Huld, R. Müller, and A. Gambardella. A new solar radiation database for estimating PV performance in Europe and Africa. *Solar Energy*, 86(6): 1803–1815, 2012.

Now, put your hand in front of the beam. It creates a shadow, robbing the receiver of those precious photons for energy conversion. Sometimes we want shadows (for patio design) and sometimes shadows are a real detriment to solar conversion (like in photovoltaic panels). It is up to the design team to learn how to predict when and where shadows will arise in a design, using the tools of *orthographic projections* and *Sun charts*.

From our little experiment with a laser pointer, we have just explored the three key geometric methods serving the goals of solar energy engineering and design! As will be demonstrated in the following sections, the concept of a "locale" for the clients is actually a combination of *place and time*. A place on the planet is connected to seasons and regional climate, to variations in positioning of the Sun on the sky dome, and to the position of the SECS north or south of the Equator. So, given the tools developed in this chapter and the next, *how can we achieve the goal of solar design?*

1. Minimize the angle of incidence between the Sun and the collector during optimal hours of solar conversion.
2. Minimize the cosine projection effect by keeping the glancing angle high in critical times of the year (relative to the client and locale).
3. Minimize or remove shading effects that rob photons from the SECS when desired (or add shading effects when desired for cooling).

SPATIAL RELATIONS

Let's start out with the Greek ABCs, shall we? Well, in this case it's the α, β, γs, but that's the basis for spatial relations in the solar field. The Sun's position at any given point in time during the day can be communicated with angles of *solar altitude* (α_s) and solar azimuth (γ_s). Note that we only used the subscript "*s*" to indicate an altitude and azimuth location for the Sun projected upon the sky dome. The Sun is a *critical point* that we track during the day. In addition to the Sun, we have our solar energy conversion system, correct? All SECS have coordinates in angles of *slope* (β) and *azimuth* (γ). To be specific, an *angle is a position* in spherical coordinates. Any projection upon the sky dome can use altitude and azimuth angles (α and γ), while any SECS can use slope and azimuth angles (β and γ) to convey orientation.

Slope (β) equivalents used in common language:

- tilt,
- pitch,
- inclination.

Azimuth (γ) equivalents used in common language:

- aspect,
- horizontal coordinate.

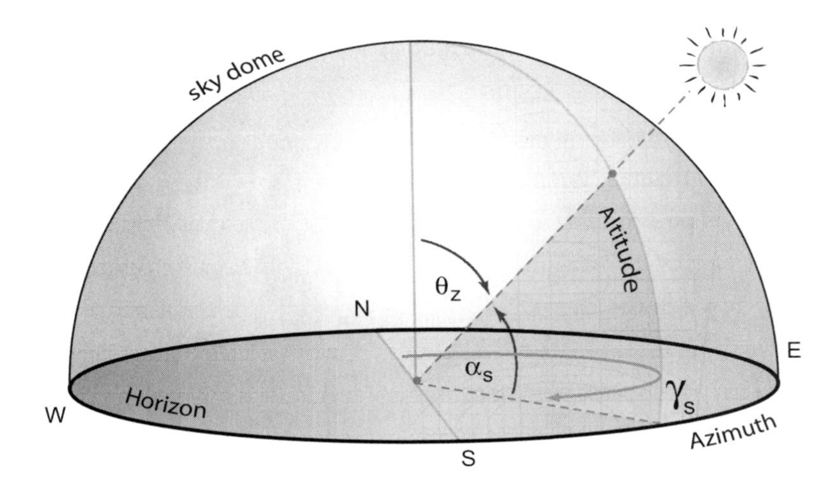

Figure 6.1: Showing the Sun's position projected onto the sky dome using angular coordinates of *solar altitude* (α_s) and solar azimuth (γ_s). In this case, North is the zero basis for the azimuth, increasing from 0 to 360° counterclockwise.

Altitude Angle (α) equivalents (and complement) used in common language:

- elevation Angle,
- elevation coordinate,
- the complement of the Altitude Angle is the zenith angle (θ_z).

Three coordinate systems are required for a spherical positioning system of a SECS. They will relate the Earth to the Sun (Earth-Sun Angles: δ, ω, ϕ), the Observer on a specific location on the surface of Earth to the Sun (Sun-Observer Angles: α_s, γ_s, θ_z; see Figure 6.1), and they will relate the orientation of the SECS aperture/receiver to the variable position of the Sun (Sun-Aperture Angles: β, (α), γ, θ; see Figure 6.2).[3]

[3] Three coordinate systems!

SPHERICAL COORDINATES

The *Celestial Sphere* is a projection of an imaginary gigantic bubble with apparent rotation (concentric and coaxial) relative to the Observer on Earth. The Earth itself is a near-spherical object, orbiting relative to the Sun in a somewhat circular elipse on an annual basis. Even when we are on the apparently flat surface of a locale spot on Earth, we have already described the hemispherical *sky dome* that one can imagine extending about our SECS, contributing the different components of light. So many spherical relations! Many of you will be familiar with Cartesian

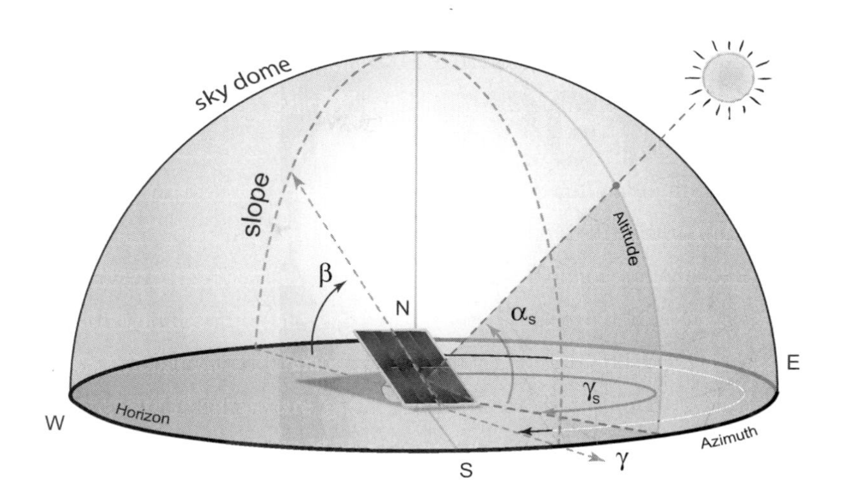

Figure 6.2: Showing the relative geometric angular coordinates of the SECS in terms of *slope* (β) and *azimuth* (γ). The coordinates of *solar altitude* (α_s) and solar azimuth (γ_s) are also shown to demonstrate the general case where $\gamma_s \neq \gamma$.

coordinates in space (x, y, z). However, when dealing with spatial relations of spherical objects like the Earth and the Celestial Sphere, we find that working with basic spherical coordinate systems makes trigonometry available to us to solve for space and time equations. As we shall see in a later vignette, the strong reason for our development of spherical trigonometry comes from early questions of relative positions of locations across the planet, such as: Where is Mecca?[4]

[4] Kryss Katsiavriades and Talaat Qureshi. The krysstal website: Spherical trigonometry, 2009. URL http://www.krysstal.com/sphertrig.html.

For spherical coordinates, we need information of radial distance, zenith angle, and azimuth, as seen in Figure 6.3. However, in solar positioning studies a radius of one (unit radius) is all we need to establish a *unit vector* and we are left with equations for only the zenith angle and azimuth (and the compliment of the zenith angle, the altitude angle). Note how the zenith angle in Figure 6.3 the generic θ angle is congruent with the solar zenith (θ_z) of the Sun, and the generic azimuth angle (γ) is congruent with the solar azimuth (γ_s).

Also, for the spherical coordinate system, the *angles are the coordinates*. So, if we were standing in a field in North Dakota (seen in Figure 6.4), looking at something tall like an enormous wind turbine, one could define a "critical point" of the position of the top of the nacelle relative to me by stating the general azimuth angle (γ, the rotation across the horizon from due South) and the general altitude angle (α, the rotation up from the horizon). Effectively, there are my x (γ) and y (α) coordinates on an *orthographic projection* of the *sky dome* onto a flat surface.

The *nacelle* part of the wind turbine is the upper housing that encloses the electrical generation part of the system (generator, gearing, drive train, and braking components).

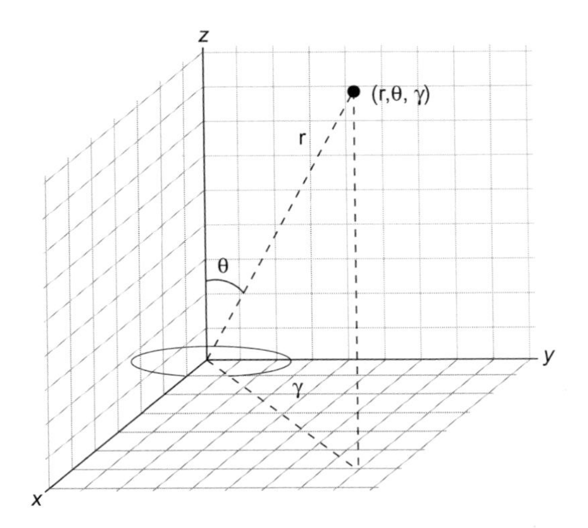

Figure 6.3: The *general* spherical coordinates of *zenith*, *x*, and *y*.

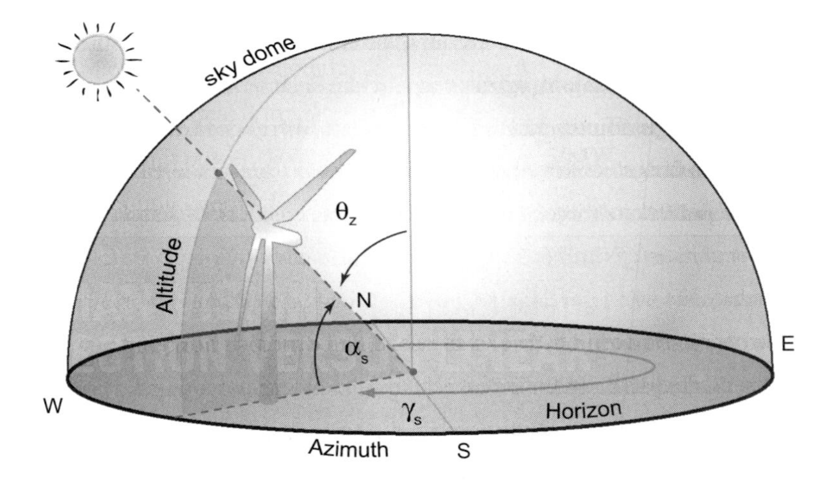

Figure 6.4: Showing the Sun's position projected onto the sky dome for a critical point of shading at the top of the wind turbine nacelle, using angular coordinates of *solar altitude* (α_s) and solar azimuth (γ_s). In this case, North is the zero basis for the azimuth, increasing from 0–360° counterclockwise.

The following equations describe the Cartesian coordinates (x, y, z) for unit vectors, followed by the equivalent functionals using the complementary altitude angle.

$$\begin{aligned} z &= \cos\theta = \sin\alpha, \\ x &= \sin\theta\cos\gamma = \cos\alpha\cos\gamma, \\ y &= \sin\theta\sin\gamma = \cos\alpha\sin\gamma. \end{aligned} \tag{6.1}$$

EARTH-SUN ANGLES

When we want to relate the relative position of the Earth with respect to the Sun, we use spherical coordinate angles again. The Earth-Sun spherical coordinate angles do not require us to specify the complete location where a collector is, because we are mainly honing in on the gross coordinates of Earth and Sun. Hence, we say that the Earth-Sun angles are independent of the Observer or Collector. However, the Earth-Sun angles are dependent on the time of year and the time of the diurnal cycle (day-night). One must be able to calculate the *declination* (δ) and *hour angles* (ω) at any time and location on Earth. Also, one must find the *latitude* (ϕ) from a resource like the web. The combination of declination and latitude angles, along with the hour angle, will later provide us with core information to calculate the solar altitude and solar azimuth angles (our Sun-Observer angles seen in Figure 6.5).

Solar declination is different from *magnetic declination*, which is the deviation of the Earth's magnetic field from true North-South.

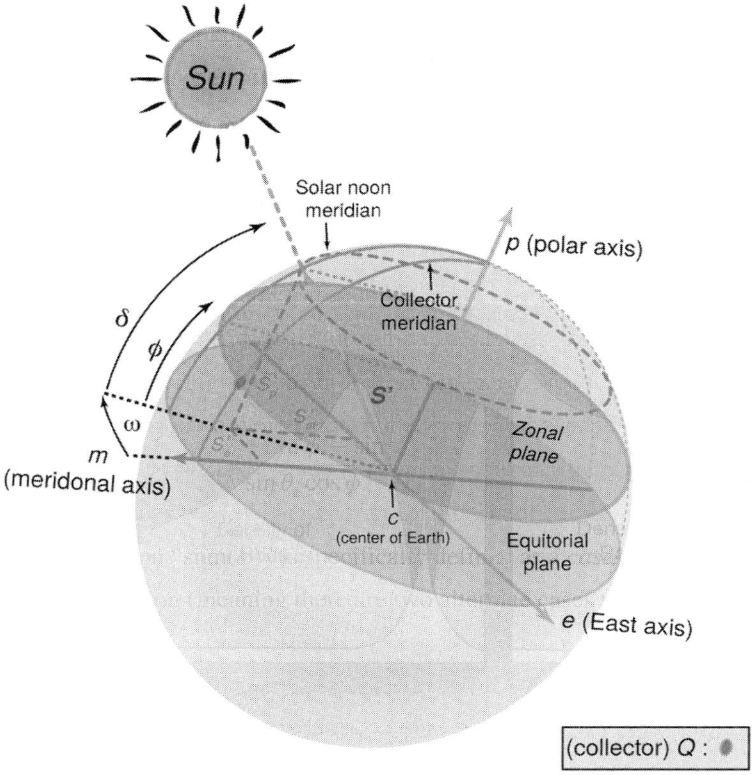

Figure 6.5: This figure shows the direction vector $\mathbf{S}'$ pointing from the center of Earth (C) to the Sun. The position of the collector (Q) on the surface of the Earth is marked, although it is only relevant to the meridonal axis, collector meridian, and the hour angle ω. $\mathbf{S}' = S'_m\mathbf{i} + S'_e\mathbf{j} + S'_p\mathbf{k}$.

Two moments, meaning two **critical points** in time, not entire days or weeks when the orbit of Earth is paused. It would be pretty weird for the Earth to pause for a day on the Solstices and the Equinoxes.

The time of year is represented by the **declination**, δ, and the input to calculate declination is the **day number**, n.

Actually, this equation approximation defines the **Northern Solstice** as $n = 172.25$, or June 21 at 6 am solar time on the Prime Meridian. The **Southern Solstice** is defined here as $n = 354.75$, or December 20th at 6 pm solar time along the Prime Meridian. When you look up the *actual* Solstices and Equinoxes online, you will find that they each vary over a few hours from year to year.

The Earth has a tilted axis, around which the planet spins in a diurnal cycle. The angle of this tilt has a maximum of 23.45° from the plane of the Earth's orbit about the Sun (the ecliptic). We call this measure of the apparent tilt of the Earth the *solar declination* (δ). Solar declination is the observed or apparent angle (due to the polar tilt) between the plane of the Earth's Equator and the plane of the ecliptic (the plane that the Earth follows to orbit about the Sun). Declination is *completely independent of your location on the surface of Earth*, and only depends on the time of year.

Now, the Earth is orbiting with that tilt relative to the Sun, and there are *two moments* in the year when the Earth is tilted far away from the Sun in one of each the Northern or Southern Hemispheres. These two moments occur when the declination has a maximum angle of $\delta = \pm 23.45°$ at either Solstice. Alternately there are *two moments* in the year when there is no apparent tilt of the Earth's axis relative to the Sun, because the tilted axis of rotation is not pointing at all toward the Sun. These alternate moments occur when the declination has a minimum angle of $\delta = 0°$ at either Equinox.

The declination can be calculated by a simple approximation found in Eq. (6.2), where n is the day of the year from January 1 through December 31.

Table 6.1: A list to calculate any day n, given the month and day i.

Month	Day
Jan	i
Feb	$31 + i$
Mar	$59 + i$
Apr	$90 + i$
May	$120 + i$
Jun	$151 + i$
Jul	$181 + i$
Aug	$212 + i$
Sep	$243 + i$
Oct	$273 + i$
Nov	$304 + i$
Dec	$334 + i$

Calculating the day (leap year not included) can be performed using Table 6.1. The first day of the year begins at 12 pm solar time on January 1 $n = 1$.

$$\delta = 23.45^\circ \sin\left(\frac{360}{365}(284 + n)\right). \tag{6.2}$$

The range of **declination** is limited to **Earth's tilt**: -23.45°(Southern Solstice) $\leq \delta \leq +23.45^\circ$ (Northern Solstice).

This equation will suffice for solar engineering of non-tracking technologies. In the case of tracking technologies, more precise algorithms have been developed, which can be found in more specialized texts.

The convention for declination is such that the Earth tilting toward the Sun in the North is valued as positive, and in the South as negative. In contrast, the declination reaches a midpoint at either of the Equinoxes. The Equinoxes are defined by a declination of 0°. Note that we do have to look up the dates (the 24 h period) within which the moment of 0° occurs for each Equinox. So the declination is a proportional measure of day length, solar altitude angle, and sunrise/sunset position on the horizon. The declination also defines the bounds of the Tropic of Cancer or the Tropic of Capricorn on a map, and explains why we have 24 h of daylight or darkness for periods of the year beyond the Arctic/Antarctic circles.

One can imagine the trend of increasing and decreasing declination each day over the course of a year as a series of Sun paths traced onto the sky dome (see Figure 6.6). The Sun traces a path high in the sky dome during the Summer Solstice, and traces a path low in the sky during the Winter Solstice. It is important to note that the length of the days is related to the declination (δ)

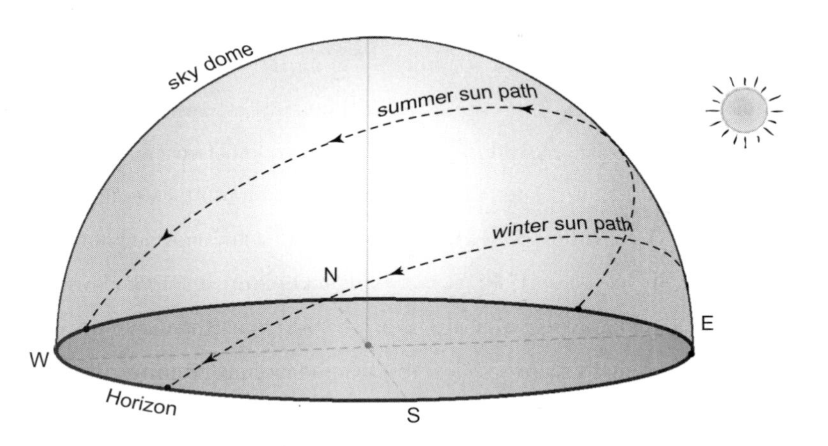

Figure 6.6: Showing the Sun's path across a day in summer and winter (Northern Hemisphere) as projected onto the sky dome.

combined with the latitude (ϕ). The point on the horizon for Sunrise can occur in the Northeast, East, or Southeast throughout the year, also directly related to the declination (δ) combined with the latitude (ϕ). The Sun also does not rise due East during the majority of the year—it rises either Northeast or Southeast depending on the declination and latitude. Correspondingly, the Sun *only* traces a path starting due East and finishing due West on the day of the Equinox.

EARTH-SUN ANGLES:

Angle of Latitude (ϕ): geographic coordinate to specify location in degrees North (+) or South (−) on the Earth's surface. Independent of longitude.

Angle of Longitude (λ): geographic coordinate to specify meridonal location in degrees East (+) or West (−) of the Prime Meridian. Independent of latitude.

Angle of Declination (δ): celestial coordinate to specify the relative location of the Sun with respect to the Equator, given the Earth's tilt through the seasons. Independent of any location on Earth.

Hour Angle (ω): celestial angular value of the Sun with respect to a meridian on Earth. The local hour angle is zero at solar noon. Prior to noon is $-15°$/hr and after noon is $+15°$/hr (coupled to longitude).

Latitude (ϕ) is independent of longitude (λ) in geospatial coordinates. In solar relations, the latitude of a location will convey information regarding the cosine projection effect, linked to the time of the year. There are a few latitude angles of interest on the planet, and they each relate to annual solar phenomena. The tropics are bounded by latitudes of $\phi = \pm 23.45°$, and represent areas where the cosine projection is low all year round, including times of the year with the Sun directly overhead during the midday hours. From the tropical latitudes ($\phi = \pm 23.45°$) through to the Arctic/Anarctic circles ($\phi = \pm 66.56°$), the cosine projection effects will be quite significant (one should observe strong seasonal variations in solar intensity). Beyond the latitudes of $\phi = \pm 66.56°$ the projection is extreme enough to guarantee times of the year with 24 h periods of full sunlight or full night.

The **cosine projection effect** is larger at greater latitudes than at the Equator, and conveys the projection of light across larger areas given the curved surface of Earth.

Interesting relation to remember: **1 min** of angular distance along a meridian on Earth ($1/60° = 0.01\overline{6}°$) is equivalent to a **nautical mile**. That physical distance is **1852 m**.

Arctic Circle: ($\phi = +66.56°$) North of this, the Sun is above/below the horizon for 24 continuous hours at least once per year (during the Solstices).

Tropic of Cancer: ($\phi = +23.45°$) During the June Solstice (summer in the Northern Hemisphere), this marks the extent of the maximum tilt of the north pole to the Sun.

Equator ($\phi = 0°$) There is no tilt of the Earth relative to the Sun at the Equinoxes. The Sun is directly overhead at solar noon during these two events.

Tropic of Capricorn: ($\phi = -23.45°$) During the December Solstice (summer in the Southern Hemisphere), this marks the extent of the maximum tilt of the sorth pole to the Sun.

Antarctic Circle: $\phi = -66.56°$ South of this, the Sun is above/below the horizon for 24 continuous hours at least once per year (during the Solstices).

In solar relations, longitude (λ) conveys information about the precession of diurnal cycles—the progression of time. With every hour of the day, there are angular changes that deviate from solar noon across the arc that is the Sun path. Those angular changes away from solar noon are represented by the *solar hour angle* (ω), in 15° increments per hour. The hour angle is also coupled to the precession of meridonal lines of longitude (λ) with respect to the beam of the Sun. Hence longitude and time are coupled. The hour angle is the way that we represent the *local* displacement of the Sun projected on the celestial sphere relative to solar noon. A ω of zero indicates that the Sun is directly above at solar noon. The sign of the hour angle is determined by occurring either before noon (the morning is *negative*) or after noon (the afternoon is *positive*). Another way of looking at it is

Solar noon is defined by a **solar hour angle** of $\omega = 0°$. A negative hour angle indicates the morning, while a positive hour angle indicates the afternoon.

Remember that the Earth rotates at 15° per hour, and 0.25° per minute. It can get confusing when your comparing spatial seconds with temporal seconds, right? That's why we stick with decimal notation in all of our references to lat./long. and angles.

the difference in angle between the local meridian of the Observer/SECS and the meridian that the beam of the Sun is intersecting at a given moment.

$$\omega = \begin{cases} -0 \text{ to } -180°, \text{ if before noon } (\textit{morning}), \\ +0 \text{ to } +180°, \text{ if after noon } (\textit{evening}). \end{cases} \tag{6.3}$$

We use Eq. (6.4) to shift hourly time into the angular representation of time as an *hour angle*. Note that here our time is evaluated in terms of hourly deviation from solar noon, in decimal notation. For example, 11h30 would be -0.5 h and 14h30 would be $+2.5$ h before and after 12h00 solar time, respectively.

$$\omega = \frac{360°}{24\text{ h}}(t_{sol} - 12\text{ h}) = \frac{15°}{\text{h}}(t_{sol} - 12\text{ h}), \tag{6.4}$$

$$t_{sol} = \omega\left(\frac{1\text{ h}}{15°}\right) + 12\text{ h}. \tag{6.5}$$

In Eq. (6.5) we are left with the *solar time* (t_{sol}) derived from the hour angle, which is typically different from the time on your clock or watch (or phone) (t_{std} or t_{sav}). Solar Energy Conversion Systems rely on solar time, which has wobbles during the year (±16 min) and shifts with respect to local longitude (±60 min). Society relies on a steady rate of time passage, uneven time zones, and occasional use of Daylight Savings (+60 min relative to standard time) for only a part of the year. We will discuss how to correct watch time into solar time shortly. For now, assume that we are already working in solar time for all scenarios.

VIGNETTE: While the study of spherical triangle extends back to the Greek mathematician Menalaus of Alexandria in the first century CE, the great emphasis on understanding the subject was developed during the 8–14th centuries CE. Why? Among other things, to help solve the query of how to locate *Mecca* from any point within the Islamic Caliphates of the Middle East, North Africa, and Spain. In the 9th century Muhammad ibn Mūsā al-Khwārizmī pioneered many elements of spherical trigonometry (you know the term "algebra" from his use of *al-jabr*, one of the two operations he used to solve quadratic equations, and the word "algorithm" stems from the Latin form of his name, *Algoritmi*), followed by Persian mathematician and astronomer Abū al-Wafā' Būzjānī in the 10th century, who discovered the Law of Sines:

Sperical Law of Sines

$$\frac{\sin(\alpha)}{\sin(a)} = \frac{\sin(\beta)}{\sin(b)} = \frac{\sin(\gamma)}{\sin(c)}. \tag{6.6}$$

The first treatise on the subject matter occurred near 1060, called *The Book of Unknown Arcs of a Sphere* by Al-Jayyani, and in the 13th century Nasīr al-Dīn al-Tūsī formalized modern spherical trigonometry as a mathematical discipline.

First Sperical Law of Cosines

$$\cos(c) = \cos(a)\cos(b) + \sin(a)\sin(b)\cos(\gamma). \tag{6.7}$$

Second Sperical Law of Cosines

$$\cos(\alpha) = -\cos(\beta)\cos(\gamma) + \sin(\beta)\sin(\gamma)\cos(a). \tag{6.8}$$

Their work led to improvements in estimating the circumference of the Earth, navigation, astronomical map making, terrestrial mapping, and to calculating the position and time of sunrise and sunset.

Wiki accessed July 31, 2012.

TIME CONVERSIONS

Recall the last chapter dealing with Meteorology, Taylor's Hypothesis connected distances between sites to periods of time for phenomena like clouds. We again connect time with spatial coordinates here, but our spatial coordinates must all be in terms of *angles*. As we noted earlier, the planetary bodies can be related in astronomy and meteorology via spherical angular relations, or spherical trigonometry.

In moving forward with our understanding of how to convert energy from the Sun to work, one should be conversant with the changes required to convert between the time on our local watch or computer and the apparent time and path of the Sun relative to the aperture or collection device. The time on our watches is reported as standard time (t_{std}; e.g., EST or MST) or daylight savings time (t_{sav}; e.g., EDT or MDT) and must be converted to solar time (t_{sol}). The approach developed here will explain the changes in time for the Sun's position relative to a point on the surface of the Earth. Once developed, the equations can also be applied to estimate the time and location of shadows. As part of a design team for SECS applications, we try to *always use solar time to plan our designs.*

The *Celestial Sphere* is a projection of an imaginary gigantic bubble with apparent rotation (concentric and coaxial) relative to the Observer on Earth (seen in Figure 6.7). The celestial sphere "rotates" in sync with the sidereal day. Imagine the need for old maritime star charts—the use of stars to reckon for time was common in navigation because of the predictable apparent rotation of the celestial sphere relative to the Observers. Now, the apparent motion of the Sun (measuring solar time) is slightly behind the rotation of the stars by about 1° per

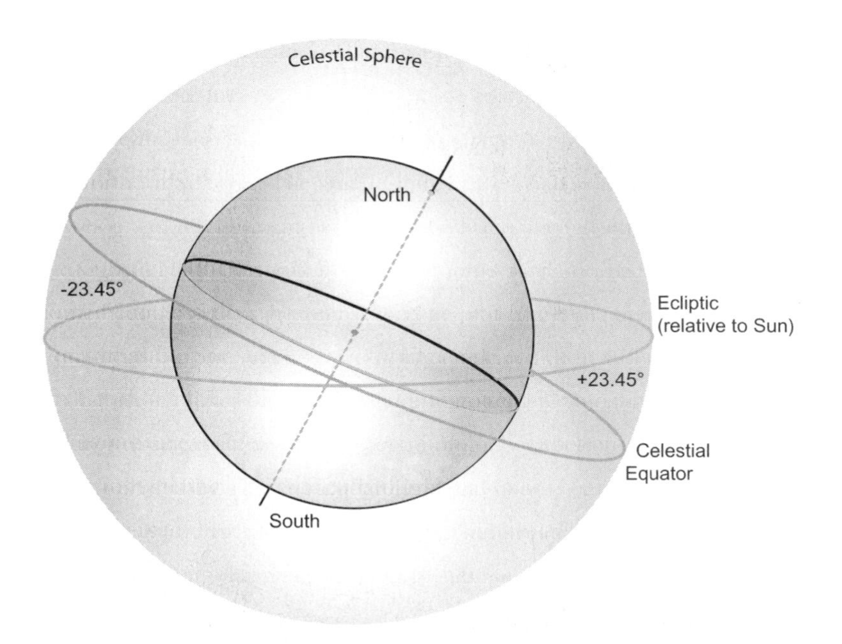

Figure 6.7: The **celestial sphere**, indicating the polar axis (pointing to the North Star, *Polaris*) and the plane of the ecliptic (the apparent annual path of the Sun in the sky).

day. As we will find out by reading further, this lag implies that the stars will appear to rise 4 min *earlier* each evening.

A *solar day* is 24 h long, and local solar noon is defined as the moment when the Sun is at its highest point in the sky. Yet in connection with the Celestial Sphere we use a fourth convention, *sidereal time*. Sidereal time is a reference to both time and apparent position of the stars (specifically the vernal Equinox), as projected onto the *celestial sphere*. This is the real time for one revolution of the Earth. A *sidereal day* is different from a solar day, in that it takes 23 h 56 min and 4.1 s for the stars to "revolve" around the Earth. In the period required for Earth to complete a rotation around its axis (a sidereal day), the Earth progresses about 1° along its orbit about the Sun. This means that even after a sidereal day, Earth must rotate an extra angular distance, *taking approximately 4 min longer than a sidereal day*. Hence SOLAR TIME refers to the apparent *time* and *position* of the Sun, as projected onto the *celestial sphere*.

This is the only reference to *sidereal time*. All other time references will be in terms of standard time, daylight savings time, or solar time.

WHY 24 H/1440 MIN/*or* 84,600 s? Because the ancient Sumerians and then Babylonians used a counting system of base 60 (sexagesimal) in the third millenium BCE (talk about prior art!), the Egyptians followed it with a 12 h day (sunrise-sunset: no night clocks), etc. The French also had an effort to make a metric time unit (base 100), but that didn't fly either (see Figure 6.8).

As we noted earlier, our notion of time is also tightly coupled with our system of *longitude*. We have used lines of longitude, or *meridians* as a reference for time and position East or West. Because time and angles are all linked together, we

Figure 6.8: A decimal clock from the French Revolution. Image by User: Rama/ Wikimedia Commons/Public Domain-Art August 2005.

cannot escape the sexagesimal (base 60) system for geographical locations. Each *standard meridian* (a major line of longitude) is spaced 15° apart, beginning with the Prime Meridian intersecting Greenwich, England. Standard meridians repeat for 360° or 24 h, although we tend to break the meridians into positive (East) and negative (West) values for each Hemisphere split by the Prime Meridian and the International Date Line. The Earth rotates at a certain speed, such that every hour, the ray from the Sun will cross one *standard meridian*. This means that we reckon the unit of one hour of time as the rotation of Earth by 15°. For the Earth to rotate and the beam of the Sun to shift 1° of longitude along the surface, it will therefore take 4 min of time.

The concept of MEAN TIME OR STANDARD TIME is used to convey time with even cycles that do not depend on the Sun or the Celestial Sphere. Today, we prefer a more cosmopolitan title (almost synonymous with GMT) for our primary world time scale, *Universal Time* (UTC; still measured relative to the Prime Meridian). This is a time system, based on *Mean Time*, according to which the length of a day is 24 h and midnight is 0 h. The "standard" is set by positioning relative to UT and your position relative to your nearby standard meridian.[5] This is our time on our regular clocks and watches, and the standard is set by the interval of the time step, rather than the precession of the heavens. We will need to correct our watches back to solar time, by considering *daylight savings*, *local time zones*, and *small shifts due to the wobble of Earth*.

[5] Because both standard time and solar time have the same acronym, we can also refer to "watch time" with solar time.

DAYLIGHT SAVINGS TIME CORRECTION

We always begin counting on the morning side of the time zone (East) for each λ_{std}. By convention, longtitudes for the Hemisphere West of the Prime Meridian are negative valued, and logntitudes East are positive valued.

Our Standard for time is established by the measured rotation of the Earth about its polar axis. In 1884, US President Chester A. Arthur (remember him?) presided over the *International Meridian Conference* in Washington DC for the selection of a globally unified zero meridian, which would act as the spatio-temporal beginning of each day. In total, 25 nations attended from all over the world. After the meeting, the basis (and home base) for measurement of relative time (and thus, *longitude*) is the *meridian* passing through Greenwich, England

and both poles. This arc is known as the *Prime Meridian*, and is the origin of the standard of time called Greenwich Mean Time (GMT). At the time, France agreed to disagree by abstaining from voting and retaining the Paris Meridian as their own Prime Meridian for several decades. We now call this the Prime Meridian for the zero hour in the "Universal Time, Coordinated," or "Universel Temps Coordonné" (UTC) standard that you see in plane tickets, meteorology reports, or clock times on the web.

Daylight Savings Time (DST) is a complication in relating clock time with the movement of the Sun. According to this concept, the Standard Time is advanced by +1 h relative to the normal shift from UTC, so in effect it is *turned on* for a period of the long days of the year (see Figure 6.9 for a visual reference).

Why, you ask? The concept of daylight savings began as an effort in Europe in 1916 (Germany/Austria) to conserve fuel used to produce electric power, from April 30 until the following October. The United States followed with the Standard Time Act in March of 1918 with the goal of saving fuel in the time of war. Now of course, the trends for energy use in the USA have changed significantly since 1918. So, in the USA (as of 2007),[6] DST was changed (not abolished) to occur

[6] Energy Policy Act of 2005, Pub. L. no. 109-58, 119 Stat 594 (2005).

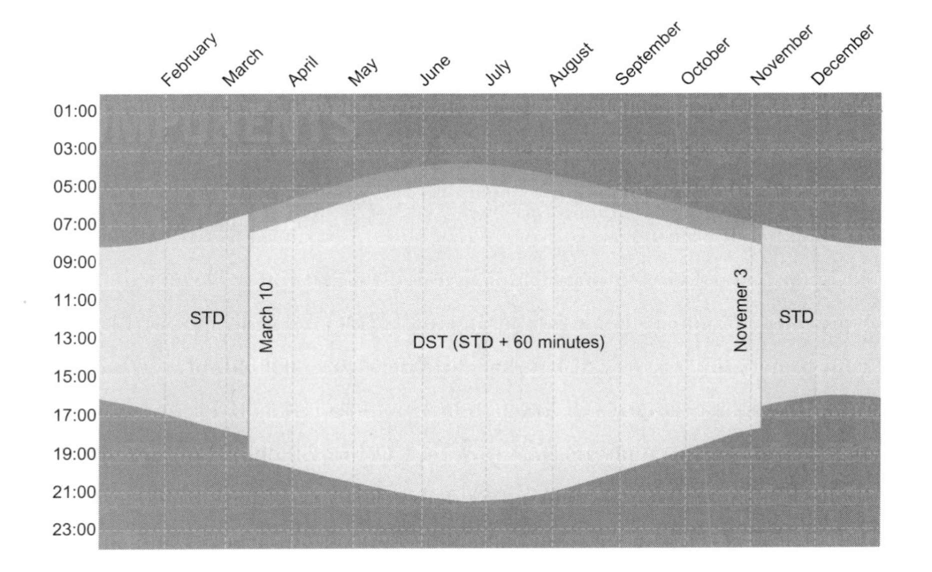

Figure 6.9: A diagram of daylight savings for GMT. Modified from original by Sualeh Fatehi/Public Domain June 2007.

from 02:00 on the second Sunday in March (was the first Sunday in April) until 02:00 on the second Sunday in November (was the last Sunday in October). DST is pejoratively termed *daylight stupid time* due to the weak evidence that the effort saves any net energy or financial cost, and the departure from using the reckoning of the Sun to work the day. Researchers in both the USA and Australia have demonstrated extremely low marginal returns for increasing the window of DST.

LONGITUDE TIME CORRECTION (t_λ)

The standard meridians are also the basis for Time Zones. The hour begins in the East and moves to the West. In order to begin correcting standard time to solar time, one must know the longitude for the standard meridian of the Observer (λ_{std}) for the local time zone, and the longitude of the Observer's location in question (λ_{loc}). The standard meridian occurs every 15°, beginning on the east or morning side of the time zone, but the relevant standard meridian is that which is linked to the Universal Time Coordinate time shift. Eastern Standard Time EST = UTC − 5 h uses the standard meridian 5 h to the West of the Prime Meridian ($\lambda_{std} = -75°$). Recall that 0.25° of longitudinal rotation will consume 1 min of time. Hence, we will need to multiply a time correction for longitudinal shifts by a factor of 4 to yield a consistent unit of minutes in time. This is shown in Eq. (6.9) as the value of the longitudinal correction (λ_{corr}) is units of minutes (temporal, not geospatial).

$$t_\lambda = 4\frac{\text{min}}{\circ}(\lambda_{std} - \lambda_{loc})\ [\text{min}], \tag{6.9}$$

where the sign convention is negative for time zones to the East of the Prime Meridian, and positive for the Hemisphere West of the Prime Meridian.

Remember that a **time zone** change of one hour is really just 15° of separation between **standard meridians**.

For example, maps place Rome, Italy at 12.5 ° East of the Prime Meridian (by convention, $\lambda_{loc} = +12.5°$). Rome is UTC + 1 h in standard time, which also means that Rome's Standard Meridian is $\lambda_{std} = +15°$ (not the Prime Meridian), as the Sun (and time) progresses from East to West. The difference from the standard longitude to the local longitude is therefore +2.5°, or (4 min/°) +10 min. Stated another way, the noon Sun passes Rome 10 min after it passes over the Standard Meridian for UTC + 1 h. State College, PA is 77.9° West of the Prime Meridian ($\lambda_{loc} = -77.9°$). State College is also UTC − 5 h in standard time, which means the Eastern time zone is $\lambda_{std} = -75°$. Much like Rome, the difference is a shift from East to West of +2.9°, or (4 min/°) +12 min. The local longitude for Philadelphia, PA happens to be $\lambda_{loc} = -75° = \lambda_{std}$, so the noon Sun passes over State College, PA (UTC − 5h) about 12 min after it passes over Philadelphia.

ANALEMMA TIME CORRECTION (E_T)

Here is a challenge: all the even steps for watch time assumes a perfect system of rotation, with no weebles or wobbles or precession of the polar axis. In reality, we know that all manner of wobbling occurs—there is measurable variability in the rotation of the Earth throughout the months of the year. This is why we add leap years and leap seconds to our calendars. So *mean time* is based on the length of an *average 24 h day*. In Greenwich, England, and along the entire Prime Meridian, standard noon (GMT/UTC + 0h) is often not the same as the time when the Sun is directly overhead (solar noon). There can be deviations from standard time along a standard meridian of up to ±16 min. The length of any one specific day can vary by about 30 s.

A much longer variation of solar intensity due to wobbling is also used to explain very long periods of climatic oscillation called Milankovich cycles. They were proposed by Serbian astronomer Milutin Milankovitch (1879–1958). Milankovich's cycles occurred in periods of 10s to 100s of thousands of years, and explain some of our glacial periods.

PHYSICAL PHENOMENA LEADING TO THE ANALEMMA AND EQUATION OF TIME:

Elliptical Orbit: Earth's orbit around the Sun describes a rather circular ellipse, but not a perfect circle (this is called *eccentricity*). This means that the Earth travels faster when close to the Sun (the Perhelion; January 3, 2010) than when it is farther away (the Aphelion; July 6, 2010). NOTE: the perhelion and aphelion do not overlap with the Solstices and Equinoxes.

Polar Tilt: Earth has a tilt to its polar axis, 23.45° away from normal to the plane of the ecliptic (the plane that the Earth follows to orbit about the Sun). This also means that the plane of the Earth's Equator is inclined relative to the ecliptic by 23.45° (see Figure 6.7). The tilt of the Earth is responsible for the variation in insolation across the latitudes, and for the seasons. This tilt also leads to a timing deviation called the *obliquity effect* at all times *but* the Equinoxes and the Solstices.

Irregular Spin & Wobble: The Earth does not spin at a uniform rate, and wobbles predictably over the years (see Figures 6.10 and 6.11).

Also, the **aphelion** occurs during the *summer* of the Northern Hemisphere. We are the furthest from the Sun, but still quite hot.

Figure 6.10: Analemma: This image was taken at the same time and location every day for one year. You are seeing the curve described by the Sun over that year. An analemma is a beautiful way to capture both the range of declination δ (along the length of the analemma) and the *Equation of Time* E_t (the expansion or width of the analemma) in a graphical format.

The variability of the hour of a day in standard time along a standard meridian (assuming the beginning of a time zone) can be corrected using the *Equation of Time* (E_t). If we look at Figure 6.10, we can see the residual deviations from

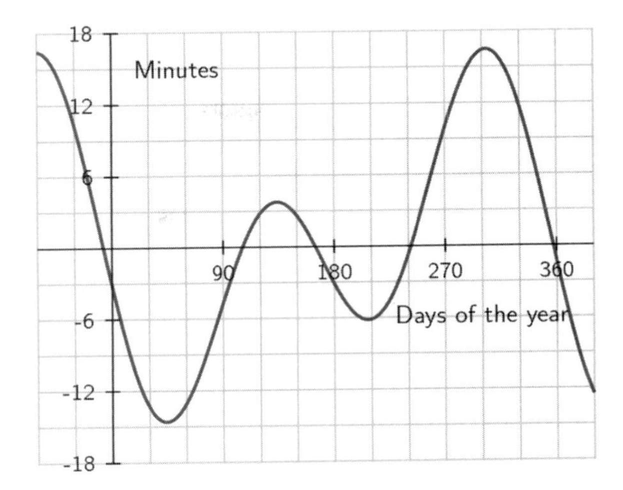

Figure 6.11: Equation of Time: Not as compelling as the Ayiomamitis' photographic analemma study, but here is another graphical depiction. Image by User:Drini/Wikimedia Commons/CC-BY-SA-3, GFDL August 2005.

solar time that are found in standard time along a given standard meridian. The photographer has used his watch to take a picture of the same point in the sky at the same time on his watch every morning. This oscillation is the graphical expression of the wobble of the Earth over the year. Our watches have even steps in time, and yet the apparent movement of the Sun-Earth orbit and diurnal spin is speeding up and slowing down in a predictable fashion. We call this image that looks like a distorted figure eight an *analemma*.

In Figure 6.11 we see the same representation laid out along a linear time line of one year. By comparing the two figures, you can see the small loop and the big loop represented in Figure 6.11. Our job in converting standard time to solar time is to remove those loops, or to flatten the loops into a single arc traced by the Sun each day, with the change in declination (δ). The Equation of Time is calculated for the number of minutes required to be added to or subtracted from standard time to adjust our clocks toward solar time. Note that the equation of time correction (E_t) must also be combined with longitudinal time correction (t_λ) for the full time correction (TC) from t_{std} to t_{sol}. Additionally, one must address whether an extra shift in -60 min is required for to convert Daylight Savings Time t_{sav} to t_{sol}.

The *Equation of Time* (E_t) correction involves first calculating a simple coefficient B for the day n (see Eq. (6.10)).[7] When plotted against the solar

[7] John A. Duffie and William A. Beckman. *Solar Engineering of Thermal Processes*. John Wiley & Sons, Inc., 3rd edition, 2006.

declination (δ), or the equivalent day number (n) the E_t is represented graphically as the *analemma*. A fit for the E_t (or analemma) function as a function of B is seen in Eq. (6.11) (arguments in degrees). Other sources may use a different numerical solution for the equation of time and B coefficient, which is older and not as accurate.[8] The answer to E_t will be in terms of minutes (no more than 16 min of correction), Refer to Table 6.2. for calculating your values of n (day number).

[8] Ari Rabl. *Active Solar Collectors and Their Applications*. Oxford University Press, 1985.

Table 6.2: Calculate any day *n*, given the month and day *i*.

Month	Day
Jan	i
Feb	$31+i$
Mar	$59+i$
Apr	$90+i$
May	$120+i$
Jun	$151+i$
Jul	$181+i$
Aug	$212+i$
Sep	$243+i$
Oct	$273+i$
Nov	$304+i$
Dec	$334+i$

$$B = (n-1)\cdot\frac{360}{365}\cdot\frac{180}{\pi} \tag{6.10}$$

$$\begin{aligned} E_t = {} & 229.2(0.000075) \\ & +229.2(0.001868\cos B - 0.032077\sin B) \\ & -229.2(0.014615\cos 2B + 0.04089\sin 2B). \end{aligned} \tag{6.11}$$

PUTTING TIME CORRECTION TOGETHER

Putting it all together, we have a method to modify local watch time (*standard time* or *daylight savings time*, DST) to *solar time*, correcting for the true time of day according to the Sun relative to the mean time. As seen in Eq. (6.12), the sum of the longitudinal correction and the equation of time is called the time correction factor (*TC*, in minutes). Note that sometimes t_λ and/or E_t will be negative values.

TC is used to correct t_{std} to t_{sol}, and is a function of the meridian of the local standard time zone (in longitude: λ_{std}), the longitude of the collector/Observer (in longitude: λ_{loc}), and the Equation of Time (E_t) that corrects for the wobbles in the Earth-Sun orbit.

$$TC = t_\lambda + E_t \text{ [min]}, \tag{6.12}$$

For the locations on the planet that use daylight savings, and for those periods when DST is in effect (turned "on," from March through November) we have to correct t_{sav} to t_{std} by subtracting 60 min, and then convert t_{std} to t_{sol} using *TC*.

$$\text{Local Solar Time } (t_{sol}) = \begin{cases} \text{Standard Time} + TC & \text{if STD,} \\ \textbf{Standard Time + TC − 60 min} & \text{if DST.} \end{cases} \tag{6.13}$$

or:

$$t_{sol} - t_{std} = 4(\lambda_{std} - \lambda_{loc}) + E_t \qquad (6.14)$$

and for DST:

$$t_{sol} - t_{dst} = t_{sol} - t_{std} + 60 \text{ min} = 4(\lambda_{std} - \lambda_{loc}) + E_t - 60 \text{ min}. \qquad (6.15)$$

Here, t_{sol} and t_{std} and t_{sav} are all in terms of normal hourly time, while the modifications to convert between solar and standard time are performed by adding or subtracting *minutes*. Answers may be divided by 60 to yield an equivalent answer in fractions of hours.

CAUTION! The equations do not tell you when DST occurs from country to country. There is a +60 min difference between March and November in the USA. Again understand that the time corrections are presented here in minutes rather than hours.

This means that the watch time must be corrected for the shift in longitude *and* the shift in the wobble of the planet (as TC), and then for periods of DST the 60 min which had been added (by policy, not for physical need of a change) must be taken back out. For most solar design projects, we will just work in solar time for simplicity. However, there are scenarios where we must be cognizant as a design team for standard/solar time conversions. When dealing with meteorological time series, the data is always reported and plotted as a shift from UTC (by hours)– such that "sunrise" being measured seems to occur at 11h00 rather than 6h00 (UTC−5 h). In other circumstances, we may be mixing data sets of local marginal electricity prices that are listed on Standard Time series, and we may even need to double check if Daylight Savings has been applied in the pricing events.

There are two very powerful tools in science and engineering: (1) **know the relative scale of things** and (2) **know how to estimate**. The rest is just refinement.

MOMENTS, HOURS, AND DAYS

How would you determine the energy incident upon the extraterrestrial surface of the Earth-Atmosphere? If we need to grab an estimate for scale, we find that all we need is bound in the average extraterrestrial irradiance (AM0), the solar constant: $G_{sc} = 1361 \text{ W/m}^2$.[9]

$$\overline{G}_{0,n} = G_{sc} = 1361 \text{ W/m}^2. \qquad (6.16)$$

If we were to estimate the irradiance of the Sun at AM0 by evaluating the intensity collected on a plane perpendicular to the *normal* radiance (the beam

[9] Actually, the average annual Solar intensity will oscillate in accord with an 11-year cycle of sunspots. Given a large number of sunspots, the solar constant will be higher (~1362 W/m²) while the value will drop to ~1360 W/m² when there are not many sunspots (changes about 0.01%). You can see NASA's Solar Radiation Climate Research Area from February 13, 2013.

shooting out from the Sun that is also perpendicular to the surface of the Sun), we would term that "normal global irradiance at AM0," or $G_{0,n}$. This would be the equivalent of an estimate at zero cosine projection error, or the angle of incidence is zero. The cosine function in Eq. (6.17) provides the variation from zero to one regarding the cyclic orbit of the Earth about the Sun.

$$G_{0,n} = G_{sc}\left[1 + 0.033\cos\left(\frac{360\,n}{365}\right)\right] \tag{6.17}$$

The value exceeds 1416 W/m² during the Australian summer, and drops near 1326 W/m² during the Canadian summer. Now why would the Australian summer be brighter than the Canadian summer (in outerspace of course)?

If we then estimate the AM0 irradiance[10] for *any* horizontal surface tangent to the curvature of some point on Earth's surface (G_0), we use Eq. (8.1).

[10] AM0 is why we have the zero in the subscript, and G is for "Global."

$$G_0 = G_{sc}\left[1 + 0.033\cos\left(\frac{360n}{365}\right)\right]\cdot[\sin\phi\sin\delta + \cos\phi\cos\delta\cos\omega]\,. \tag{6.18}$$

By analysis, we will see that the equation is actually multiplying Eq. (6.17) with Eq. (6.26). We shall observe soon that the sine of the altitude angle ($\sin\alpha_s$) is equivalent to the cosine of the zenith angle ($\sin\alpha_s = \cos\theta_z$). We have already performed the analysis showing that the altitude angle is dependent upon the Earth-Sun angles: ϕ, δ, ω in turn.

Peeking ahead: the **solar altitude angle** is α_s. The relation to find α_s is: $\sin\alpha_s = [\sin\phi\sin\delta + \cos\phi\cos\delta\cos\omega]$.

The notation for *irradiation* (energy density, in J/m²) is broken down into hourly values and daily values. Hourly values use the coefficient I, while daily values use the coefficient H. We will see that our time values also need to be converted into both radians (1 rad $= \pi$) and degrees.

Again, $G_0 = G_{sc}\cdot\sin\alpha_s$.

If I wish to obtain hourly values (which have the scale of J/m²), then I require two separate hour angle instants (ω_1 and ω_2). Again, my analysis of the equation shows that we are still using a variant with the same form as Eq. (6.18).

You should divide any answers for I_0 (or H_0) by 10^6 J/MJ to convert the results into smaller useful numerical values, in units of MJ/m² (much easier to read and check).

$$I_0 = \frac{12\cdot 3600}{\pi}\cdot G_{0,n}\cdot\left[\frac{\pi}{180}(\omega_2 - \omega_1)\sin\phi\sin\delta + \cos\phi\cos\delta\,(\sin\omega_2 - \sin\omega_1)\right]. \tag{6.19}$$

For *irradiation*, units of seconds are required to convert a Watt to a Joule (s ·J/s = J). Also the first term is a conversion of time in 12 h of exposure on a single Hemisphere per 1 rad (π).

$$\text{considering}: 12\frac{\text{h}}{\text{d}} \cdot 3600\frac{\text{s}}{\text{h}} = 43200 \text{ s per day}, \tag{6.20}$$

$$\textbf{Unit check}: \frac{\text{s}}{\text{d}} \times \frac{\text{d}}{\text{rad}} \times \frac{\text{J}}{\text{s}} \times \frac{\text{rad}}{^\circ} \times {}^\circ = \text{J}.$$

FINDING SUNSET, DAY LENGTH, AND H_0

If I wish to obtain daily values (which have the scale of J/m^2), then I require the extent of the Sun's arc from noon until *sunset* (which I will double, because the morning and evening are mirror images). This hour angle is positive valued as it occurs *after solar noon* and is the opposite sign of the sunrise angle (ω_{sr}). The hour angle of geometric sunset (we assume there are no mountains blocking our view in space) can be obtained by setting the sine of the altitude angle to zero, and solving for the special case of $\cos\omega_{ss}$ (which simplifies into tangent functions). By referring to Eq. (6.26), we set the solar altitude angle α_s to zero (making $\sin(\alpha_s)$ equal to zero as well), and solve for the hour angle, we arrive at Eq. (6.23).

$\tan(x) = \frac{\sin(x)}{\cos(x)}$.

You can use Equation 6.5 to convert hour angles to hours of time.

$$0 = [\sin\phi \sin\delta + \cos\phi \cos\delta \cos\omega], \tag{6.21}$$

$$\cos\omega = \frac{-\sin\phi \sin\delta}{\cos\phi \cos\delta}, \tag{6.22}$$

Simplifying by the trigonometric relation, we have the final form for the hour angle of sunset in Eq. (6.23), and again $\omega_{ss} = -\omega_{sr}$.

$$\omega_{ss} = \cos^{-1}(-\tan\phi \tan\delta). \tag{6.23}$$

Because of our convention for the hour angles, where negative values occur before solar noon, while positive values occur after solar noon, calculations are done for sunset to yield only positive values for the length of a half-day. The hours within a day (represented in degrees) are therefore going to be equal to double the

So what about the *blue hour* during twilight, when the Sun is below the horizon, yet the sky is not yet fully visible as night? This is a representation of a totally diffuse sky light (no beam component), and the irradiance is too low for most SECSs. But it is a beautiful sight.

half-day hour angle, or $2 \times \omega_{ss}$. Converted from hour angle units of degrees into hours, we would express the day length as a modification of Eq. (6.5).

$$\text{Day Length} = 2 \times \omega_{ss}\left(\frac{1h}{15^\circ}\right) \tag{6.24}$$

Finally we insert the contribution of the hours of the day into Eq. (6.25) to arrive at the MJ/m^2 of irradition upon the extraterrestrial Earth surface (Air Mass 0) in a single day. You will note that the 2× factor is still included in this relation, and the equation still maintains the basic form of the sine of the altitude angle ($\sin\alpha_s = [\sin\phi\sin\delta + \cos\phi\cos\delta\cos\omega]$).

Divide H_0 by 10^6 J/MJ to convert your results into units of MJ/m^2.

$$H_0 = \frac{12\cdot 3600}{\pi}\cdot G_{0,n}\cdot 2\left[\frac{\pi}{180}(\omega_{ss})\sin\phi\sin\delta + \cos\phi\cos\delta(\sin\omega_{ss})\right]. \tag{6.25}$$

Now, what about gathering together an average energy for the month? The average daily energy density (in MJ/m^2) for a month has a summary value, $\overline{H}_0$. Aside from taking the mean of the daily values in a month, Klein developed a simplified approach to approximate the average day in a month, using a single critical day that bears the value of the mean.[11] We can see a list of these key days in Table 6.3. By examining the table, we notice that the mean value of the daily energy density moves *toward the beginning of the month in Solstice months*. In essence, this shift in the days near the Solstices is a temporal manifestation of the slow-down and speed-up of the daily change in the declination. The rate of change in declination is fastest during the Equinoxes, and slowest during the Solstices.

[11] S.A. Klein. Calculation of monthly average insolation on tilted surfaces. *Solar Energy*, 19: 325–329, 1977.

The cosine of a 6° angle of incidence is ~0.995.

Putting it into another perspective, the reason why the long Summer days (or the long Winter nights) seemed so long when you were a kid playing outside, is precisely because there are not significant daily changes in the declination about the times of the Solstices! Also, when school starts in the Northern Hemisphere and it seems as though the daylight is just evaporates with the coming of Fall, well that is because the change in declination is actually happening quite fast. We can see this graphically by looking at Figure 6.12, where we have created a window of 6° of change in declination to bin our days. How many days hover within six degrees of the same declination about the Solstices? Over 80 days, or

Month	Avg Energy Day (AM0)	n
Jan	17	17
Feb	16	47
Mar	16	75
Apr	15	105
May	15	135
Jun	11	162
Jul	17	198
Aug	16	228
Sep	15	258
Oct	15	288
Nov	14	318
Dec	10	344

Table 6.3: A list of the energy density days where the value of the day approximates the mean value for the month.

approximately two and a half months! Meanwhile, the change of $\Delta\delta = 6°$ moves along at a monthly rate for the rest of the year. As Kalogirou and others have noted, we could take advantage of this oscillation and in simple systems with low concentration mechanisms and adjust the tilt of our collectors on a bi-annual basis to increase SECS performance.[12] So, in recap, the average energy for a day in the months of the Solstices will be relatively far removed from the midpoint of the month due to the fact that the maximum for the year occurs in that month, the maxima of the Solstice months occur near the end of the month, and the curve of the variability in the intensity is relatively flat over the months surrounding the Solstices (see Table 6.4).

[12] Soteris A. Kalogirou. *Solar Energy Engineering: Processes and Systems.* Academic Press, 2011.

SUN-OBSERVER ANGLES

The aperture/receiver of our solar energy conversion system is also where we will place our hypothetical Observer. We will now describe the angular coordinates to describe the position of the Sun relative to an Observer. Time, as figured in angular relations of the Earth-Sun relationship, requires knowledge of our

Remember, **declination** is dependent on the time of the year, but is *independent of location on the planet!*

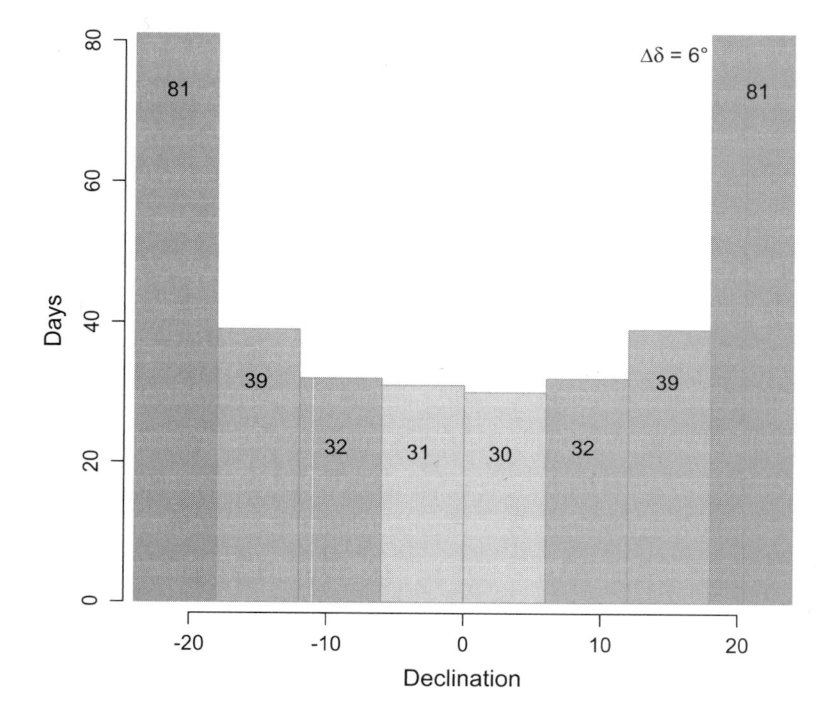

Figure 6.12: Histogram of days spent in a 6° window. Indicates long periods of Winter nights and Summer days about the Solstices.

Table 6.4: Reminder of the various angles used in the text.

FOR HORIZONTAL SURFACE ONLY		
Earth-Sun angles	ϕ, λ	latitude, longitude
	ω	hour angle
	δ	declination
Sun-Observer angles	α_s	solar altitude
	γ_s	azimuth angle
	θ_z	zenith angle

Observer's position on Earth, which we know in terms of ϕ (latitude) and λ (longitude). Time also requires the solar position relative to the Observer on Earth. We have already learned of the required coordinates of solar altitude (α_s) and solar azimuth (γ_s). With each passing day, the relative change of the solar noon altitude angle also represents changes in the declination (δ). The declination rate of change is slowest surrounding the Solstices, and will change the fastest from day to day near the Equinoxes.

$\gamma \neq \gamma_s$: NOT THE SAME AS THE SOLAR AZIMUTH.

$\theta \neq \theta_z$: NOT THE SAME AS THE ZENITH ANGLE.

One must be able to calculate the *solar altitude* (α_s) and *solar azimuth* (γ_s) at any time and location on Earth (see Figure 6.13). This can be done with knowledge of the latitude (ϕ, the knowledge of how far North/South the locale is), declination (δ, which implies knowing the day of the year independent of location), and hour angle (ω, which implies knowing the longitude and time of day).

The Solar Altitude (α_s) measures the angle between the central ray from the Sun (beam radiation), and a horizontal plane containing the Observer. As we noted in the beginning of this chapter, the solar altitude angle is essentially a *glancing angle*, in compliment with the zenith angle, which is essentially a special *angle of incidence*. Recall that the subscript "*s*" is only used to indicate that we choose the elevation angle relative observation of the Sun projected on the *sky dome*. This will become important in evaluating the altitude angles of other objects projected onto the sky dome, like buildings, overhangs, wing walls, and arrays of solar receivers.

$$\sin \alpha_s = [\sin \phi \sin \delta + \cos \phi \cos \delta \cos \omega] . \tag{6.26}$$

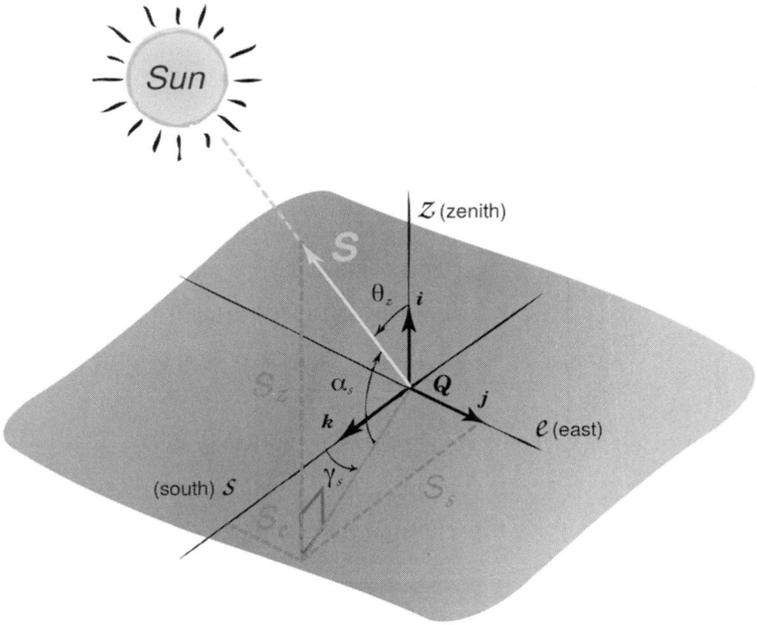

Figure 6.13: This figure shows Convention I, where the vector **S** is pointing from the surface of Earth (with the collector positioned horizontally at *Q*) to the Sun, where the zero basis is directed toward the Equator (±180°). This follows the vectorial form of $\mathbf{S} = S_z\mathbf{i} + S_e\mathbf{j} + S_s\mathbf{k}$. Note that the alternate Convention II γ_s is where the zero basis is directed to the North (0–360° clockwise).

Recall our discussion of the laser pointer for the cosine projection effect.

We can break down Eq. (6.26) into contributing parts. Notice how the form of the equation takes on the general form $\sin(\phi)\sin(\delta)$ plus $\cos(\phi)\cos(\delta)\cos(\omega)$. You can exploit this pattern to commit the basic equation to memory for calculations of extraterrestrial irradiance (G_0) and hourly/daily irradiation (I_0, H_0), along with a few helpful pointers to recall the relative contributions of each Earth-Sun angle (ϕ, δ, ω).

First, the sine of the *solar altitude angle* is proportional to the *latitude* (ϕ). For those of us in solar energy, the latitude conveys the intensity of cosine projection effect that an Observer on the surface of Earth is exposed to. Specifically, during the Equinox there is no apparent tilt of the Earth relative to the Sun. When the latitude is at an extreme (trending away from the Equator and toward the poles), the cosine projection effect is high and incident photons are spread out over a very large area. When the latitude approaches zero, the same number of incident photons is much more concentrated within a smaller surface area because the cosine projection effect is smaller. For the purposes of Eq. (6.26), the sine and cosine of latitude angle convey the specific solar altitude angles during the days of the Equinoxes at solar noon, given a location north or south of the Equator on the surface of the Earth.

Next, the sine of the *solar altitude angle* is proportional to the *declination* (δ). The declination is a specific reference to days of the year as they deviate from the Equinoxes, given a half orbit of the Earth about the Sun.[13] A high positive declination angle (near $+23.45°$) will indicate that the Northern Solstice is occurring and the northern pole of the Earth's spinning axis is at an extreme tilt toward the Sun. If an Observer is at a higher latitude they will receive an extra boost of solar energy from the favorable tilt toward the Sun. Of course, the intensity is reversed for the Southern Hemisphere during the Northern Solstice. For the purposes of Eq. (6.26), the sine and cosine of the declination angle convey the increase or decrease of the solar altitude angles at solar noon, as they work each day away from the Equinox and toward the critical times of the Northern and Southern Solstices.

[13] Because the sequence of the change in declination from June 21 to December 21 is a mirror of the change in declination from December 21 to June 21, we are only referring to a half orbit.

The latitude and declination parameters contribute to the resulting sine of the altitude angle ($\sin(\alpha_s)$) in Eq. (6.26) at solar noon only. We would like to be able to represent how high the Sun is in the sky dome for moments other than solar noon. Hence, we finally see how the hour angle (ω) contributes to the last functional contribution in Eq. (6.26). The cosine of zero is equal to one, and thus the hour angle is defined as zero degrees at solar noon. So if we only wished to plot or calculate the solar altitude angle at solar noon for a given location north or south of the Equator (latitude, ϕ) and for each day of the year (declination, δ), then Eq. (6.26) would simplify by allowing the cosine of the hour angle to be equal to one ($\omega = 0°$). Again, recall that the solar sunrise can only occur on the planet at an hour angle of $\omega = -90°$ during the two days of the Equinoxes. Otherwise the sunrise occurs at whatever time (and corresponding hour angle) that the solar altitude angle is equal to zero (as will be demonstrated later in Eq. (6.23)). So we can have hour angles for sunset (ω_{ss}) greater than 90° in the summer months (longer than 12 h days), and hour angles for sunset less than 90° in the winter months (days shorter than 12 h).

Sun-Observer Angles:
Solar Altitude (α_s).
Solar Azimuth (γ_s).
Solar Zenith (θ_z).

The zenith angle (θ_z) is the geometric complement of the solar altitude angle: $\sin(\theta_z) \equiv \cos(\pi/2 - \theta_z) = \cos\alpha_s$. This can be seen in the inverse cosine of the same function in Eq. (6.26). We direct your attention to the use of θ here, as the general concept of the angular deviation of the Sun's ray from the *normal* projection of a surface is called the *angle of incidence*, θ. In effect, the zenith angle represents the angle of incidence for a *horizontal surface*.

Recall: **normal** means perpendicular to a surface.

$$\theta_z = \cos^{-1}[\sin\phi \sin\delta + \cos\phi \cos\delta \cos\omega]. \quad (6.27)$$

The Solar Azimuth (γ_s) measures the rotational angle along the horizontal plane. The choice of the geometric basis (the reference point for the zero angle) for the *solar azimuth* will create slightly different equations used for solar path and shading calculations from text to text and online. Two conventions are described that are commonly used in solar calculations. The angle between the direction of the Equator (facing South for the Northern Hemisphere, and the North for the Southern Hemisphere) and the meridonal projection of the Sun's central beam (the Sun's meridian) is described as Convention I. The angle varies from 0° at the equatorial-pointing coordinate axis to ±180°. East (earlier than noon) is *negative* and West (later than noon) is *positive* in this basis. This azimuthal convention has the same directionality and sign as the *hour angle* (centered about solar noon), making Convention I slightly easier to learn and apply in practice. Convention I is a well-used standard drawn from the original text of M. Iqbal.[14] We use it in Eq. (6.28) and with the subsequent relations. Convention I is also used in the advanced design tools for solar energy, TRNSYS (UW-Madison: Transient eNergy Simulation Software), PVSyst, and SAM (NREL: *System Advisor Model*).

[14] Muhammad Iqbal. *An Introduction to Solar Radiation.* Academic Press, 1983.

SUN-OBSERVER ANGLES:

Solar Altitude α_s; Also called the *elevation angle*. Measures the angle between the normal vector from the Sun and the horizontal plane. Could also be thought of as the vertical angle between the *critical point* of the Sun (s) and horizontal plane, leading to the subscript "s."

Solar Azimuth γ_s; This is the projected rotation across the horizontal plane (the azimuthal rotation) that the Sun will have as it passes across the sky dome. Thus, the solar azimuth can be thought of as the azimuth angle between the critical point of the Sun (s) and the origin point at 0°.

Convention I: (easier to learn with the hour angle and longitude) Pointing directly to the Equator is $\gamma_s = 0°$ (the truly solar plane of reference). This means that East $= -90°$ (before solar noon) and West $= +90°$ (after solar noon).

Convention II: (used by meteorologists) Pointing to the North is 0°, while East $= +90°$, South $= +180°$, and West $= +270°$. The convention does not necessarily have a solar logic, but is common to multiple fields and software tools. It has also been determined to be a standard in the solar literature.

Zenith Angle θ_z; This is the angle between the normal of the horizontal (the vector pointing to the zenith of the sky dome) and the projection of the Sun's central beam.

$$\gamma_s = \text{sign}(\omega) \left| \cos^{-1} \frac{\cos\theta_z \sin\phi - \sin\delta}{\sin\theta_z \cos\phi} \right|, \tag{6.28}$$

where the function "sign(ω)" is specifically defined as a *cases* form of positive and negative notation (meaning there are two alternate cases to choose from).

$$\gamma_s = \begin{cases} +\cos^{-1} \frac{\cos\theta_z \sin\phi - \sin\delta}{\sin\theta_z \cos\phi} & \text{if } \omega > 0, \\ -\cos^{-1} \frac{\cos\theta_z \sin\phi - \sin\delta}{\sin\theta_z \cos\phi} & \text{if } \omega < 0. \end{cases} \tag{6.29}$$

There are also textbooks and useful software that use the azimuthal convention of meteorology and astronomy,[15] measuring *clockwise only* along the horizontal plane from a North-pointing 0° coordinate axis. We have titled this Convention II, and this azimuthal basis uses 360° for the meridonal projection of the Sun's central beam. Convention II is just as valid as the first, and is the accepted standard for solar conventions. It is relevant to be aware of the difference, in that the North basis is used in common tools for simulation and education such as the Sun path tool at the University of Oregon, the basic PV simulation tool PVWATTS, and in the interactive software of *PVEducation.org*.[16] Again, we begin with Convention I, but explore using both Convention I and II in our Sun path charts and shading analyses.

[15] William B. Stine and Michael Geyer. *Power From The Sun*. William B. Stine and Michael Geyer, 2001. Retrieved January 17, 2009, from http://www.powerfromthesun.net/book.htm.

[16] Christiana Honsberg and Stuart Bowden. Pvcdrom, 2009. URL http://www.pveducation.org/pvcdrom. Site information collected on January 27, 2009.

COLLECTOR-SUN ANGLES:

Slope, β: Also called the tilt. Measures the angle between the plane of the aperture and the horizontal. $\beta = 90°$ is a vertical tilt (like a wall), and a $\beta = 0°$ has no slope (like a table), with a normal pointing straight up to the *zenith* axis of the sky dome. For $\beta < 90°$, the aperture will have an intermediate downward-facing tilt.

Collector Azimuth, γ: This is the E–W rotation that an aperture will have for a given tilt. This is a form of azimuth angle determined with respect to the collector deviation from due South/North (with respect to the Equator) (could be fixed in place, could be tracking in time with the Sun's azimuth). Of course, if $\beta = 90°$, γ is meaningless.

Angle of Incidence, θ: We have seen this variety of angle on numerous occasions. This angle is critically important, and it is a tough one, *so learn it and never mess it up*. The angle of incidence measures the angle between the normal of the aperture (the vector perpendicular to the collector plane—in general cases, not horizontal) and the projection of the Sun's central beam to the collector surface. In a sense, this is a special case of a zenith angle, where the zenith is referenced to the aperture as a normal projection.

COLLECTOR-SUN ANGLES

Last of all, we need to convey the relative angles of the tilted collectors. Typically, a collector is oriented in any number of general angles that point up into the sky (see Figure 6.14). The symbols for the aperture orientation in the SECS are of course quite similar to those used when relating the coordinates of the *Sun-Observer* relationships. The slope, surface azimuth of the collector (typically just called the azimuth), and the angle of incidence are key orientations that we need to hold in our minds.

REALLY IMPORTANT! The most general case of the angle of incidence (θ) can be seen below in Eq. (6.30).

$$\begin{aligned}\cos\theta = {} & \sin\phi\sin\delta\cos\beta - \cos\phi\sin\delta\sin\beta\cos\gamma \\ & + \cos\phi\cos\delta\cos\beta\cos\omega + \sin\phi\cos\delta\sin\beta\cos\gamma\cos\omega \\ & + \cos\delta\sin\beta\sin\omega\sin\gamma.\end{aligned} \tag{6.30}$$

FAIRLY IMPORTANT: There are several special cases that are worth mentioning. These are specific scenarios where the equation for the angle of incidence can be simplified:

1. If $\beta = 0°$ and $\theta = \theta_z$: Eq. (6.30) reduces to Eq. (6.27). This is the case for a horizontal collecting surface.
2. If $\beta = 90°$, then all components of Eq. (6.30) with $\cos\beta$ are equal to zero, while components with $\sin\beta = 1$, and Eq. (6.30) will reduce to:

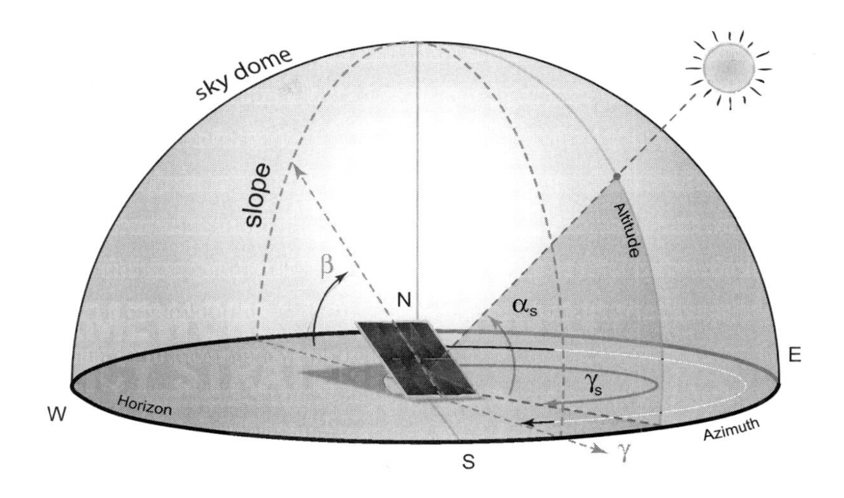

Figure 6.14: Showing the relative geometric angular coordinates of the SECS in terms of *slope* (β) and *azimuth* (γ). The coordinates of *solar altitude* (α_s) and solar azimuth (γ_s) are also shown to demonstrate the general case where $\gamma_s \neq \gamma$.

$$\cos\theta = -\cos\phi\sin\delta\cos\gamma + \sin\phi\cos\delta\cos\gamma\cos\omega + \cos\delta\sin\omega\sin\gamma. \tag{6.31}$$

This is the case for a vertical collecting surface like a solar wall, or just the side of a building.

Fixed collector orientation is not a sensitive parameter for SECS design. Save money on supporting structures and engineering costs by adapting to the surroundings you are provided.

3. If $\gamma = 0°$ (south orientation) with arbitrary tilt: Eq. (6.30) reduces to:

$$\cos\theta = \sin(\phi - \beta)\sin\delta + \cos(\phi - \beta)\cos\delta\cos\omega. \tag{6.32}$$

4. If $\gamma = 180°$ (north orientation) with arbitrary tilt: Eq. (6.30) reduces to:

$$\cos\theta = \sin(\phi + \beta)\sin\delta + \cos(\phi + \beta)\cos\delta\cos\omega. \tag{6.33}$$

A COMMENT ON OPTIMAL TILT

If we recall the tools that we are developing to achieve the goal of solar design, we can adjust our system design to accomplish the following:

1. Minimize the angle of incidence (θ) during optimal hours of solar conversion.

2. Minimize the cosine projection effect by keeping the glancing angle high in critical times of the year.
3. Minimize or remove shading effects.

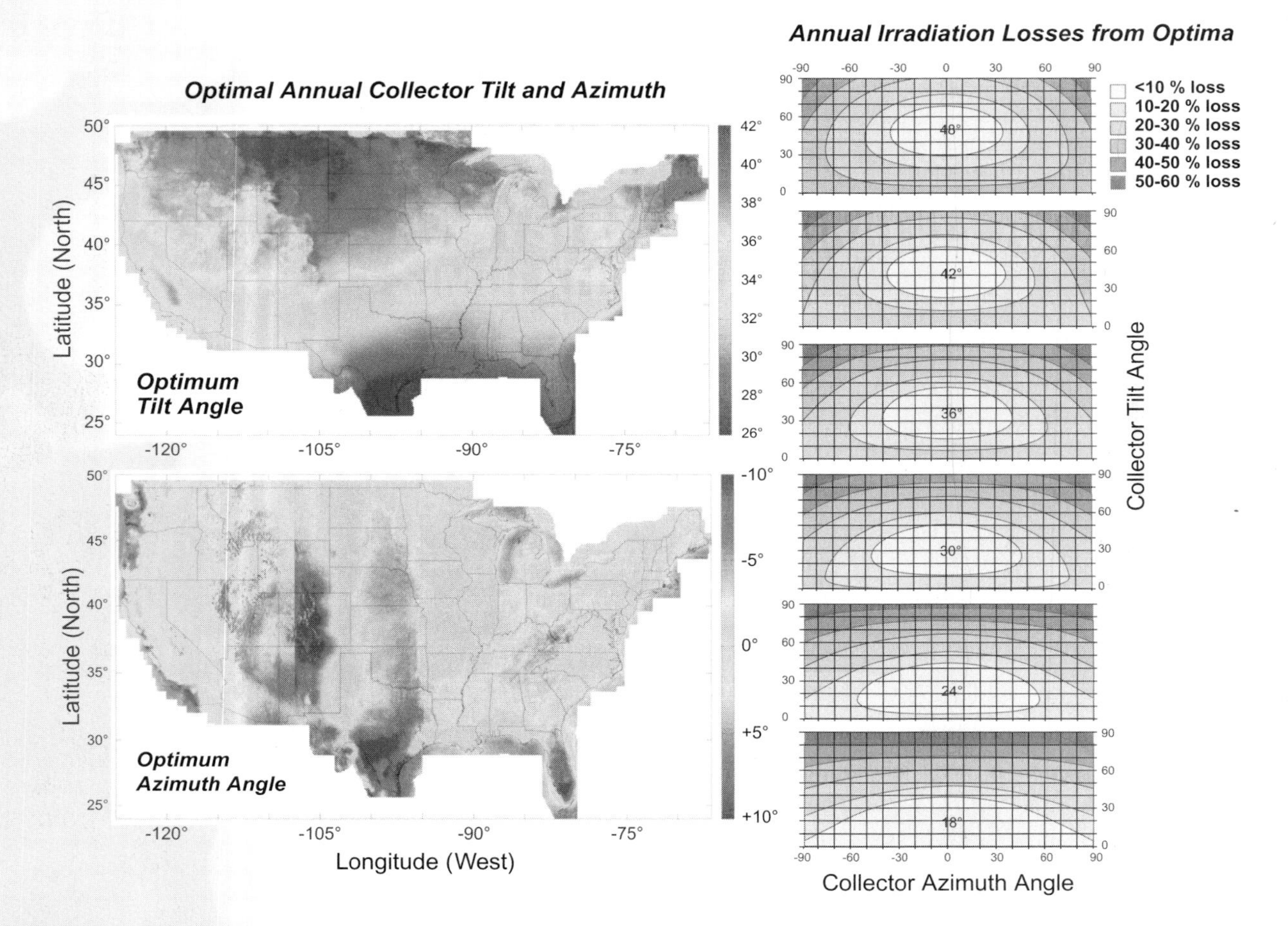

Figure 6.15: (Left) Map display of optimal annual collector orientations in the USA (Lave & Kleissl, 2011), along with (Right) the respective sensitivity (according to annual losses) of those optima in 6° tilt increments (Christensen & Barker, 2001).

Assuming a fixed-tilt and azimuth collector, we will have to select an optimal tilt (β) and azimuth (γ) for the collector system. In doing so, we will accomplish task 2 in this list. The cosine projection effect is a result of the relative tilt of the Earth from the ecliptic plane of orbit, leading to stronger seasonal projections as one progresses in latitude (ϕ) away from the Equator. As we display in Figure 6.15, researchers have investigated the optimal average annual orientations for collectors using dynamic energy simulation tools.[17]

[17] M. Lave and J. Kleissl. Optimum fixed orientations and benefits of tracking for capturing solar radiation in the continental united states. *Renewable Energy*, 36: 1145–1152, 2011.

However, one must note that these optima are generally *not sensitive parameters in SECS design*, even less so the closer the locale is to the Equator. Also displayed in Figure 6.15 are the relative annual losses for deviations from each optimal tilt and azimuth, in six degree increments for an "optimal" β assessment.[18] We observe that major changes in tilt and azimuth can be retained in many scenarios, without losing more than 10–15% of the total annual solar budget. For this reason, we recommend that beginning designers are tolerant of an abundance of possible orientations during the integrative design process. Chances are good that a trade-off in annual collection may also contribute to lower system costs by simplifying structural mounting requirements.

ROBOT MONKEY MINDSTRETCH

ROBOT MONKEY DOES SPHERICAL DERIVATION!

OK, so go take a good look at Figure 6.13. We will initially define a vector **S** pointing toward the Sun from an Observer[19] located at point Q on the surface of Earth. This initial equation relates the Observer/collector to the Sun by linear coordinates of *zenith* (pointing normal to the horizontal), *east*, and *south* (or *north*). See Eq. (6.34) and subsequent trigonometric relations of the angles α_s, θ_z, and γ_s (see Figure 6.14).

$$\mathbf{S} = S_z\mathbf{i} + S_e\mathbf{j} + S_s\mathbf{k}, \tag{6.34}$$

where **i**, **j**, and **k** are unit vectors along the z (zenith), s (south), and e (east) axes, respectively. The direction cosines of **S** relative to the z, e, and s axes are

[18] Craig B. Chistensen and Greg M. Barker. Effects of tilt and azimuth on annual incident solar radiation for United States locations. In *Proceedings of Solar Forum 2001: Solar Energy: The Power to Choose*, April 21–25 2001.

[19] Let's call him Quincy, though he doesn't look much like Jack Klugman from the 1970s…

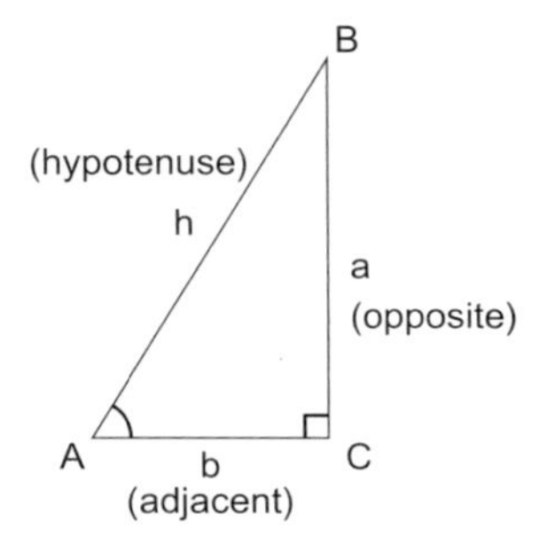

Figure 6.16: Trig reminder: Time to stimulate the gray cells again!

$$\sin A = a/h,$$
$$\cos A = b/h,$$
$$\cos B = a/h$$
$$\Rightarrow \sin(\theta_z) = \cos(\pi/2 - \theta_z)$$
$$= \cos(90^\circ - \theta_z) = \cos(\alpha_s).$$
$$\therefore \sin\alpha_s = \cos\theta_z.$$

The radius of Earth is about 6378 km, compared to the average distance from Earth to the Sun of about 150 million km. The real translation means moving an object from the center of Earth to the surface, a difference of 0.004%. Hence, we will call that translation *negligible* due to the sheer difference in scales.

[20] William B. Stine and Michael Geyer. Power from the Sun. William B. Stine and Michael Geyer, 2001. Retrieved January 17, 2009, from http://www.powerfromthesun.net/book.htm.

S_z, S_e and S_s, respectively. These may be written in terms of solar altitude and azimuth as:

$$\begin{aligned} S_z &= \sin\alpha_s, \\ S_e &= \cos\alpha_s \sin\gamma_s, \\ S_s &= \cos\alpha_s \cos\gamma_s. \end{aligned} \tag{6.35}$$

Now take some more time to absorb Figure 6.6. We will now define a second vector **S**′ pointing toward the Sun from the center of the Earth. Equation (6.36) relates the Earth to the Sun by linear coordinates of *m* (the meridonal axis, pointing to the meridian of the Observer/collector), *e* (east), and the *p* (polar axis, pointing to the north star Polaris).[20]

$$\mathbf{S}' = S'_m\mathbf{i} + S'_e\mathbf{j} + S'_p\mathbf{k}, \tag{6.36}$$

$$\begin{aligned} S'_m &= \cos\delta\cos\omega, \\ S'_e &= \cos\delta\sin\omega, \\ S'_p &= \sin\delta. \end{aligned} \tag{6.37}$$

Mixing the Systems of Angles: These two sets of coordinates are connected by a common rotation about the east axis (e) through the latitude angle (ϕ), and a negligible translation along the Earth radius QC. The difference between the direction vectors **S** and **S**′. The following form displaying the rotational relation between the direction vectors **S** and **S**′ can be expressed in matrix notation as in Eq. (6.38).

$$\begin{vmatrix} S_z \\ S_e \\ S_s \end{vmatrix} = \begin{vmatrix} \cos\phi & 0 & \sin\phi \\ 0 & 1 & 0 \\ -\sin\phi & 0 & \cos\phi \end{vmatrix} \cdot \begin{vmatrix} S'_m \\ S'_e \\ S'_p \end{vmatrix}. \tag{6.38}$$

The solution of which is:

$$\begin{aligned} S_z &= S'_m\cos\phi + S'_p\sin\phi, \\ S_e &= S'_e, \\ S_s &= S'_p\cos\phi - S'_m\sin\phi. \end{aligned} \tag{6.39}$$

Expanded to:

$$S_z = \cos\phi\cos\delta\cos\omega + \sin\phi\sin\delta,$$
$$S_e = \cos\delta\sin\omega, \tag{6.40}$$
$$S_s = \cos\phi\sin\delta - \sin\phi\cos\delta\cos\omega.$$

You will notice that we now have the component coordinates for $[S_z, S_e, S_s, S'_m, S'_p, S'_e]$ on both sides of the equation in Eq. (6.40). Substituting Eqs. (6.35) and (6.37) into Eq. (6.40), we get the following results:

$$\sin\alpha_s = \sin\phi\sin\delta + \cos\phi\cos\delta\cos\omega, \tag{6.41}$$

$$\cos\alpha_s\sin\gamma_s = -\cos\delta\sin\omega,$$

$$\cos\alpha_s\cos\gamma_s = \cos\phi\sin\delta - \sin\phi\cos\delta\cos\omega. \tag{6.42}$$

The **zenith angle** is an important equation. Remember it.

Alternat ively, the compliment of α_s is θ_z, and the compliment of the sine is the cosine, so Eq. (13.7) can also be expressed as (6.27):

$$\cos\theta_z = \sin\phi\sin\delta + \cos\phi\cos\delta\cos\omega. \tag{6.43}$$

WE'RE ALMOST THERE! Eq. (13.7) can be converted into our first *usable equation* for the altitude angle (α_s, in degrees):

$$\alpha_s = \sin^{-1}[\sin\phi\sin\delta + \cos\phi\cos\delta\cos\omega]. \tag{6.44}$$

WE'RE REALLY ALMOST THERE! By an act of Herculean derivation found in books on spherical trigonometry and astronomy, Eq. (6.42) can be converted into our second *usable equation* (Eq. (6.45)) for the solar azimuth angle (γ_s, in degrees). I will not do that here,

$$\gamma_s = \text{sign}(\omega)\left|\cos^{-1}\frac{\cos\theta_z\sin\phi - \sin\delta}{\sin\theta_z\cos\phi}\right| \tag{6.45}$$

NOTE: The function "sign(ω)" is specifically defined as a *cases* form of positive and negative notation (meaning there are two alternate cases to choose from).

Brain Rest: Well done! Go get a tasty snack and relax.

$$\gamma_s = \begin{cases} +\cos^{-1}\frac{\cos\theta_z\sin\phi - \sin\delta}{\sin\theta_z\cos\phi} & \text{if } \omega > 0, \\ -\cos^{-1}\frac{\cos\theta_z\sin\phi - \sin\delta}{\sin\theta_z\cos\phi} & \text{if } \omega < 0. \end{cases}$$

Congratulations!

Robot Monkey grants you one glowing banana for getting through the mindstretch

PROBLEMS

1. Assign definitions, correct symbols, and sign convention to the following terms:
   - Angle of Incidence.
   - Solar Azimuth.
   - Collector Azimuth.
   - Solar Altitude.
   - Zenith Angle.
   - Declination.
   - Hour Angle.

2. Calculate the solar declination (δ) for the spring and fall Equinoxes and the summer and winter Solstices. [Extra] estimate the error between the calculation and the defined declinations for each event.

3. Calculate the sunrise and sunset times (standard time) for four days in Paris: the spring and fall Equinoxes and the summer and winter Solstices, given 49°N latitude (ϕ) and 2°E longitude (λ). Do not use daylight savings in any of these solutions.

4. Calculate the day lengths for four days in Paris: the spring and fall Equinoxes and the summer and winter Solstices at 49°N latitude (ϕ) and 2°E longitude (λ). Do not use daylight savings in any of these solutions.

5. Determine the solar altitude and azimuth angles at 14h00 local time (watch time) for Fargo, North Dakota on August 13. Assume daylight savings is in effect (DST).

6. Calculate three conditions for the day of June 9, at 10h30 solar time in Athens, Greece: (a) the solar zenith and azimuth angles, (b) the sunrise and sunset times, and (c) the day length.

7. Repeated calculations for two days and two times: March 15 and September 15, each at 10h00 and 14h30 solar time. Solve for the following two conditions given the locale of the borrough of Brookly in New York City, NY.: (a) the solar azimuth (γ_s) and altitude (α_s) angles at the appointed times, and (b) the sunrise and sunset times.

8. What is the solar time in Denver, Colorado, on June 9 at 10h30 Mountain Standard Time (MST)?

9. Calculate the time in Philadelphia, PA at 15h00 on May 13, moving from solar to local time (DST).

10. Calculate the time in Chicago, IL at 15h00 on May 13, moving from solar to local time (DST).

11. Correct for time from local (standard time) to solar in Paris, France at 11am on February 23. Then calculate the appropriate hour angle.

12. A flat plate collector in Pittsburgh, PA, is tilted at 34° from horizontal and pointed 5° West of South. Calculate the angle of incidence (θ) on the collector at 10:30am and 2:30 pm solar times for both March 15 and September 15.

RECOMMENDED ADDITIONAL READING

- An Introduction to Solar Radiation.[21]
- Heavenly Mathematics: The Forgotten Art of Spherical Trigonometry.[22]
- KryssTal: Spherical Trigonometry.[23]
- Power from the Sun.[24]
- Solar Engineering of Thermal Processes.[25]
- Solar Energy Engineering: Processes and Systems.[26]

[21] Muhammad Iqbal. *An Introduction to Solar Radiation*. Academic Press, 1983.

[22] G. Van Brummelen. *Heavenly Mathematics: The Forgotten Art of Spherical Trigonometry*. Princeton University Press, 2013.

[23] Kryss Katsiavriades and Talaat Qureshi.The krysstal website: Spherical trigonometry, 2009. URL http://www.krysstal.com/sphertrig.html.

[24] William B. Stine and Michael Geyer. Power From The Sun. William B. Stine and Michael Geyer, 2001. Retrieved January 17, 2009.

[25] John A. Duffie and William A. Beckman. *Solar Engineering of Thermal Processes*. John Wiley & Sons, Inc., 3rd edition, 2006.

[26] Soteris A. Kalogirou. *Solar Energy Engineering: Processes and Systems*. Academic Press, 2011.

07

APPLYING THE ANGLES TO SHADOWS AND TRACKING

ALL OF THE ANGLES that we have learned about in the past chapter can be combined and applied to dynamic systems in the real world, we have already stated that these angular space-time relations are used in tools for maps, GIS, and global positioning systems (as satellites and in your mobile phones). The two key dynamic applications of angular relations in SECS are for estimation of shading across an entire year and for each hour of the day, and two assist in our design and control of solar dynamic tracking systems.

SHADING ESTIMATION

As we have learned from the history of solar energy, *solar access* or unobstructed solar gains are essential for several SECS, often during key times of the year. And in contrast, there are some SECS that would like to use appropriate shading planning as a control solution to regulate solar gains and losses, say daylighting and optocaloric gains inside of a home. This should emphasize that shading is important, either as a control mechanism or as an obstruction. Although we have saved the section on shading until the end, this is the culmination of the spatio-temporal relations that we have been developing all through the chapter. Shading analysis is key for an effective project design of your SECS.

While the diffuse component of irradiance (G_d) is significant during the day in any location (keeps things warm too), the direct component of light (G_b) is generally sought out for site assessment and design. Coincidentally, the direct beam of irradiance is also a great proxy for the vector between the Sun's normal ray ($G_{b,n}$) and the *critical points* of shading that stand between the Sun and the collecting surface of the receiver or aperture. This vector can be plotted for all times of the year in a *Sun path*, as we demonstrated earlier. But the shadowing from obstructions can also be plotted in terms of *altitude angle* and *azimuth angle* using a polar or orthographic projection with a Sun path overlay.

Let's proceed with the example of a building blocking the Sun's beam for a window of another building. We will first review the application of plotting spherical data onto a flat projection.

PROJECTIONS OF SPHERICAL DATA

Most of our work surfaces are now flat (video screens, paper, maps, tables, pizza boxes). If we wish to visually convert my observations of what is happening on the sky dome onto a flat surface, there are a few choices to project the data onto a two-dimensional medium. One could either take a panoramic photo of the surroundings or use a reflective fish-eye surface and take a picture of the sky and horizon captured by the convex reflective surface. In the former case, a panoramic photo converts the hemisphere of sky-horizon data onto an *orthographic projection*. We do the same thing in many flat map projections of the globe. In the latter case, we convert the hemisphere of sky-horizon data into a *polar projection* onto a polar chart. We will show how one can use calculations or graphical methods to establish times of solar availability or shadowing for a given point of solar collection.

The *Sun path* describes the apparent arc of the Sun across the sky (during one day) relative to an Earth-bound Observer at a given latitude and time (see Figure 7.1). There are excellent tools online to assist with plotting Sun paths onto *orthographic projections* or *polar projections* (as *Sun charts*), either through Scilab code for this text, or from the University of Oregon Solar Radiation

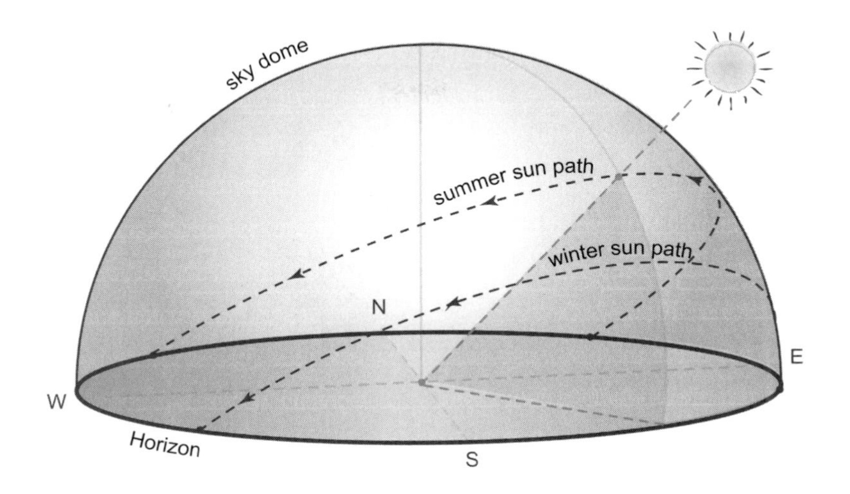

Figure 7.1: Showing the Sun's path across a day in Summer and Winter (Northern Hemisphere) as projected onto the sky dome using angular coordinates of *solar altitude* (α_s) and solar azimuth (γ_s).

Monitoring Laboratory (http://solardat.uoregon.edu/SunChartProgram.php). However the Sun path is only one layer of information that can be plotted on a projection chart. Recall that our solar polar coordinate system is composed of *altitude* values (α, no subscript) of elevation angle up from the horizontal plane, and *azimuthal* values (γ, no subscript) of rotation along the horizontal plane. No subscript on the coordinates indicates a generic point. We add subscripts by convention to indicate the angle from our point of origin as the Observer or Collector relative to *critical points* of interest. Hence, when we are reporting the position of the Sun we use α_s, γ_s.

Things one can plot:

- the path of the Sun for any day of the year, and
- the position of an object obstructing the Sun-Collector relation.

The key point to keep in mind for an orthographic projection is that the most distortion occurs at the top of the plot, because we are stretching the lateral and vertical parts of the zenith of the sky the most in an orthographic projection. In contrast, vertical lines near the horizon will not have much distortion at all. In Figure 7.2, we have used the Sun chart tool from the University of Oregon Solar

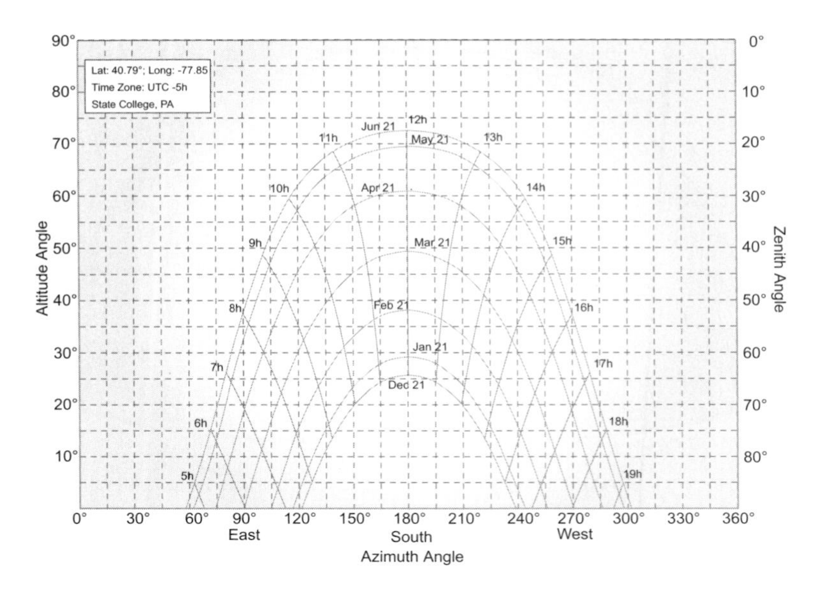

Figure 7.2: The orthographic projection of the Sun's path across the sky dome, in coordinates of azimuth and altitude angle. North is the zero basis for the azimuth here, increasing from 0°–360° clockwise. The figure was adapted from the software at the University of Oregon Solar Radiation Monitoring Laboratory (http://solardat.uoregon.edu/SunChartProgram.php).

Radiation Monitoring Laboratory to create an orthographic projection (Cartesian coordinates) of half the year in State College, PA ($\phi = +41°$, $\lambda = -78°$). Note that this plot uses the azimuthal Convention II, a standard in meteorology. We see that the Sun rises each day in the East (to the left) and sets in the West (to the right), as if we were facing South with a large curved sheet of paper in front of us. We also note that the largest arc in the chart is at the top, and is labeled as June 21, the Northern Solstice. The smallest arc in the chart is at the bottom, showing the least hours of daylight, and occurring on December 21, the Southern Solstice.

The significant distortion in an orthographic projection near the zenith of the sky dome becomes very apparent when one plots the Sun path for a latitude near the Equator, as the Sun reaches an altitude angle of 90° near the Equinoxes. Such severe distortion makes the plot challenging to read for most people, in particular clients! For this reason, a polar projection may be a better way to communicate the effects of shading near the Equator.

Using the same location and the same half-year of State College, PA, we can explore what a polar projection looks like, and examine the difference from an orthographic projection Figure 7.3. We see that the arc of the Sun path is

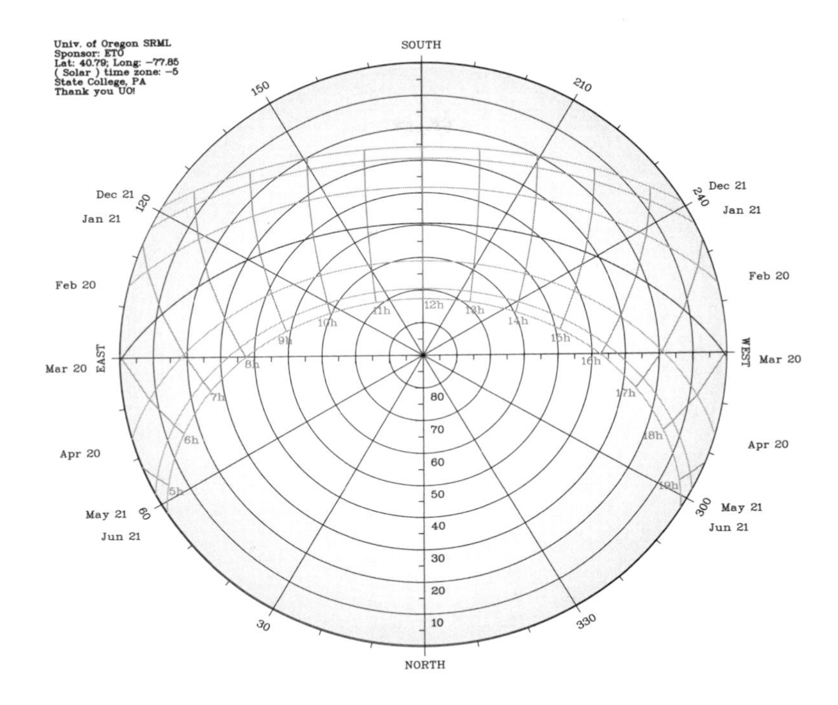

Figure 7.3: The 2-D polar projection of the Sun's path across the sky dome, in polar coordinates of azimuth and altitude angle. The figure was also adapted from the software at the University of Oregon Solar Radiation Monitoring Laboratory (http://solardat.uoregon.edu/SunChartProgram.php).

plotted such that the Sun still rises in the East (left) and sets in the West (right), and there are arcs curving through the plot each month. However, the arc for December 21 is now at the *top* of the chart, while the arc for the longest day on June 21 is at the *bottom* of the arcs. This is because we are effectively lying on the ground with our head pointed South, and holding that large piece of paper straight up to the sky. Now, most distortion in a polar projection occurs around the perimeter of the plot, because we are stretching the lateral and vertical parts of the *horizon* in the sky dome rather than the zenith. In such cases "vertical" lines like the sides of a building will follow the radii of the plots, and "horizontal" lines like the roof line of a commercial building will follow the circular arcs of the altitude angles.

Now, if I have the skills to access Trimble *SketchUp* (open software) or Autodesk's software *Ecotect* ($$ software for architecture and architectural engineering), I have access to pseudo three-dimensional modeling tools for shading and energy balance assessment. As an extension of SketchUp, furthermore, I can use GIS software for 3-D analysis as well. In ESRI's *ArcGIS* software suite, there is a tool called *Spatial Analyst* that has all of the described geometries embedded for shading analysis. Several add-ons have been developed for ArcGIS to assess solar potential in urban environments. For the open source GIS variant called *GRASS*, there is the powerful extension *r.sun* developed by researchers at PVGIS. PVGIS in itself is a treasure for large-scale estimations of solar potential in national regions of Europe and Africa.

SHADING ESTIMATION: CRITICAL POINT AND PLOTTING

For every solar energy conversion system we need to be able to predict whether major shadowing is occurring throughout the year. There is a method to assess the shading using 2-D projections that we will describe here. First, let us think of a single point, like the center of a window, or the spot on a roof where you will install PV panels, that *may or may not be* obstructed by a nearby tall building during the winter months. If the building is an obstruction, how many hours of the day will it be in the way? Will the shading be just in the early morning hours, when the intensity of the Sun is negligible anyway, or right during midday, when most electricity production will occur?

We begin by making a sketch of the scenario, both in cross-section (facing the surface of interest) and from plan view (map view). For those familiar with the tool, we could just use SketchUp and design the inter-building setup in 3-D to make this assessment. For now though, we use the simple tools. Our goal in this process is to plot all the time, throughout the year, that a shadow exists on our point of interest. It is very important to keep in mind that *we are not plotting what the shadow looks like*. Yes, shadows on a building or flat PV module are going to look like straight lines, or will be rectilinear in shape, but the plot of the times where shading occurs is not necessarily rectangular. The orthographic and polar projections distort straight lines. This is a common mistake (particularly in orthographic projections) for learners beginning the process.

YOUR JOB IS TO:

1. Calculate the angles as critical coordinates from origin to shading point;
2. Tabulate the γ, α coordinates for each point pair;
3. Plot the γ, α coordinates on top of a Sun chart in a given projection;
4. Shade in the correct zone for which shadowing effects are occurring;
5. Interpret the results for the hours of the day and months of the year, within which a collector will be shaded; and
6. Input your shading results into your simulation software for project design (e.g., System Advisor Model from NREL).

The origin point of interest may in fact be one of multiple *critical origin points*, like the corner of the PV array to be installed. Moving outward from the origin point, we look for a few *critical points* of potential shading occurring from the building in the distance. Good choices would be the upper and lower corners of the building and some midpoint along the roof line (if the building were a simple box). We will now proceed by calculating the azimuth angles (γ) and altitude angles (α) from the critical origin point to each of the critical shading points. This process will create a small table of γ, α coordinates, just like x, y coordinates in a Cartesian system. We will then plot those points on top of one of the projections with the Sun paths, and shade in the bounded zone of shadow effects. From analyzing our shaded box, we will be able to determine both the months of the year when shading occurs, and the hours of the day each month that the shading will fall on the critical origin point of interest.

WING WALL

There are three general forms that one can experiment with to understand the role of trigonometric calculations, angular plotting, and shading in orthographic or polar projections. The first is the *wing wall*, where the shading on a surface obscures either the morning Sun (to the East) or the evening Sun (to the West). We present a simplified form in Figure 7.4, where the critical origin point is centered on a vertical wall with a southwest azimuth. As mentioned earlier, it is often helpful to break this projection diagram into two sketches: a cross-section

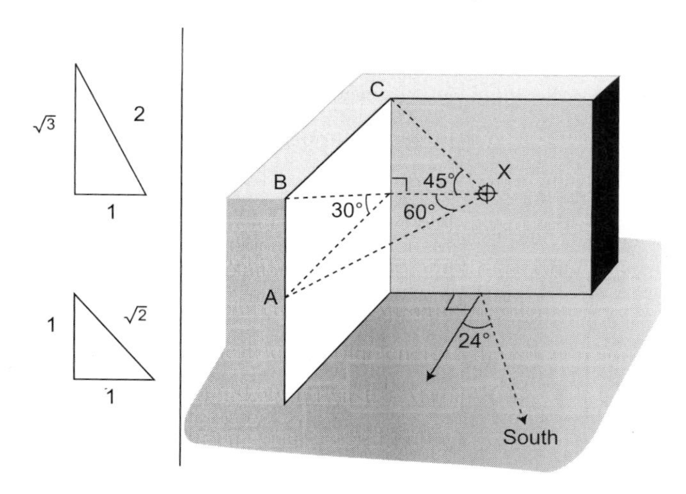

Figure 7.4: A simple projection of a wing wall to the West of a southwest oriented wall. The critical origin point is *X*, and the critical points of potential shading are *A*, *B*, and *C*. We do not consider points lower than *X*, which would be below the horizon relative to the beam of the Sun.

view (facing the wall with point *X*) and plan view (top, map view), labeling the relevant points and angles in each (some points will overlap). There will also be one specific angular coordinate pair that will not plot directly in the planes of the cross-section or the plan view. Looking at Figure 7.4, that coordinate pair is $(\gamma_{X-B}, \alpha_{X-B})$, which extends diagonally through the body of the projection. Given the trivial angles represented in the figure, α_{X-B} is really the only angle that must be calculated using trigonometry. By visual analysis, $30° < \alpha_{X-B} < \alpha_{X-C} = 45°$, which means we should expect a value to fall between 30° and 45°. We use the inverse tangent relation of "opposite over adjacent" and calculate an altitude angle of 35.3°:

$$\alpha_{X-B} = \tan^{-1}\left(\frac{\sqrt{1}}{\sqrt{2}}\right) = 35.3°. \tag{7.1}$$

The azimuth angle for the same points can be calculated using the knowledge that the entire vertical face for the critical origin point is oriented as $\gamma = +24°$ from due South. Hence, the building is facing into the afternoon Sun, so we must add 24° to the measured azimuthal rotation of points $X-B$, which is 30° from inspecting the complement of the 60° angle noted in Figure 7.4:

$$\gamma_{X-B} = 30° + 24° = 54°. \tag{7.2}$$

Finally, the remaining azimuthal coordinate for γ_{X-C} must be calculated in the same manner, by adding the positive rotation of 24° to the 90° rotation already expressed by the measured azimuthal relation:

$$\gamma_{X-C} = 90^\circ + 24^\circ = 114^\circ. \tag{7.3}$$

The resulting pairs of γ, α can be seen in Table 7.1. Those pairs can now be plotted in an orthographic or polar projection, and shaded accordingly. Note that the connections between the coordinates for $X - B$ and $X - C$ will form a soft arc (bending upwards) in an orthographic projection, due to the distortion of the data near the zenith of the sky dome. The arc will also extend beyond point $X - C$, as the presence of the origin wall will block any Sun access for azimuth's greater than 90°. The shading fill between the points will occur *down* and toward the *West* (see Figure 7.5).

If the wing wall had been to the east of the origin point, then the shading fill would occur *down* and toward the *East*.

Table 7.1: A table of altitude angles and azimuth angles for critical points in the wing wall example.

Points	$\gamma_{orig,shad}$ (°)	$\alpha_{orig,shad}$ (°)
X, *A*	54	0
X, *B*	54	35.3
X, *C*	114	45

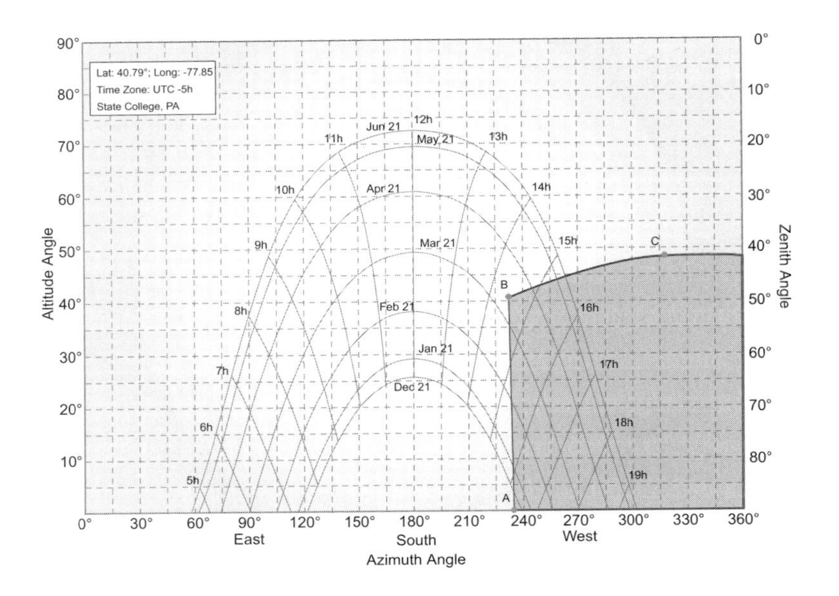

Figure 7.5: The orthographic projection of the Sun's path across the sky dome, with the shading zone defined by points *A*, *B*, and *C*. The figure was adapted from the software at the University of Oregon Solar Radiation Monitoring Laboratory (http://solardat.uoregon.edu/SunChartProgram.php).

If we analyze the shaded region intersecting with the monthly arcs of the Sun paths, we can interpret the hours of the day for which point X will be obscured from beam irradiation, and the complementary hours of the day for which point X will observe full irradiation on a clear day.

AWNING

The next general shading structure is an obstruction from above, as from an awning or soffit extending out over the vertical wall of a building. Again, we present a simplified form in Figure 7.6, where the critical origin point is centered on a vertical wall for a different southwest azimuth. The critical points of shading are now identified as A–E, with D and E flush with the origin wall.

Now we can identify two specific angular coordinate pairs that will not plot directly in the planes of the cross-section or the plan view. Looking at Figure 7.6, that coordinate pairs are γ_{X-A}, α_{X-A} and γ_{X-C}, α_{X-C}, which extend

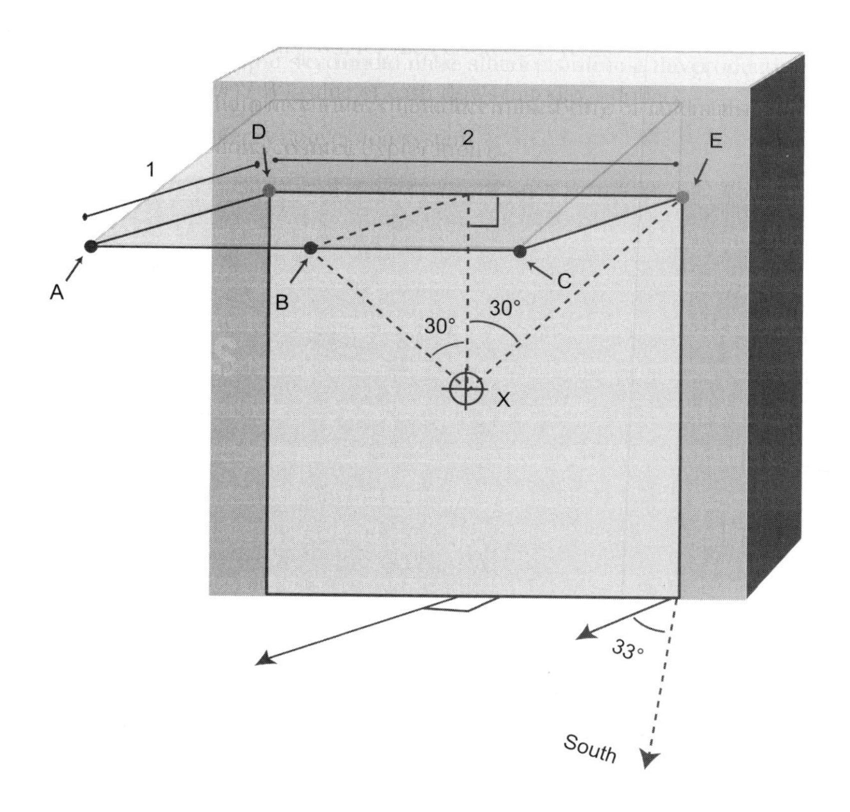

Figure 7.6: A simple projection of an awning over a southeast oriented wall. The critical origin point is X, and the critical points of potential shading are A, B, C, D, and E.

diagonally to the southwest and southeast through the body of the projection, respectively. Given the angles represented in the figure, $\alpha_{X-A}=\alpha_{X-C}$, and they must be calculated using trigonometry. By visual analysis, $\alpha_{X-B}=\alpha_{X-D}=\alpha_{X-E}=60°$, the complement of the presented angles of 30° (these are the altitude angles, rotating up from horizontal). Also, $\alpha_{X-C}<\alpha_{X-E}=60°$, which means we should expect a value to be less than 60°. We use the inverse tangent relation of "opposite over adjacent" and calculate an altitude angle of 50.8°:

$$\alpha_{X-A} = \alpha_{X-C} = \tan^{-1}\left(\frac{\sqrt{3}}{\sqrt{2}}\right) = 50.8°. \quad (7.4)$$

Unlike the altitude angles, the azimuth angles for the same points will be different from each other. They can be calculated using the knowledge that the entire vertical face for the critical origin point is oriented as $\gamma=+33°$ from due South. Hence, the building is again facing into the afternoon Sun, so we must add 33° to the measured azimuthal rotation of points $X-A$ and $X-C$, which are ±45°, noted in Figure 7.4:

$$\gamma_{X-A} = +45° + 33° = 78°, \quad (7.5)$$

$$\gamma_{X-C} = -45° + 33° = -12°. \quad (7.6)$$

The resulting points are presented in Table 7.2 and plotted upon an orthographic projection in Figure 7.7. The points will connect via three arcs, resembling the bottom of the "bat signal" (wings of the bat). As the shadowing effects for an awning occur when the Sun is high in the sky, the affected shadowing times will be shaded upward.

Points	$\gamma_{orig,shad}$ (°)	$\alpha_{orig,shad}$ (°)
X, A	78	50.8
X, B	33	60
X, C	−12	50.8
X, D	123	60
X, E	−57	60

Table 7.2: A table of altitude angles and azimuth angles for critical points in the awning example.

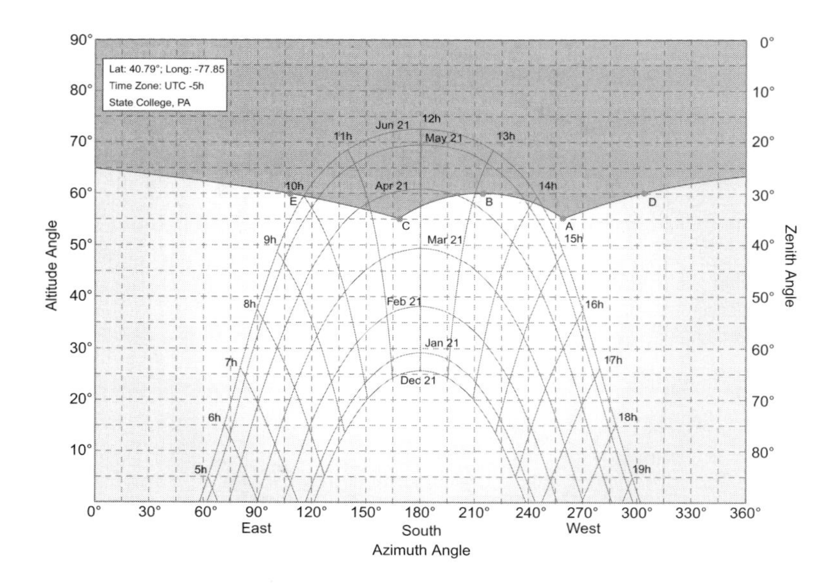

Figure 7.7: The orthographic projection of the Sun's path across the sky dome, with the shading zone for this awning defined by points *A*, *B*, *C*, *D*, and *E*. The figure was adapted from the software at the University of Oregon Solar Radiation Monitoring Laboratory (http://solardat.uoregon.edu/SunChartProgram.php).

ARRAY PACKING

As final scenario we describe the shading setting for an object blocking another in the horizontal plane. This could be a situation where a tree is obscuring a solar hot water panel, or where a building in an urban setting is shadowing another neighboring structure. Alternatively, for utility-scale PV installations or concentrating solar power (CSP) thermal applications, we need to ask how far apart must one space the collectors in an array to remove shadow losses during the whole year? (see Figure 7.7).

Let's create a scenario for a fixed-axis array of PV modules that are facing a few degrees East of South, just outside of Philadelphia. We will specify the size of a single PVmodule: 1 m wide by 1.25 m in length (top to bottom). Hence a row of the array in Figure 7.8 stacks two PV modules from base to top (2.5 m dimension), and each array row is 8 modules wide, or 8 m across in Table 7.3, we have calculated the critical points, where the distance d has been set to 3 m. Using a sun chart, we can observe that seasonal shading (and thus power losses) will occur if this array were set in Philadelphia, PA. If they are mounted on a horizontal ground, but with a fixed tilt of $\beta = 35°$, how far apart must one

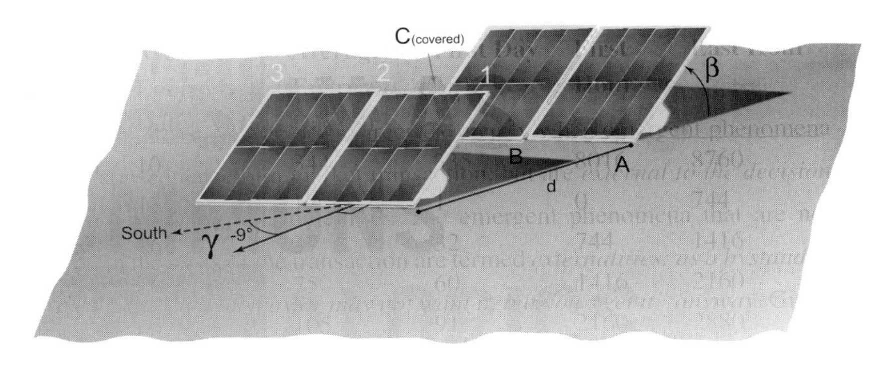

Figure 7.8: A simple array facing 9° southwest with unknown spacing *d*. The critical origin points are multiple this time, found at the base of the rear panels. The critical points of potential shading are found at the top of the front panels.

Table 7.3: A table of altitude angles and azimuth angles for critical points in Figure 7.8. Here, the distance d is set at 3 m and the length of two PV modules from base to top is 2.5 m. The entire row is 8 modules wide, or 8 m across. The array tilt and azimuth are already specified in the Figure 7.8.

Points	$\gamma_{orig,shad}$ (°)	$\alpha_{orig,shad}$ (°)
1, *A*	−9	56
2, *A*	67.5	19
3, *A*	74	10
1, *B*	−85.5	19
2, *B*	−9	56
3, *B*	67.5	19
1, *C*	−92	10
2, *C*	−85.5	19
3, *C*	−9	56

space them to *never have any shading on any of the rear panels*? This becomes quite important as a problem when you are *area-constrained*, as you would like to pack in as many panels as possible in a limited space, but also minimize the degree of shadowing occurring on the panels. PV modules are assembled as cells connected in a series circuit (like batteries attached from top to bottom) to increase the total voltage of the module. Hence, a shadow on the corner cell of a "dumb" panel (one without reverse bypass diodes or microelectronics) could reduce the current flow to the entire module. Note that in some tracking systems for PV, the corner cells are in fact removed to avoid this shading loss altogether.

A CLOSING NOTE ON SHADING: Keep in mind that a compliment of being an expert in solar design will be the access to design criteria where we actually *desire* shadowing effects. In complement with architects and landscape architects, we can identify systems for which a cool microclimate, shaded from solar gains, would be highly desirable. Our human comfort systems can have such a range of acceptable conditions for keeping cool on a hot day, and sometimes the shade of a tree, or the canopy of an awning in a patio can be just the right strategy to avoid using intensive engineered electronic technologies like a heat pump air conditioner. Keep in mind that you are solving for patterns in coupled systems, and listen to your integrated team for solutions that increase solar utility *and* deliver a comfortable environment to your clients.

TRACKING SYSTEMS

Increasingly, there are systems design cases where we wish to not only avoid shading, but we also wish to design a system that tracks the Sun. There are two primary methods to do so: the first is to use a sensor with a feedback loop to direct the collection surface or aperture toward the brightest region of the sky. Sometimes this is the Sun, but clouds can interfere with the Sun, and can even create a lensing effect to deliver temporary brighter conditions.

The traditional way to track the Sun (using computers) was to program an algorithm that would seek to minimize the *angle of incidence* (θ) between the vector normal to the Sun (providing $G_{b,n}$ irradiance) and the vector normal to the collector surface. As we shall see, there are trade-offs to different tracking methods, often at the expense of shading among an array of collectors. First, let us recall the most general equation for the angle of incidence, θ in Eq. (7.7) (no subscripts, so it refers to any orientation of the collector relative to the Sun):

$$\begin{aligned}\cos\theta = {} & \sin\phi\sin\delta\cos\beta - \cos\phi\sin\delta\sin\beta\cos\gamma \\ & + \cos\phi\cos\delta\cos\omega\cos\beta + \sin\phi\cos\delta\sin\beta\cos\gamma\cos\omega \\ & + \cos\delta\sin\omega\sin\beta\sin\gamma.\end{aligned} \tag{7.7}$$

Now we can identify several common scenarios for minimizing θ with corresponding criteria. The following scenarios have been described in detail in several texts, and we summarize their algorithms and merits/detractions here.[1] A key point to bring into play is the role of concentration. If the collector systems are only susceptible to low concentration, then precision tracking is a limited trade-off. The more that one incorporates concentration of light into the systems design, the more one must be aware of the limitations of tracking systems and their respective trade-offs:

[1] John A. Duffie and William A. Beckman. *Solar Engineering of Thermal Processes*. John Wiley & Sons, Inc., 3rd edition, 2006.

Fixed Azimuth due N-S, Periodic Tilt Adjustment: In this first case, we assume the system will only be analyzed for solar noon of each day, and then the tilt (β) will be inclined up/down with a crank to meet the best conditions for the angle of incidence minimization ($\theta=0$ at solar noon). Equation (13.16) is then simplified to

$$\cos\theta = \sin^2(\delta) + \cos^2(\delta)\cos(\omega), \tag{7.8}$$

where the slope of the surface will be adjusted to

$$\beta = |\phi - \delta|. \tag{7.9}$$

But do we need to adjust this system daily? We have observed that we can be tolerant of several degrees of change in the declination angle while maintaining a near-zero angle of incidence. In fact, one can accept up to 8° of error in deviation from the actual angle of declination and still achieve $\cos(8°) = 0.99 \approx 1$. The implication is: a periodically adjusted tilted collector only needs seasonal adjustment, or even bi-annual adjustment. The trade-off is that this is not a good system for a periodic array of solar modules stacked North to South. Recalling our work on shading assessment, optimal spacing between panels is dependent on a fixed tilt, so this arrangement would lead to seasonal shading challenges, particularly in the Winter months when the Sun is low in the sky.

Inclined N-S Axis, E-W Tracking: This is the first of several single-axis continuous daily tracking methods. In this case, the modules are mounted on a beam or track that is oriented with an azimuth pointing to the Equator. The tilt of the track (*not the modules*) is at an optimum angle near the angle of the local latitude, but often reduced to some degree to local climate effects.[2,3] Then, the modules track the Sun from East to West rotating on the axis of the tilted track. The strategy is to minimize theta by setting $\cos\theta \approx \cos\delta$ (in a polar tilt scenario, $\cos\theta = \cos\delta$).

In this case, the true tilts and azimuths of the collector are continuously varying throughout the day, much like a dual-axis tracking system. The collector β is found from the steepest descent for the panel, and the corresponding γ is aligned with the path of that steepest descent[4]:

$$\beta = \tan^{-1}\left(\frac{\tan\phi}{\cos\gamma}\right). \tag{7.10}$$

The variable collector azimuth angle has a special algorithm with cases:

$$\gamma = \tan^{-1}\left(\frac{\sin\theta_z \sin\gamma_s}{\cos\theta' \sin\phi}\right) + 180 \cdot C_1 C_2. \tag{7.11}$$

Notice the denominator has a special angle "theta prime" (θ') that is different from any other values we have seen before, where:

$$\cos\theta' = \cos\theta_z \cos\phi + \sin\theta_z \sin\phi \sin\gamma_s \tag{7.12}$$

and

$$C_1 = \begin{cases} 0 & \text{if } \tan^{-1}\left(\frac{\sin\theta_z \sin\gamma_s}{\cos\theta' \sin\phi}\right)\gamma_s \geq 0, \\ +1 & \text{otherwise} \end{cases}$$

and

$$C_2 = \begin{cases} +1 & \text{if } \gamma_s \geq 0, \\ -1 & \text{if } \gamma_s < 0. \end{cases}$$

The value in this particular approach is that one is creating a *pseudo-two axis* variation for the collector, in that each module is changing in terms of tilt and azimuth throughout the day. The performance in the system is maximized near the Equinoxes, while the cosine project effect will be the largest at each of the solstices. Specifically, the Equinox days can perform slightly better than the Summer Solstice. In fact, calculations from researchers in Europe have found that this configuration can offer a potential increase of 12–50%

[2] For example, in State College, PA the $\phi = 41°$, but the optimal tilt for the year is $\beta \sim 33°$.

[3] Craig B. Chistensen and Greg M. Barker. Effects of tilt and azimuth on annual incident solar radiation for United States locations. In *Proceedings of Solar Forum 2001: Solar Energy: The Power to Choose*, April 21–25 2001; M. Lave and J. Kleissl. Optimum fixed orientations and benefits of tracking for capturing solar radiation in the continental united states. *Renewable Energy*, 36:1145–1152, 2011; and T. Huld, R. Müller, and A. Gambardella. A new solar radiation database for estimating PV performance in Europe and Africa. *Solar Energy*, 86(6):1803–1815, 2012.

[4] John A. Duffie and William A. Beckman. *Solar Engineering of Thermal Processes*. John Wiley & Sons, Inc., 3rd edition, 2006.

annual power production relative to that of a flat plate PV array.[5] However, shadowing effects are particularly challenging to address in this configuration, and these systems tend to have modules assembled on the tracks with gaps in each corner to minimize shadowing losses.[6]

Vertical Axis, Azimuthal Tracking with Fixed Optimal Tilt: This case is similar in performance to the inclined track method, in that the approach has been shown to offer a potential increase of 11–55% annual power production relative to that of a flat plate PV array.[7] The base of each system is a vertical beam upon which modules are mounted with the optimal fixed tilt for the year ($\beta = \beta_{opt}$, given latitude and local climate conditions). The module then rotates clockwise along the azimuthal plane from sunrise to sunset on the central beam ($\gamma = \gamma_s$). The angle of incidence is minimized by making the collector azimuth and the solar azimuth angles equal:

$$\cos\theta = \cos\theta_z \cos\beta + \sin\theta_z \sin\beta. \quad (7.13)$$

Horizontal N-S Axis, E-W tracking: For this case, we take the inclined N-S axis and drop it to a horizontal beam with E-W tracking of the array. This particular configuration has a characteristically "peaky" power profile over the day:

$$\cos\theta = \sqrt{\sin^2\alpha_s + \cos^2\delta \sin^2\omega}. \quad (7.14)$$

The tilt of the surfaces also varies through the day, and can be obtained as:

$$\beta = \tan^{-1}\left(\tan\theta_z |\cos(\gamma - \gamma_s)|\right). \quad (7.15)$$

While the collector azimuth is either −90° before solar noon, or +90° after solar noon.

The advantage of this particular array class for tracking is that one obtains very small shadowing effects. Essentially, shadowing will only occur at the early and late hours of the day. A lesser drawback is that the power is produced in a tighter curve about solar noon comparted to the inclined N-S axis:

Horizontal E-W Axis, N-S Tracking: Our final case seems unusual: we will rotate the horizontal axis of the tracking beam to be along an E-W azimuth instead

[5] T. Huld, M. Šúri, T. Cebecauer, and E. D. Dunlop. Comparison of electricity yield from fixed and sun-tracking PV systems in Europe, 2008. URL http://re.jrc.ec.europa.eu/pvgis/.

[6] Soteris A. Kalogirou. *Solar Energy Engineering: Processes and Systems.* Academic Press, 2011.

[7] T. Huld, M. Šúri, T. Cebecauer, and E. D. Dunlop. Comparison of electricity yield from fixed and sun-tracking pv systems in europe, 2008. URL http://re.jrc.ec.europa.eu/pvgis/.

of N-S. Then, we "rock" the array, tracking from N-S-N or S-N-S, depending on the time of year and location.[8] As with the previous class of single-axis tracking with a horizontal track, the effects of shadowing are not large:

[8] John A. Duffie and William A. Beckman. *Solar Engineering of Thermal Processes.* John Wiley & Sons, Inc., 3rd edition, 2006; Soteris A. Kalogirou. *Solar Energy Engineering: Processes and Systems.* Academic Press, 2011.

$$\cos\theta = \sqrt{1 - \cos^2\delta \sin^2\omega}. \tag{7.16}$$

Again, the tilt of the surfaces varies through the day, and can be obtained as:

$$\beta = \tan^{-1}\left(\tan\theta_z |\cos(\gamma)|\right). \tag{7.17}$$

While the collector array azimuth (γ) will be facing either due South or due North, depending on whether the Sun is in the northern half of the sky dome or the southern half. One would also expect the tracking system to remain static and only rock toward the Equator and back near the Equinoxes, when the Sun rises due East and sets due West.

An interesting result of this strategy is that the profile of power generated over the day is unusually flat (not peaky) across the hours of the day. The roughly square profile could be useful for solar thermal applications where a level thermal character for power production is called for across the day.[9] However, in the months near the Winter Solstice, the cosine projection effect is strongly influencing the system performance on otherwise horizontal arrays, and leads to significantly decreased daily performance.

[9] Soteris A. Kalogirou. *Solar Energy Engineering: Processes and Systems.* Academic Press, 2011.

So there you have it. Multiple trade-offs to solve for in a pattern of coupled problems dealing with tracking decisions for a solar energy conversion system. Some solutions will automatically remove tracking as an option for your client. *The author only knows of three rotating houses: one near Pittsburgh, one in San Diego, and one in Vauban, Germany, just outside of Freiburg. Tracking houses are tricky!* For those solutions in utility-scale electric power, or using tracking concentration to heat a fluid for power or water/air heating linked to industrial process heating, these questions will be necessary in the integrative design process. Sometimes you will be constrained by the physical limits of the land, and sometimes you will think about taking advantage of natural hill slopes to create a positive systems outcome! Good luck in your systems designs.

PROBLEMS

1. Plot an orthographic projection of a southeasterly tall building interfering with a first floor window of interest, seen in Figure 7.9 (problem adapted from Kalogirou). Assume the locale is in Raleigh, NC.[10]
2. Plot a polar projection of the same scenario (using the same critical points).

NOTE that both problems can be plotted on templates using software from the University of Oregon Solar Radiation Monitoring Laboratory (http://solardat.uoregon.edu/SunChartProgram.php).

[10] Soteris A. Kalogirou. *Solar Energy Engineering: Processes and Systems.* Academic Press, 2011.

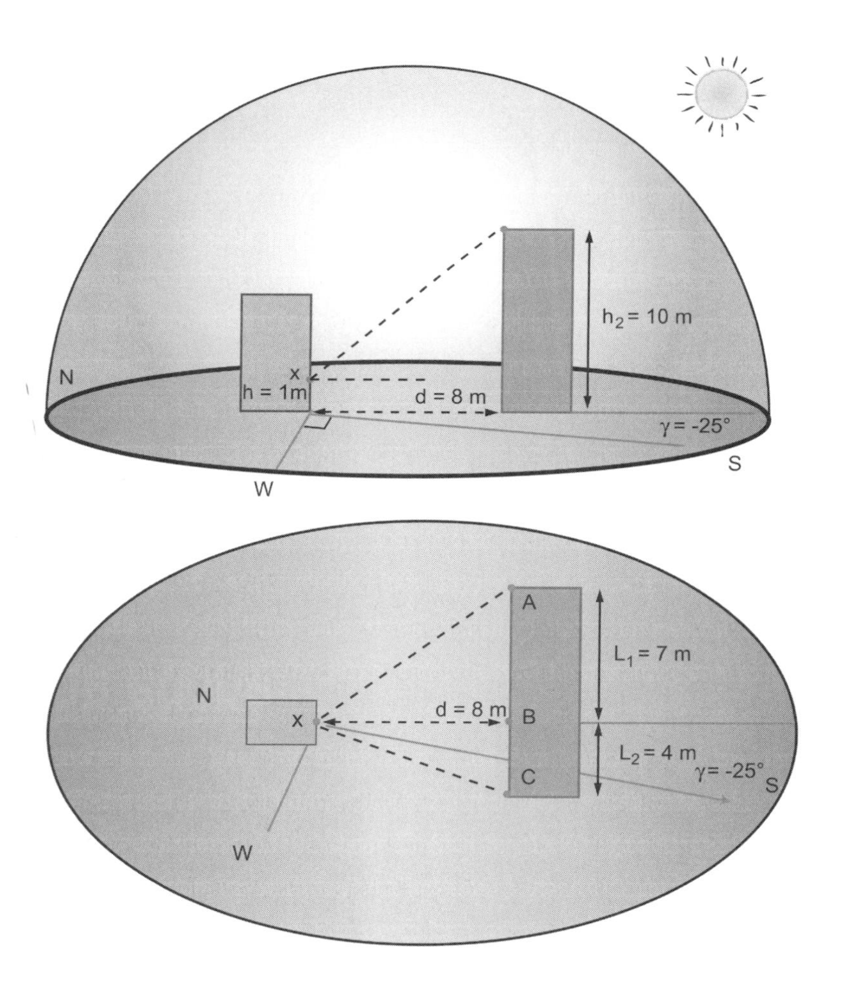

Figure 7.9: A demonstration of shade offered from a taller building to the southeast of a point *X* on the South wall. The critical origin point is *X*, and the critical points of potential shading are *A*, *B*, and *C*.

RECOMMENDED ADDITIONAL READING

- Effects of tilt and azimuth on annual incident solar radiation for United States locations.[11]
- Optimum fixed orientations and benefits of tracking for capturing solar radiation in the continental United States.[12]
- A new solar radiation database for estimating PV performance in Europe and Africa.[13]
- Solar Engineering of Thermal Processes.[14]
- Solar Energy Engineering and Processes and Systems.[15]

[11] Craig B. Chistensen and Greg M. Barker. Effects of tilt and azimuth on annual incident solar radiation for United States locations. In *Proceedings of Solar Forum 2001: Solar Energy: The Power to Choose*, April 21–25 2001.

[12] M. Lave and J. Kleissl. Optimum fixed orientations and benefits of tracking for capturing solar radiation in the continental United States. *Renewable Energy*, 36:1145–1152, 2011.

[13] T. Huld, R. Müller, and A. Gambardella. A new solar radiation database for estimating PV performance in Europe and Africa. *Solar Energy*, 86(6):1803–1815, 2012.

[14] John A. Duffie and William A. Beckman. *Solar Engineering of Thermal Processes.* John Wiley & Sons, Inc., 3rd edition, 2006.

[15] Soteris A. Kalogirou. *Solar Energy Engineering: Processes and Systems.* Academic Press, 2011.

MEASURE AND ESTIMATION OF THE SOLAR RESOURCE

08

Optimization of collector orientation for any solar process that meets seasonally varying energy demands, such as space heating, must ultimately be done taking into account the time dependence of these demands. The surface orientation leading to maximum output of a solar energy system may be quite different from the orientation leading to maximum incident energy.

John Duffie and William Beckman, Solar Engineering of Thermal Processes (2006)

OUR ENVIRONMENT surrounds us, both indoors and outside. We are a part of our environment, engaged in the exchange of energy and mass in flows across systems boundaries.[1] When we want to understand the patterns emergent in our environment, particularly the interplay among the Earth, collector, sky, and Sun, we need *measurables*—qualities of the environment that can be measured and recorded as time passes (also called empirical observations). Keep in mind, measurable parameters for SECS are coupled within a larger pattern—they are a part of the Earth-Sky-Sun system. Coupled parameters in a system mean that the pattern of our habitable environment cannot be pulled apart easily on Earth. By measuring one emergent property of the system (particularly an energy flux like irradiance), we are providing some additional information regarding other properties of the pattern in the system (like temperature, wind speed, rain).

[1] Annie Leonard. *The Story of Stuff: How Our Obsession with Stuff is Trashing the Planet, Our Communities, and our Health—and a Vision for Change*. Simon and Schuster, 2010.

Key concept: *nature* is not "outside," and there is no "away" to throw things. We are a part of our environment.

As Lord Kelvin said, "If you cannot measure it, you cannot improve it."

Again, as a result of coupled environmental parameters, by measuring phenomena like irradiance (W/m^2), temperature (°C), humidity or dew point (T_{dp}, °C), and pressure (mbar) within our environment, we can use the resulting information to *estimate* many other emergent properties of the larger system that we are a part of. When estimating coupled but less certain parameters in a system using parameters measured from the environment, we develop what is called an *empirical model.*

Get it? Measure some accessible parameters of the environment, and *estimate* the coupled behavior of the other parameters in a *model.*

As an example of an environmental measurable that we are familiar with, let's first turn to the scenario of accounting for a hot day. Ah, *temperature*, now we are back to a familiar territory! Temperature is an emergent property of matter (gas, liquids, solids) as a consequence of atomic movement (vibration, twisting, or jetting around). Temperature can also be viewed as a summary statistic, the conjugate of *heat* (related by the *heat capacity*), or the singular measure describing the full distribution of *thermal energy* in a parcel of air or in a solid (see the Appendix on energy distributions for more discussion).

In solid materials, the two principal types of thermal energy are the *vibrational energy* of atoms about their mean lattice positions (including organopolymers), and *kinetic energy* of conducting electrons (in metals). Electrons contribute a relatively small amount of kinetic energy (as conducting electrons) in semiconductors. As a solid absorbs energy, its temperature rises and its internal energy is increased. The principal type of thermal energy in gases is kinetic energy (jetting around in space). We explore these details to point out that society has accepted temperature as a useful measure and has incorporated it into everyday language. However, the underlying science of temperature is just as challenging and interesting as the science behind irradiance.

So how do we account for a hot day? The simplified answer is that we put a thermometer[2] outside and *measure* the caloric quality of the air. If we additionally measure the dew point conditions,[3] we can *estimate* the comfort level outside for humans that have acclimated to a regional climate condition, or the likelihood of rain.

Why do we measure energy? We measure so that we can reveal insight into the underlying pattern of the environment. From our insight, we propose models and iterate with more detailed and thorough measures in a process of learning,

[2] Thermometers can be formed from liquids that change volume with increases in thermal energy, or by changes in other properties including bimetallic electrical resistance.

[3] The *dew point* is the temperature at which a given parcel of air will become saturated with water vapor if cooled at constant pressure and water content.

additionally revealing our societal perceptions of the environment (biased or not). If we are creative and intend to apply our model back onto the pattern of the environment, we begin the process of design.

MEASURING LIGHT

Why do we need to measure light conditions (as irradiance; W/m^2) for solar energy conversion systems design and operation? Consider that light is a form of energy just like thermal energy, and we measure thermal energy throughout society to better inform the technologies that we use. *Everything* in solar assessment is meteorological and climatological in nature. Hence, by understanding the manner in which meteorological information is parsed and processed, we form stronger design teams to deliver measures or expected values of variance and uncertainty for the solar resource within different bands of time under consideration.

In the earlier chapters we described the interplay and exchange of radiant energy between surfaces, we described the complex nature of the sky interacting with sunlight, and then we focused on trigonometric relations for space and time. We also worked to use those angular relations to help address the goals of Solar Design. In doing so, we often simplified our model of light from the Sun, and imagined it as a single beam (normal to the surface of the Sun) of monochromatic light. The real solar resource has a broad spectrum, is scattered off of many surfaces in non-linear fashion, and has a dynamic character. The real solar resource is *location dependent* (which is why we maximize the solar utility for a client in their locale). So how can we build a suite of concepts together to justify measuring the solar resource for better solar design?

Goal of Solar Design:

- Maximize the solar utility,
- for the client,
- in a given locale.

We know that there are several parameters that contribute to the character of light, such as *bands* (spectra) and *components* (directionality). Let's return to the analogy that the light can be thought of as a *pump* rather than a fuel. Again, this is because photons are converted as a dynamic flow of energy. So we measure solar energy to get proper estimates of the behavior of our solar pump. We can extend our analogy in several distinct ways. The following list provides key parameters that describe or quantify the character of light, coupled with an analogous question related to our pump analogy:

a. **Intensity** How *big* is the pump at any given time? (no "pumping" at night without supplemental storage).
b. **Bands** What *color* is my pump at any given time? The solar resource has variable contributions of UV/Vis/IR (shortwave) during the day.
c. **Components** What *direction* is it pumping now? The solar resource can be divided into beam, diffuse sky reflectance, and ground reflectance (at the most simple level) relative to a tilted collector.
d. **Intermittency** *When* and *where* will the pump stop? Intermittency brings in the concept of risk assessment in project design.
e. **Network Distribution** What *portfolio* of sites might one need to make a *super-pump*? Solar exists across regions, as such it is a decentralized resource that can be captured in a *portfolio*.
f. **Climate Regime** How is a *Fall* pump different from a *Summer* pump? Climate regimes (spatio-temporal) impact the *component* and *band* contributions to the resource.

Now think about how we could evaluate the quality of a sunny or overcast day. How do we account for the puffy cumulus clouds in the sky on a bright day, and account for smog in an urban environment? The answer is *measurement*. In order to estimate the character of light, we need to measure the light.

In our description of angular relations, we took time to introduce a few concepts of light direction from the sky dome and the ground. In review, the center beam of light from the Sun projected onto the receiving surface of a collector is termed *Direct* or *Beam* irradiance (G_b), while the light sources in the shortwave band other than *Direct* light are termed *Diffuse* (G_d).[4] The intensity of *Beam* and *Diffuse* irradiance is due to the degree of light *scattering* by molecules and particles (in addition to the time of year and latitude discussed in the last chapter). The atmospheric physics of light reflection and refraction, transmission, and absorption can be considered as an emergent systems properties, which we can expand upon as components of light.

[4] If we point our collection surface toward the Sun, we align with the **Direct Normal Irradiance** (DNI, $G_{b,n}$), or the beam/vector that emerges *normal* (perpendicular) to the surface of the Sun. DNI is *not equivalent* to **beam irradiance** ($G_{b,n} \neq G_b$) when our collection surface is oriented otherwise.

HORIZONTAL SURFACES ABOVE THE SKY DOME: We have already covered three basic calculations of power density and energy density for the time steps of seconds, hours, and days with the following three Eqs. (8.1), (8.2), and (8.3). Each of the three equations represents the irradiance or irradiation upon a *horizontal surface* sitting upon the extraterrestrial rim of the Earth's atmosphere, as calculated by geometric relations proportional to the cosine of the zenith angle for a given latitude and time of the year (declination and hour angle).

AM0 Irradiance (W/m^2):

$$G_0 = G_{sc}\left[1 + 0.033\cos\left(\frac{360n}{365}\right)\right] \cdot [\sin\phi\sin\delta + \cos\phi\cos\delta\cos\omega]\,. \tag{8.1}$$

Calculated values, not measured values—but useful for understanding the trends of solar intensity in a locale over the year.

Hourly AM0 Irradiation (MJ/m^2):

$$I_0 = \frac{12 \cdot 3600}{\pi} \cdot G_{0,n} \cdot \left[\frac{\pi}{180}(\omega_2 - \omega_1)\sin\phi\sin\delta + \cos\phi\cos\delta\,(\sin\omega_2 - \sin\omega_1)\right]. \tag{8.2}$$

Daily AM0 Irradiation (MJ/m^2):

$$H_0 = \frac{12 \cdot 3600}{\pi} \cdot G_{0,n} \cdot 2\left[\frac{\pi}{180}(\omega_{ss})\sin\phi\sin\delta + \cos\phi\cos\delta(\sin\omega_{ss})\right]. \tag{8.3}$$

Answers for I_0 and H_0 must be divided by 10^6J/MJ to convert the units of J/M^2 to MJ/m^2, which are easier units to read and check.

HORIZONTAL AND TILTED SURFACES UNDER THE SKY DOME: So what do we do for surfaces that are *underneath the atmosphere?* for surfaces that actually interact with the sky and clouds, and rain, and even smog? Furthermore, what do we do as a design team to understand the solar resource incident upon a *tilted surface*, a collector with a *non-horizontal plane of orientation?*

My answer to you, the modern entrepreneurial team who will change the world: *measure the solar resource!* Measure irradiance on a horizontal surface, measure it on a tilted surface or a tracking surface. Measure irradiance as a whole ensemble of bands and components *or* measure it as spectral signals at each wavelength and for decoupled beam and diffuse components. If you wanted to know the way a new fuel behaves in an engine, you would certainly measure the characteristics of the chemistry before and after combustion. You wouldn't just hold a beaker of biofuel up to the light and assess with your eyes: *well, it seems good enough to burn?* No, you use modern technology to measure the characteristics of the resource, so that your team can get the best utility out of the energy conversion process for your client in that given locale of interest! Once again quoting Lord Kelvin (William Thomson, 1st Baron Kelvin from the 19th century): "If you cannot measure it, you cannot improve it."

And remember: *do not trust your eyes for solar measurement*. In a following section we will explore the value and great detraction of using our sense of vision to assess a solar resource. Can you think of some reasons why the eye is better at some things, but could be quite horrible in assessing the solar resource accurately over the year?

MEASUREMENT TERMINOLOGY:

- BANDS: A convenient term for groups of wavelengths of light (e.g., IR band, UV band, longwave band).

- COMPONENTS: A term for the groups of physical orientations and scattering of light (e.g., diffuse component, beam component).

- SHORTWAVE RADIATION: Includes UV/Visible/Infrared bands of radiation. Solar shortwave radiation is often regarded as being bounded from 290 to 2500 nm (the shortwave band).

- LONGWAVE RADIATION: In the past, it has been termed thermal radiation, or radiation emitted from the Earth as it "glows" around 300 K. Of course, the Sun produces this band as well. The intensity of longwave radiation is significantly lower than shortwave radiation.

- APERTURE: This is a reminder. As a collector, even the measurement device can be viewed as an aperture. Typically, a collector is oriented in any number of general angles that point up into the sky (but not always, when connected to a reflecting device).

- PYRANOMETER: Instrument used to collect total shortwave radiation (beam + diffuse). Pyranometers are *almost always* oriented horizontally.

- PYRHELIOMETER: Instrument used to measure only the direct/beam component of shortwave radiation. The pyrheliometer has a two-axis tracking system. Pyranometers are normally calibrated against pyrheliometers.

- SUNSHINE DURATION: Periods for which $G_{b,n}$ exceed 120 W/m^2.

METEOROLOGICAL YEARS: COLLAGE OF DATA

Sometimes we need to develop a project design using less information, or with an approximation of weather information based on the expected conditions in a locale. We do have historical records of solar conditions in many locations across the world. From these records, solar scientists and engineers were able to form an approximation of hourly irradiation conditions (and other weather data) that could be used to model long-term performance of systems like solar hot water panels and homes—derived from NSRDB (National Solar Radiation Data Base) data spanning 1961–1990 and 1991–2005. Beginning in the late 1970s the *Typical Meteorological Year* (TMY) was developed for the USA and surrounding regions as a design tool to be applied to long-term systems design and estimations.

See dynamic maps of the resources at the NREL GIS link.

The TMY data are not a data set of the year to date, but rather is synthetic data based on long averages and mosaicking days together. Estimates are evaluated relative to multi-decadal averages of weather conditions at specific locations. The two common databases used in computer simulation programs such as TRNSYS, Energy+, and SAM (System Advisor Model) are TMY_2, collected from the period of 1961–1990, and the TMY_3 data for over 1400 stations, spanning 1976–2010 or 1991–2010.[5] TMY_3 data sets are recommended by NREL for use instead of TMY_2 data, as they are based on more recent and accurate data.

[5] NSRDB link: the National Solar Radiation DataBase.

[6] F. Vignola, AMcMahan, and C. Grover. *Solar Resource Assessment and Forecasting,* chapter "Statistical analysis of a solar radiation data set—characteristics and requirements for a P50, P90, and P99 evaluation." Elsevier, 2013.

The TMY data are very useful, but are to be used with *care*.[6] These data are reasonable estimates for first-order simulations evaluating the effect of *regional climate* on the long-term performance of systems, such as an *average June* or an *average January* in Williamsport, PA. They are *not* reasonable markers of *extreme local weather events* or outliers from a normal data distribution. Extreme events are low probability (long tail) phenomena that have extremely high impact on energy use and costs (e.g., heat waves, hurricanes, tornadoes that insurance companies are concerned with).

TMY data is not appropriate for project design that cannot tolerate uncertainties in the hourly data set >10%. Wilcox and collaborators have divided the locations into Class I, II, and III sites, and reported a range of uncertainties for the TMY_3 modeled data ranging from 8% (optimal) >25% in locations with sub-optimal input data.[7] While the data have a large root mean square (RMS) error (a wide spread of data), they were found to have a smaller bias error (not too high or too low). TMY data is also less appropriate for project designs where sub-hourly data is called for, or where tracking and concentration are critical factors in the deployed system. For further reading we highly recommend reviewing the "National Solar Radiation Database 1991–2010 Update: User's Manual."

[7] Stephen Wilcox. National solar radiation database 1991–2010 update: User's manual. Technical Report NREL/TP-5500-54824, National Renewble Energy Laboratory, Golden, CO, USA, August 2012. Contract No. DE-AC36-08GO28308.

HUMAN VISION: LOGARITHMIC DETECTION VERSUS LINEAR

Let's start with our native solar detection system, the human eye. Sight and visual perception is an unusual system. Our vision has evolved to be advantageous to humans when seeking to *minimize risk* in dark environments, such as avoiding a lion prowling through the forest in the evening. Our visual system also adapts with intensely bright signals, trying to maintain a stream of useful information even without sunglasses.

Our eyes were designed to provide dense information, to minimize risk in our environment.

The human visual system also has distinct limitations, in that our visual senses are linked to a cognitive system which can extrapolate small signals into valuable information, or scale very intense signals into accessible information. The goal of sight is *information* about the world around us, not the amount of light delivering that information to use (which we think of as *power*).

The eye has two primary classes of large molecules, called *rods* and *cones*. They are both located in the *retina*. The two molecular systems have adapted for full daylight (cones), and for dim lighting conditions (rods). Rods have developed to absorb certain photons of lower energy and wavelengths of light that are longer. The system of cones (in fact multiple kinds of cones) has developed to absorb separate bands (wavelengths) of light that we subsequently interpret as *color*.

In both systems the maximum band of absorption is limited, and does not include the ultraviolet or infrared regions of the shortwave band that comprise about 50% of solar irradiance. The consequence of this design is that our eyes do not detect a large range of solar wavelengths—they are limited in the photons that can be accepted. Additionally, the response factor of each of those receptors is not linear either. In terms of measurement, we seek reproducible information that is linear in response, making the rods and cones system one of the weaknesses to trusting the eye to be used as a quantitative solar measurement tool.

Notice in Figure 8.1 that our rod and cone receptors are distributed across the back of the eye. Yet the two systems do not share the same spatial distribution. We see that rods are distributed broadly across the *retina*, with

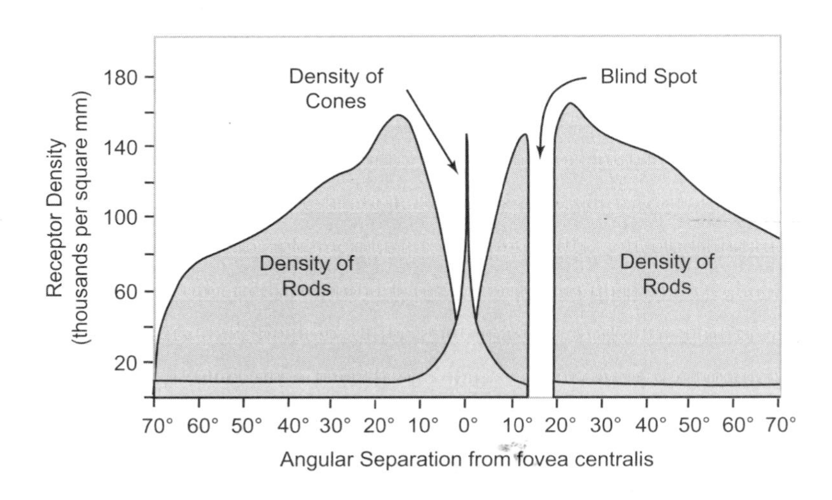

Figure 8.1: Schematic of the distribution of receptors within the eye. Figure adapted from Osertberg, G. "Topography of the Layer of Rod and Cones in the Human Retina," *Acta Opthalmologica*, Supplement, Vol. 6, 1–103, 1935.

the exception of the *fovea centralis*. As a contrast, cones have evolved to be distributed mainly within the *fovea centralis*, which is the focal point of our optic lens system.

We will discuss lenses again in the next section with the Campbell-Stokes sunshine reader.

Conveniently, the lens system in the eye allows us to think about concentrating solar systems and the limitations of a lens. A lens directs light onto a focus, but as a consequence can *only collect light from the direction that the lens is pointing*. This implies that (for most systems) a concentration process calls for tracking of bright light sources to achieve greater performance.

This property of concentration means that the cone system of the eye will mainly detect light for the direction that the eyes are pointing. The big implication is that our color detection system that we enjoy so much for high-definition visual experiences is relatively *poor* at sensing diffuse, or scattered light from zones of the sky or ground at which the eye is not pointing. Mark down one more detraction for the eye as a quantitative solar measurement tool.

Resuming our discussion of the rod system, we note that rods are distributed everywhere *except* the focal point (and our blind spot). This distribution means that rods will detect *diffuse light* entering the eye from all angles. Rods are not color sensitive, so they just detect longer wavelength shortwave photons. Examples would include ambient starlight, dim night lights, or during the twilight. A stargazer recommendation: if you want to see more at night while looking for constellations, or while riding a bike or running at night, try defocusing your vision to allow peripheral light to be detected.

The optical implications for the human eye are that rods are not part of a concentrating system, and will detect diffuse light better at the expense of color discrimination. However rods only present a useful signal to the brain at low levels of light. Mark down yet another detraction for the eye as a quantitative solar detector.

Our eyes have adapted to provide high-quality *information* to the brain, not maximal solar *power*.

If we expand our exploration of the eye SECS beyond the absorber portion, we note an additional macroscopic feedback measure that controls light acceptance. *The goal of the human eye is to provide the brain with the best information*, not power, and the human iris helps to achieve that goal. The *iris* for the eye is a feedback system meant to open up a wide aperture when the light is dim, and squeeze into a small aperture when the light is intense. No matter what your rods and cones are automatically doing as a photochemical response to incoming photons, the iris is constantly adapting to maximize the signal of visual information to the brain. In contrast, for an ideal solar measurement system, we would not want an adaptive iris system, because it again detracts from our goal of linear detection of irradiance changes across the day.

Of course, going into the larger system of the face, one could also consider the eyelids and eyebrows, as they block or shade much of the bright light to your eyes. Considering human behavior (humans are a part of the visual system, right?), we recognize that very few people actually "look up" to sense the light in the sky. Instead, we tend to look to the horizon, meaning our lenses are not trained vertically upward where the signal would be strongest, but rather along a horizontal plane. In contrast, we will show that solar detectors are often mounted flat, to point up to receive the entire sky dome of light. In total, one arrives at the conclusion that the eye cannot inform us quantitatively of the amount and changes in irradiance during the day.

This is all a reminder to the solar design team that the eye is designed to provide you with *information* sufficient to avoid bumping into bad things in extreme lighting conditions, or to process subtle and diverse streams of data from a high-definition imaging device. We call all of this information, but it is not the useful information for assessing the solar resource quantitatively. A linear detector like a photovoltaic device or a thermopile is required to accurately measure the radiant flux from the Sun, such that we may design a useful system for society.

GROSS MEASURE: SUNSHINE DURATION

Beyond the human eye, but related to vision in a general context, *we can define what sunshine means*. The World Meteorological Organization has specified that *sunshine duration* (SD) are the periods for which the direct normal irradiance (DNI, $G_{b,n}$) exceeds 120 W/m^2.

No *really*, sunshine has a definition!

> The term "sunshine" is associated with the brightness of the solar disk exceeding the background of diffuse sky light, or, as is better observed by the human eye, with the appearance of shadows behind illuminated objects. As such, the term is related more to visual radiation than to energy radiated at other wavelengths, although both aspects are inseparable. In practice, however, the first definition was established directly by the relatively simple Campbell-Stokes sunshine recorder…, which detects sunshine if the beam of solar energy concentrated by a special lens is able to burn a special dark paper card. This recorder was already introduced in meteorological stations in 1880 and is still used in many networks.
> –WMO (2008).[8]

[8] *Guide to Meteorological Instruments and Methods of Observation*, Chapter 7–8. World Meteorological Organization, 7th edition, 2008.

As noted by the WMO report, sunshine is the ability of irradiance to create a thermal burn in paper using a large lens, called the Campbell-Stokes recorder. This type of measurement is called an *operationally* defined, because the threshold is linked to a physical piece of equipment. Seen in Figure 8.2, the lens is a large glass sphere which will concentrate DNI from any time of day. As a

Figure 8.2: This is a typical Campbell-Stokes sunshine recorder. Image by User: Vfarboleya/Wikimedia Commons/CC-BY-SA 3.0. August 2005. Feb 2006.

consequence of the physics of the type of lens,[9] one can only collect the direct normal irradiance (DNI), making the SD measure a *simple but useful way to evaluate the periods of the day when the DNI exceeds a threshold.*

[9] ...and sharing the same limitations of the cones in the eye.

PYRANOMETER: GLOBAL IRRADIANCE MEASUREMENTS

Pyranometers act as solar energy *transducers*, meaning they can collect irradiance signals and transform them into electrical information signals (see Figure 8.3). Pyranometers produce a voltage in response to incident solar radiation. Provided that a pyranometer uses a thermopile (thermoelectric detector) the device acts as an effective "integrator" of all *components* and *bands* of light. In the case of a glass enclosure, even a thermopile detector will operate only in the shortwave band. Pyranometers based on photodiodes are used only for shortwave global radiation measurements. If we wished to measure only the *direct* component of solar radiation measured normal to the Sun's surface (DNI), we would use a *pyrheliometer*, which uses a thermopile detector.

Excellent instrument descriptions and photos can be found at the University of Oregon Solar Radiation Monitoring Laboratory. Please take time to explore this resource.

Researchers traditionally mount pyranometers in a horizontal orientation for high accuracy data collection. The measure is called Global Horizontal Irradiance (GHI). However, GHI only accounts for light originated from the sky dome, not reflected light from the ground (proportional to the albedo, or ground reflectance). Also, most collectors are *not horizontal*, and so the horizontal information must be translated to tilted orientations without knowledge of true ground reflectance. Recalling that we just mentioned that the pyranometer acts to effectively *integrate* components of light across a selected band, practitioners are now mounting pyranometers in the *plane of the array* (POA) in addition to horizontal measurements. The POA mounted devices will then additionally incorporate the changes in albedo and total ground reflectance. The trade-off is that tilted pyranometers will introduce some new errors in thermal losses to convection and cosine projection errors.

POA pyranometers measure **global tilted irradiance** and incorporate albedo effects without component models.

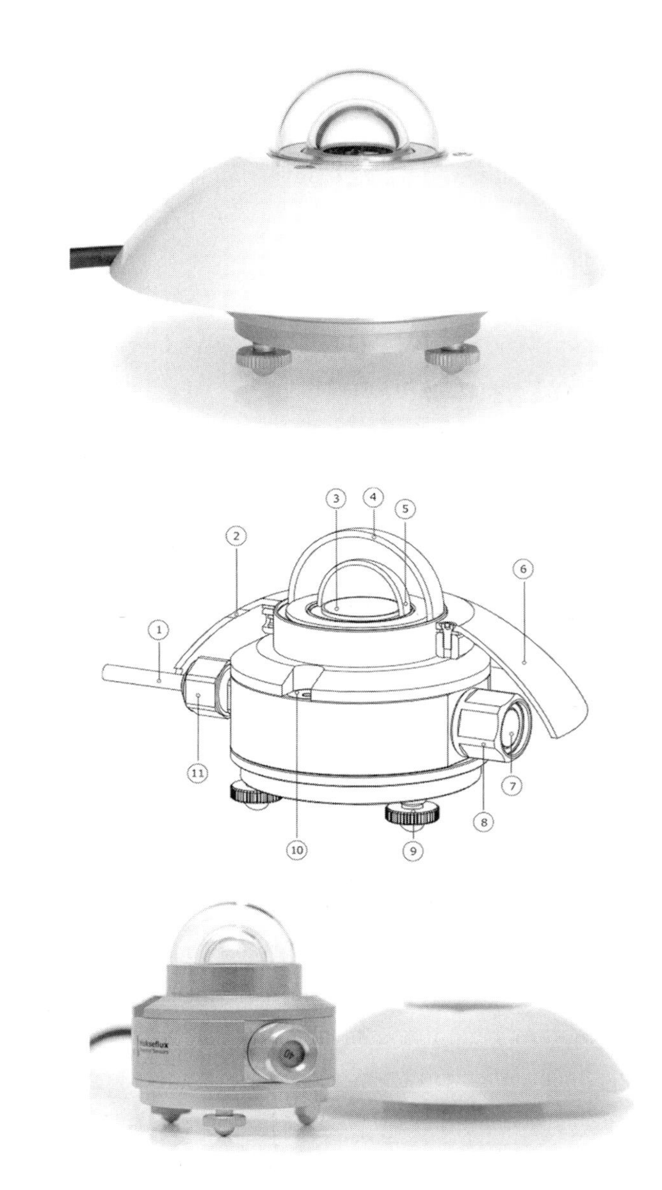

Figure 8.3: This is a typical pyranometer for research. The detail of components includes: 1. signal cable, 2. hole to fix radiation screen, 3. thermopile sensor (black body), 4. glass dome (selective surface), 5. glass dome (selective surface), 6. radiation screen (white diffuse reflector), 7. humidity indicator, 8. desiccant (moisture is bad),9. leveling feet, 10. level (bubble), 11. cable fastening nut. Referenced from Hukseflux Thermal Sensors SR11 Manual.

> A THERMOPILE was so-named as a "pile" of thermocouples. This is a series of bimetallic junctions which deliver an electric signal when exposed to a thermal gradient.

Research-grade pyranometers use a film of opaque material to collect thermal energy. The thermal energy then diffuses into a thermopile. However, we should note that metals (in general) are very good reflectors, making them also very poor *absorbers*. So how do we get a material that functions on thermal gradients to make use of the radiation from the Sun? The key is in the absorber material: Parson's black is a paint with very low reflectance across shortwave and longwave bands of light (~300–50,000 nm; making it an effective *blackbody*). However, if covered by glass (a *selective surface*), the "window" of light acceptance from the Sun is about 300–2800 nm. This system assembly forms a shortwave (band) global (component) pyranometer. Now imagine, if we develop a thermopile with a thin coating of a black absorber, but replace the glass with a material that is transparent in the longwave band (many organopolymers/plastics), we will have created a longwave (band) global (component) pyranometer.

Inexpensive pyranometers can use *photodiodes*. Keep in mind that all *photodiodes are photovoltaics*. They are semiconductor films that directly convert shortwave band radiation into electrical signals (no thermal conversion step necessary). While the cutoff for a silicon photodiode is <1100 nm, the integrated power response is fairly comparable to that of a Parson's black-coated thermopile detector. However, they do not perform as well (relative to thermopile detectors) near sunrise and sunset due to a *cosine response error*. In the morning and evenings, for small glancing angles created by low solar altitude (α_s), some of the radiation incident on the detector is reflected, that

The **cosine response error** is due to the cosine projection effect.

Robot Monkey Warning: the following sections are dense with relations and equations which explain how one parses solar measurements into useful tools for SECS simulation software. Those who need to learn this material for engineering applications should proceed, while those using this text for an overview of the art in design may wish to come back when the situation calls for it.

reflected light from a horizontal surface produces a reading less than it should be. Some correction can be made for this using a black cylinder casing and a small white plastic diffuser cover (with a low reflectance at low angles to minimize the cosine error). While pyranometers can be manufactured from parts, professional monitoring in major solar installations requires calibrated instrumentation to support vetted data collection.

DIFFUSE AND DIRECT NORMAL MEASURES

Please visit U Oregon SRML for more detailed examples of instrumentation.

An **RSP** measures both GHI and DHI, and can calculate DNI and G_b.

We will describe two more typical instruments used to collect irradiance data. The first is a *Rotating Shadow Band Pyranometer* (RSP), which measures GHI and DHI followed by calculation of G_b and DNI (Direct Normal Irradiance). The second is a *pyrheliometer*, which measures DNI ($G_{b,n}$), and can calculate G_b.

The RSP is a modified pyranometer will first measure Global Horizontal Irradiance conditions with no obstructions, followed by a rotating curved black band that intentionally obstructs the brightest cone of sunlight for the second measurement (only seconds afterward) of Diffuse Horizontal Irradiance. As we can see in Eq. (8.4), the relation among sky dome components for a horizontal collecting surface is trivial.

$$G_b = G - G_d. \tag{8.4}$$

However, given a beam component for a horizontal surface, how does one then calculate DNI? We can review Figure 8.4 to see the simple geometric relation in Eq. (8.5) for horizontal surfaces. Thus, by having an estimate of G_b from the difference in Eq. (8.4), we can calculate DNI (also labeled $G_{b,n}$) seen in Eq. (8.6). Bear in mind, as the integrated time step increases beyond an instantaneous rate (one minute → one hour → one day) error is introduced due to the natural variation of the real solar signal as it progresses through the day. Hence the beam value will only be an approximation.

$$G_b = G_{b,n} \cdot \cos\theta_z, \tag{8.5}$$

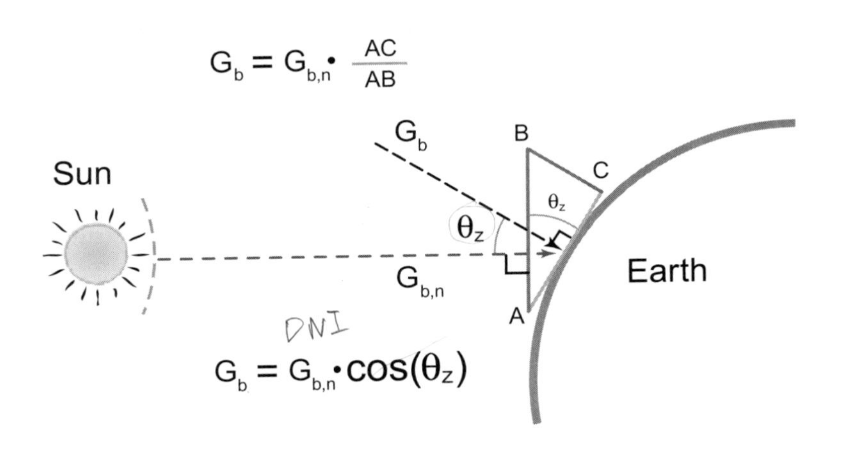

Figure 8.4: A geometric analysis of the dot product projection from DNI to G_b.

$$G_{b,n} = DNI = \frac{G_b}{\cos\theta_z}. \tag{8.6}$$

The *second device* is a *pyrheliometer*, which is a tubular device mounted upon a two-axis tracking system. In a pyrheliometer the long tube is coated black on the inside, with baffles to decrease the acceptance of light from large angles of incidence (typically $\theta < 3°$), and a thermopile is mounted at the base. A tracking pyrheliometer measures DNI ($G_{b,n}$) (and some circumsolar irradiance). From that DNI value the device can calculate G_b using Eq. (8.5).

A **pyrheliometer** measures DNI, and can calculate G_b or $G_{b,t}$.

SATELLITE MEASURES OF IRRADIATION

Ground measurements are complemented and extended using *remote sensing* from satellite imaging. We now even have recreational access to remote sensed images through Google Earth and Microsoft Bing Maps, made possible through advances in satellite imaging. Remote sensing spacecraft began in the 1960s for the purposes of making either low-resolution images of weather patterns or high-resolution images for military spy satellites. We have since expanded our use of satellite imaging to document climate changes and weather patterns in periodic images (hourly or 15 min) as well as a continuous stream of spectral data.[10] Sequences of images are collected from geosynchronous satellites stationed above the Equator (a satellite with a period of orbit matching the Earth's rotation period, hovering above the same location each day), and once again applying *Taylor's Hypothesis* that the change in measurement over time results from the lateral change in the conditions across region of the meteorological event, the sequence of images conveys events that translate downstream later in time for forecasting purposes. There are also Sun-synchronous orbits that also have geocentric orbit, but are positioned such that the satellite ascends or descends over the assigned latitude (ϕ) for the same local mean solar time, thus delivering consistent lighting throughout the year.[11]

US geosynchronous weather satellites are called *GOES* (GOES-East and GOES-West), operated by the National Oceanic and Atmospheric Administration (NOAA) within the jurisdiction of the Department of the Interior. The intergovernmental organization EUMETSAT operates the *Meteosat* weather satellites for Europe and Africa. In Japan, the JMA operates the GMS system, Russian satellites are managed by the Roskomgidromet, India by the ISRO (Indian Space Research Organization), and in China by the CMA (China Meteorological Administration). The satellites orbit 200–800 km above ground, with precise altitude control required for high-resolution remote sensing.

The challenges in remote sensing from space are numerous, yet the value of the data to solar energy design is significant. Satellites are able to measure *radiosity* from a horizontal surface in specific bands. In the shortwave (or visible) band, the measure follows the light reflected from the Earth's surface and the materials in the atmosphere. The difference between the incoming extraterrestrial irradiance and the reflected irradiance should provide an estimate of the terrestrial irradiance. In the longwave (or infrared) band, the measure follows the radiance emitted from the Earth and the atmosphere, collectively.[12]

Given the horizontal nature of the measurements, we are still left with the challenge of how to quantify irradiance on a *tilted surface*.

[10] Théo Pirard. *Solar Energy at Urban Scale*, Chapter 1: The Odyssey of Remote Sensing from Space: Half a Century of Satellites for Earth Observations, pages 1–12. ISTE Ltd. and John Wiley & Sons, 2012.

[11] Théo Pirard. *Solar Energy at Urban Scale*, Chapter 1: The Odyssey of Remote Sensing from Space: Half a Century of Satellites for Earth Observations, pages 1–12. ISTE Ltd. and John Wiley & Sons, 2012.

Radiosity: $J = \rho G + \varepsilon E_b$.

[12] Jesùs Polo, Luis F. Zarzalejo, and Lourdes Ramírez. *Modeling Solar Radiation at the Earth's Surface: Recent Advances*, Chapter 18: Solar Radiation Derived from Satellite Images, pages 449–461. Springer, 2008.

Despite the appearance of component-based data sets in purchased weather data sets, the design team should be aware that the majority of satellite data and historical ground measurements are typically produced from measures of global horizontal irradiance, and perhaps diffuse horizontal irradiance (subsequently integrated into units of irradiation at the data logger). It is simply cost-prohibitive to deploy a full meteorological surface radiation station at every potential location for a SECS. Yet decades of historical observation and successful empirical correlations by solar scientists and engineers have provided us with strategies hourly, daily, and monthly average day data. The main tools we need are the equations for hourly and daily extraterrestrial irradiance (Air Mass Zero, or *AM0*) and the integrated energy density (irradiation: MJ/m^2) gathered from a horizontally mounted pyranometer, which you learned of in the last chapter. We shall also find that we can infer more than just the *components* of light from the ratios of measured irradiation to AM0 calculated irradiation—we can describe the fractions of days in a given month that lighting conditions will be clear or overcast/cloudy.

EMPIRICAL CORRELATIONS FOR COMPONENTS

As you may have noticed in our discussion of pyranometers and seen in Figure 8.5, even in the best of scenarios we will not be able to measure every component of light (scattered and unscattered) with one instrument to make estimations on the contributions of each component to the total *irradiation* incident on the aperture of interest. We have already learned about estimating the components of light incident on a horizontal surface from the clear sky models of Bird and Gueymard's SMARTS.[13] When limited by the expense of measurement equipment, scientists and systems design teams over the past 50+ years have developed methods to approximate irradiation conditions on multiple surfaces of any orientation, given the information collected for a horizontal surface directed up to the sky dome.

[13] R. E. Bird and R. L. Hulstrom. Simplified clear sky model for direct and diffuse insolation on horizontal surfaces. Technical Report SERI/TR-642–761, Solar Energy Research Institute, Golden, CO, USA, 1981. URL http://rredc.nrel.gov/solar/models/clearsky/.

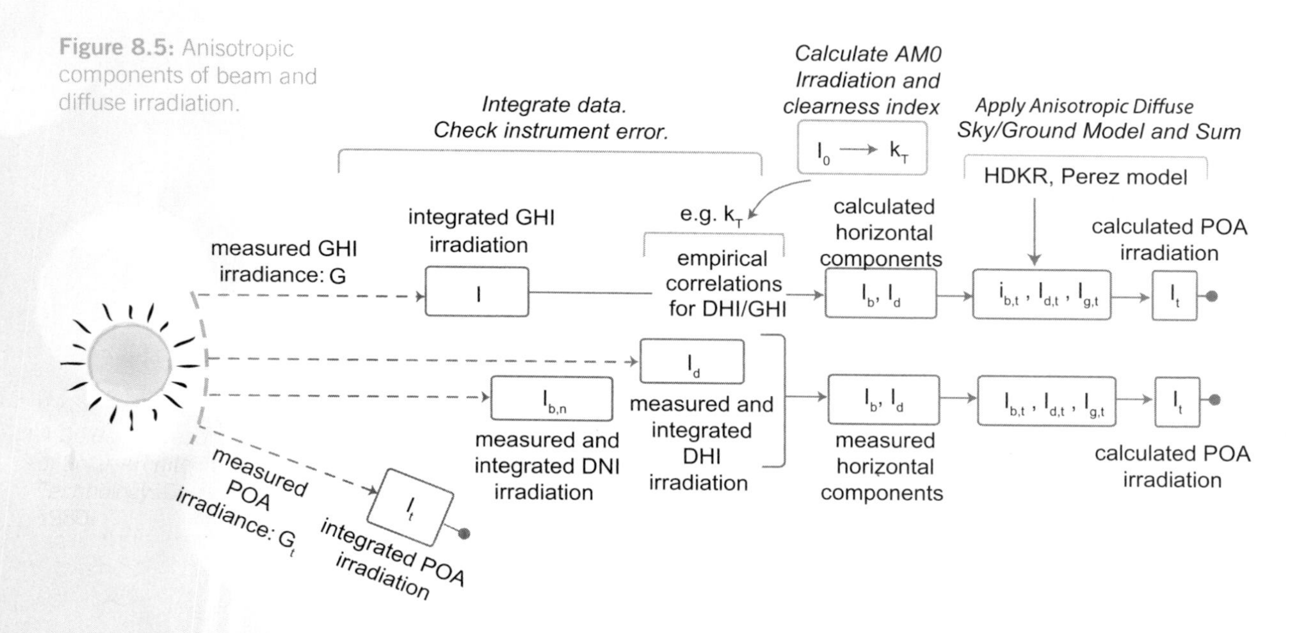

Figure 8.5: Anisotropic components of beam and diffuse irradiation.

These methods are called *empirical correlations*. The first of our correlations was already introduced for estimating beam irradiance on a horizontal surface.

Tilted collectors are far more important in SECS design, as the tilt and azimuth of the collector (γ, β) can minimize the solar cosine projection effects due to the tilt of the Earth. We have diagrammed the sky dome and ground as sources for emitted (beam) and reflected (sky diffuse and ground diffuse) light previously (see Figure 8.6). In Eq. (8.7), we show those contributions as three

On an inclined surface, $G_{g,t}$ increases, relative to a horizontal collector.

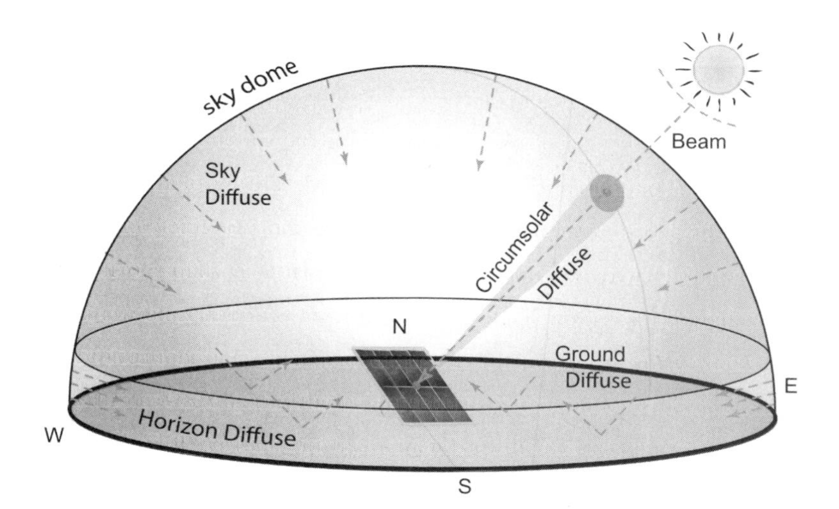

Figure 8.6: Flowchart of light processing options for hourly data sets, from measurement to summed POA global irradiation.

major directional components of shortwave light sources. The beam ($G_{b,t}$), sky diffuse ($G_{d,t}$), and ground diffuse ($G_{g,t}$) light sources incident upon the tilted collector must now be estimated from *empirical correlations* and *isotropic* or *anisotropic models* of light source components, if they have not been integrated using a POA pyranometer.

$$\begin{aligned} \text{Tilted Plane of Array} &= \text{beam} + \text{diffuse, sky} + \text{diffuse, ground} \\ G_t &= G_{b,t} + G_{d,t} + G_{g,t}. \end{aligned} \tag{8.7}$$

The beam component contributions to tilted surfaces ($G_{b,t}$), can be solved using Eq. (8.8). If we only have the beam irradiance on a horizontal surface, we can apply a simple geometric ratio of beam irradiance on a tilted surface relative to beam irradiance on a horizontal surface: the *beam irradiance tilt factor*, R_b. Notice in Eq. (8.9) that the DNI portion of the equation ($G_{b,n}$) cancels out, leaving the tilt factor proportional to the ratios of the cosines, seen again in Eq. (8.11).

$$G_{b,t} = G_{b,n} \cdot \cos\theta. \tag{8.8}$$

$$R_b = \frac{G_{b,t}}{G_b} = \frac{\cos\theta}{\cos\theta_z}. \tag{8.9}$$

Notice how this tilted beam irradiance $G_{b,t}$ is proportional to the cosine of the *angle of incidence*, θ, not the zenith angle.

Thus, Eq. (8.8) can be rewritten:

$$G_{b,t} = G_b \cdot R_b. \tag{8.10}$$

ISOTROPIC DIFFUSE SKY MODEL

In the 1960s, the *isotropic sky model* was developed to approximate the diffuse sky on a tilted surface, complimented by an estimate for diffuse light from the ground.

$$\begin{aligned} G_t &= G_{b,t} + G_{d,t} + G_{g,t}, \\ G_{tilted} &= G_b \frac{\cos\theta}{\cos\theta_z} + G_d(F_{surface-sky}) + G\rho_g(F_{surface-ground}), \end{aligned} \tag{8.11}$$

where,

Surface: the aperture.

$$F_{surface-sky} = \frac{1+\cos\beta}{2}, \tag{8.12}$$

ρ_g: collective reflectance of the ground. Reduces the irradiance G by a value between 0 and 1.

$$F_{surface-ground} = \frac{1-\cos\beta}{2}, \tag{8.13}$$

Memory device: looking *up* is a *plus* in solar energy!

$$G_{d,t} = G_d \cdot \left(\frac{1+\cos\beta}{2}\right), \tag{8.14}$$

where the fraction proportional to the collector tilt is termed the diffuse sky irradiance tilt factor for an *isotropic sky* model.

Memory device: looking *down* is so *negative* for solar energy.

$$G_{g,t} = \rho_g(G_b + G_d) \cdot \left(\frac{1-\cos\beta}{2}\right), \tag{8.15}$$

where the reflectance of the ground is termed the *albedo* (a fractional value from 0 to 1), and is multiplied by the GHI (not G_t) and the diffuse ground irradiance tilt factor for an isotropic sky model. The albedo is almost never measured in practice,[14] and instead proxy values have been used of $\rho_g \sim 0.3$ in the Summer and $\rho_g \sim 0.7$ in the snowy Winter.

[14] Boo. Dissatisfied experimentalist.

Again, all three components presume that the site has useful information regarding the *horizontal* irradiance conditions over the day. In the 50 years since those initial isotropic models, *anisotropic sky models* were introduced and refined by Hay, Davies, Klucher, and then Reindl, cumulatively abbreviated the *HDKR model*.[15] A separate approach to estimating the anisotropic sky irradiance on tilted surfaces was developed by Richard Perez from SUNY-Albany. The Perez model is now used as a standard in simulation software such as TRNSYS, Energy+, and SAM, and has been used extensively to estimate the geographically extensive solar resource potential in the USA.

[15] John A. Duffie and William A. Beckman. *Solar Engineering of Thermal Processes.* John Wiley & Sons, Inc., 3rd edition, 2006.

PEREZ ANISOTROPIC MODEL

This is an *anisotropic diffuse* model that takes into consideration the real observations of subcomponents of diffuse light. The Perez model[16] adds the

[16] Richard Perez is a Research Professor in the Atmospheric Sciences Research Center in SUNY-Albany. He has a great web site at http://www.asrc.cestm.albany.edu/perez/.

circumsolar diffuse component and the *horizon diffuse component* to the diffuse, sky component of the isotropic model.

$$G_{d,t} = G_{iso} + G_{cir} + G_{hor} + \text{diffuse, ground.} \tag{8.16}$$

Notice how the beam component is not mentioned—it doesn't change from the original relation. Also the ground diffuse is added directly into the equation here.

$$\begin{aligned} G_{d,t} = & \left[G_d(1 - F_1) \cdot \frac{1 + \cos\beta}{2}\right] \\ & + \left[G_d(F_1) \cdot \frac{\cos\theta}{\cos\theta_z}\right] + \left[G_d(F_2) \cdot \sin\beta\right] \\ & + \left[G_d\rho \cdot \frac{1 - \cos\beta}{2}\right]. \end{aligned} \tag{8.17}$$

For the purpose of this text we won't delve into the shape factors (F) in this model. Further reading on the model use can be found in the original literature.[17] We will just observe that $F_{surface-sky}$ is *reduced* by a proportion of F_1 (circumsolar radiance), and F_2 can either increase or decrease the contribution of horizon radiance.

[17] R. Perez, R. Stewart, R. Arbogast, R. Seals, and J. Scott. An anisotropic hourly diffuse radiation model for sloping surfaces: Description, performance validation, site dependency evaluation. *Solar Energy*, 36(6): 481–497, 1986. Perez, Ineichen, and Seals. Modeling daylight availability and irradiance components from direct and global irradiance [17]. *Solar Energy* J, 44(5): 271–289, 1990.

CLEARNESS INDICES AND CLIMATE REGIMES

The most common measurement that researchers have available is Global Horizontal Irradiance (GHI), which can be collected either from a ground station or from a satellite measurement. As seen in Figure 8.5, by integrating the sampled GHI over periods of time, we acquire *global irradiation* measures for a horizontal surface. As a comparison we may calculate the irradiation conditions for two hypothetical scenarios: (1) for extraterrestrial solar irradiation above the sky dome (AM0 conditions), and (2) for solar conditions underneath the sky dome, excluding emergent cloud phenomena (clear sky conditions). We have discussed calculations for irradiation with "no atmosphere" (infinite transparency) and calculations for irradiation with a clear sky earlier.

Hence, highly used empirical correlations in the solar field employ the relations of the various *clearness indices* to *diffuse sky ratios*. The clearness index is a ratio of *measured irradiation* in a locale relative to the extraterrestrial irradiation *calculated* (AM0) at the given locale. The value can be commonly evaluated for a given day of interest K_T, or for an hour of interest on specific day k_T.

Notice how the day uses a capital "K," while the hour uses the lower case "k." The subscript "T" in this case is historic and refers to *Total* irradiation.

Researchers have also explored the *clear sky index* (K_c and k_c), where the ratio denominator is replaced by the calculated values of clear sky irradiation (H_c) for the time period of interest. Ianetz and Kudish confirm that the two measures have a high correlation with each other ($R^2 > 0.99$ for linear regressions of each month) for data sets in Israel. The magnitude of the clearness index (K_T) encompasses parameters of clear sky transparency with the cloud content in the sky. In contrast, the clear sky index marginalizes the diurnal cycle of airmass by removing the clear atmosphere from the equation. The remainder $(1 - K_c)$ marginalizes the contribution of cloudiness. In such a case, a preferred metric of $(1 - K_c)$ has been proposed as an indicator of cloudiness for the day.[18]

[18] Amiran Ianetz and Avraham Kudish. *Modeling Solar Radiation at the Earth's Surface: Recent Advances*, Chapter 4: A Method for Determining the Solar Global Irradiation on a Clear Day, pages 93–113. Springer, 2008.

We have already demonstrated the use of components on a horizontal surface to further estimate the component contributions of irradiation to tilted surfaces using anisotropic diffuse light models. Thus knowledge of methods to acquire estimates of horizontal components is of value to early design steps. Clearness indices have been explored extensively over the past 50 years as tools for empirical correlation that allow engineers to estimate the components of light incident upon a horizontal surface from minimal measurements of GHI alone.

Consider the spatio-temporal scales of light-related pheneomena: synoptic scale weather that occurs over large regions, meso/microscale weather that occurs over much smaller area, and diurnal (daily) events occurring at scales between the synoptic and mesoscale phenomena. As we keep mentioning, time and space are linked for the solar resource! *Synoptic scales* of meteorology (also called cyclonic meteorology) refer to spatial distances on the order of 1000 km and time scales of 30–90 days. Now what two time periods do we know like this?[19] So, for regional spatial scales (>1000 km) we seek out clearness indicators that reflect the time scales of months. The solar industry uses the average day measurement for the month ($\overline{H}$, an average measure of a day's energy in the month) to assess this scale, and the months are divided into seasons in the

[19] Answer: **Months** and **Seasons**.

Month	Average Energy Date (AMo)	Average Energy Day (n)	First Day (n_1)	First Hour	Last Hour
December	10	344	335	8016	8760
January	17	17	1	0	744
February	16	47	32	744	1416
March	16	75	60	1416	2160
April	15	105	91	2160	2880
May	15	135	121	2880	3624
June	11	162	152	3624	4344
July	17	198	182	4344	5088
August	16	228	213	5088	5832
September	15	258	244	5832	6552
October	15	288	274	6552	7296
November	14	318	305	7296	8016

Table 8.1: The first day (n_1) of each month, where the data has been separated into the Northern Hemisphere, mid-latitude fingerprints of:

Winter, (December–February)

Spring,

Summer, and

Fall (September–November), followed by the recommended days for average daily irradiation in that month (n).

mid-latitudes according to Table 8.1, where the thermal behavior of the region tends to lag the Solstices by 20–25 days. Thus, the middle of the "Winter fingerprint" in PA or WI is January, not December.

The average day's *measured irradiation* for a given month ($\overline{H}$ in MJ/m^2) can be evaluated by integrating the daily irradiation over the month and dividing by the number of days. As we mentioned in the chapter dealing with solar geometry, the average daily irradiation for extraterrestrial irradiation for a month has a summary value, $\overline{H}_0$. Aside from taking the mean of the daily values in a month, Klein recommended a simplified approach to approximate the average day in a month, using a single critical day that bears the value of the mean.[20] Thus, the average day's *calculated extraterrestrial irradiation* for a given month ($\overline{H}_0$, also in MJ/m^2) can be evaluated by using the singular daily value for the critical day of the month (Average Energy Day) found in Table 8.1.

Although we do not use the average energy day measures very much in practice ($\overline{H}$) because hourly data is readily available in most locations on the planet, the average values are great learning tools to provide estimates for changes in irradiation or clearness indices across the year.

[20] S. A. Klein. Calculation of monthly average insolation on tilted surfaces. *Solar Energy*, 19: 325–329, 1977.

$$\overline{H} = \frac{1}{n_2 - n_1} \sum_{i=n_1}^{n_2} H_i. \tag{8.18}$$

For $K_T \rightarrow 0.75$–1: atmosphere is clear. For $K_T \rightarrow 0$: atmosphere is cloudy. However, this measure incorporates both light scattering and light absorption.

The method for generating the daily clearness index is the same as for the average day of the month, with the exception that one only integrates the irradiation for the singular day in question. It should make sense that one day's clearness index would fall within the spread of data making up the monthly clearness index mean.

$$K_T = \frac{H}{H_0}. \tag{8.19}$$

It has subsequently been determined that the spread of monthly irradiation data retains unique fingerprints from region to region (e.g., the tropics had different correlations than the temperate USA, India was different than much of Africa, etc.) This work was followed by Hawas and Muneer for India and Lloyd for the UK, among others.[21] The spread of the data across the days of the month is different for different regions. From *Solar Radiation and Daylight Models* (Chapter 3, page 123):

> For example, for Indian locations, during a month for which $\overline{K}_T = 0.7, K_T \leq 0.73$ for 70% fractional time, and $K_T \leq 0.68$ for 20% of the time. In contrast the corresponding figures for the US are, respectively, 56% and 30%. It is therefore evident that the distributions for the Indian locations are flatter. This means that for the Indian sub-continent the daily clearness index K_T varies in a narrower range.
> —T. Muneer (2004)

However, remember from meteorology that there are certain common phenomena specific to spatio-temporal regions called *climate regimes*. To repeat our studies from earlier, each climatic regime should be treated effectively as a *different geographical place* throughout the year. Thus we should expect that daily and diurnal clearness indices are strongly coupled to the climate regime under study! In relation to solar energy design, the light interacting character of the sky dome changes with climate regime shifts as well as the conditions of air temperature. In the geographical regions in the mid-latitudes

[21] M. Hawas and T. Muneer. Generalized monthly K_t-curves for India. *Energy Conv. Mgmt.*, 24: 185, 1985; P. B. Lloyd. A study of some empirical relations described by Liu and Jordan. Report 333, Solar Energy Unit, University College, Cardiff, July 1982; T. Muneer. *Solar Radiation and Daylight Models*. Elsevier Butterworth-Heinemann, Jordan Hill, Oxford, 2nd edition, 2004; and Joaquin Tovar-Pescador. *Modeling Solar Radiation at the Earth's Surface: Recent Advances*, Chapter 3: Modelling the Statistical Properties of Solar Radiation and Proposal of a Technique Based on Boltzmann Statistics, pages 55–91. Springer, 2008.

(North and South Hemispheres), the tendency is for *four climate regimes* (i.e., *seasons*). In regions affected by monsoonal swings or areas closer to the oceans and tropics, there may be two or three climate regimes. We have found that *clearness indices* and their resulting correlations to diffuse sky models on a tilted surface are in fact *emergent properties of climate regimes*. In fact, what was being confirmed by the data collection from engineers of the past decades was the concept of *climate regime* in meteorology. For synoptic scale meteorological phenomena (data on the order of months and days), there are expected regional commonalities found in measures such as the daily clearness indices within similar climatic regimes. Of course, differing climate regimes will have different synoptic characteristics, leading to different distributions of irradiance data.

- $k_T = \frac{I}{I_0}$: hourly clearness index. Ratio against energy density for extraterrestrial solar in one hour.
- $K_T = \frac{H}{H_0}$: daily clearness index.
- $\overline{K}_T = \frac{\overline{H}}{\overline{H}_0}$: average day of the month clearness index.

Now we revisit a concept from the chapter on the Sky. The prolific and creative 20th century British scientist Geoffrey I. Taylor (1886–1975) was a noted expert on a broad range of topics from math and physics, including the study of fluid dynamics. According to Taylor's hypothesis, we assume that a change in irradiance measurement over time from a pyranometer results from the lateral change of sky and cloud conditions across the area of a meteorological event. Taylor's hypothesis permits irradiance observations as a time series over fixed locales to be converted as a translation across space relative to propagation speed (advection) of the corresponding cloud pattern.[22] Again, this means that time and spatial scales are linked—as long as the advective wind speed is much greater than the time scale of the evolving meteorological event being investigated, as is often the case.

Taylor's hypothesis: a series of changes in time for a fixed place, like you observing a thunderstorm passing over your house, is due to the passage of an unchanging spatial pattern over that locale, like the dark storm cloud passing over your house then your neighbors house then the next house.

[22] G.I. Taylor. The spectrum of turbulence. *Proc Roy Soc Lond*, 164: 476–490, 1938.

And finally, we transition to data from the *mesoscale meteorological phenomena*. The *hourly clearness index* is used extensively to estimate diffuse fractions of light from the sky dome incident upon a tilted surface, when only a horizontal measurement is available. As seen in Eq. (8.20), the ratio has the same form as the prior two, using hourly measured irradiation data relative to calculated extraterrestrial irradiation data.

$$k_T = \frac{I}{I_0}. \tag{8.20}$$

DIFFUSE FRACTION

The diffuse fraction for the average day of the month (a ratio of average monthly diffuse values to average monthly global irradiation values) has been related via empirical correlation as a function of the average day month clearness index ($\overline{K}_T$). As seen in Eq. (8.21), the correlations differ for mid-latitude climate regimes (e.g., Massachusetts or the UK) versus desert and tropical climate regimes (e.g., India) with high turbidities (particles scattering light: aerosols), including high fractions of diffuse skies during monsoon months. Generally, we find the range of values to be $0.3 < \overline{K}_T < 0.7$.[23]

[23] T. Muneer. *Solar Radiation and Daylight Models.* Elsevier Butterworth-Heinemann, Jordan Hill, Oxford, 2nd edition, 2004.

$$\frac{\overline{DHI}_{d,midlat}}{\overline{GHI}_{d,midlat}} = \frac{\overline{H}_d}{\overline{H}} = f(\overline{K}_T) = 1.00 - 1.13 \cdot \overline{K}_T. \tag{8.21}$$

(Page, 1977)

Keep in mind that a *fraction* is not a percentage, and in our case for a cumulative distribution, it is a decimal value between 0 and 1.

$$\frac{\overline{DHI}_{d,trop}}{\overline{GHI}_{d,trop}} = \frac{\overline{H}_d}{\overline{H}} = f(\overline{K}_T) = 1.35 - 1.61 \cdot \overline{K}_T. \tag{8.22}$$

(Hawas & Muneer, 1984)

Again, K_T distributions are *not universal*—they are regional and empirically derived, as the distributions reflect regional meteorological conditions for a given prevailing air mass. That being said, we provide a general empirical correlation from Muneer that has been used for Canada, India, the USA, and the UK.[24]

[24] T. Muneer. *Solar Radiation and Daylight Models.* Elsevier Butterworth-Heinemann, Jordan Hill, Oxford, 2nd edition, 2004.

$$\frac{H_d}{H} = f(K_T) = \begin{cases} 0.98, & \text{if } K_T < 0.2, \\ 0.962 + 0.779K_T - 4.375K_T^2 + 2.716K_T^3, & \text{if } K_T \geq 0.2. \end{cases} \tag{8.23}$$

Similarly, in lieu of a preferred regression for a locale, the hourly relation was recommended by Muneer:

Refer back to Figure 8.5 to identify where the process of using an hourly correlation to determine DHI fits within the great protocol to estimate G_t from GHI.

$$\frac{I_d}{I} = f(k_T) = 1.006 - 0.3170k_T + 3.1241k_T^2 - 12.7616k_T^3 + 9.7166k_T^4. \tag{8.24}$$

Through simple arithmetic we can back out estimates of the two horizontal components:

$$I_d = I \cdot \frac{I_d}{I}, \tag{8.25}$$

$$I_b = I - I_d. \tag{8.26}$$

Once the horizontal components are known, the progression is to use the beam irradiance tilt factor (R_b)[25] and the anisotropic diffuse irradiance models to calculate irradiance on a tilted surface. For an hourly estimation or irradiation on a tilted surface, the midpoint of the hour is typically used (e.g., 10:30a for the hour from 10 to 11a).

[25] $R_b = \frac{\cos\theta}{\cos\theta_z}$.

We have established the link between GHI and DHI via the clearness indices ($\overline{K}_T, K_T$, and k_T). We have also indicated that the clear sky indices ($\overline{K}_c, K_c$, and k_c) would be of similar use given proper empirical study for the region or locale of interest. Although limited by lower precision, estimation of irradiation components on a horizontal surface using empirical data correlation can be useful for rapid assessment of the resource potential in a region, and the resulting components of diffuse light over a day. The only tools we need are the equations for hourly and daily extraterrestrial irradiance (Air Mass Zero, or *AMo*) and the integrated irradiation (J/m^2) gathered from a horizontally mounted pyranometer.

Separate from hourly data sets, the monthly average daily radiation intensity is also important for the design of solar energy conversion systems. The average day of the month is a summary statistic, describing the *quality of light* in terms of the frequency of various measures of radiation intensity. Information regarding the magnitude of beam irradiation on a tilted surface ($\overline{H}_{b,t}$ is useful for rapid assessment of the potential of a locale/site for concentrating solar conversion systems, and the magnitude of the POA global irradiation ($\overline{H}_t$) is crucial for estimating system sizing of any system. Beyond the summary value, knowledge of the variation of $\overline{H}_t$ is desirable to facilitate system sizing. A small variance suggests that there will be fewer days for which the final deployed system functions as undersized, underutilized, or oversized for integrative system functioning.

WHEN EMPIRICAL CORRELATIONS ARE NOT APPROPRIATE

After a lengthy discussion of the utility of empirical correlations in long-term estimations of systems performance, we should also assess the conditions for which the correlations are ill suited, and may even fail to serve our needs as a solar design team. *Uncertainty* propagates in our design solutions requiring multiple physical orientations.

The **Gaussian** (or *normal*) distribution is a continuous probability distribution, that is symmetric and bell shaped, defined by the mean (μ) and standard deviation (σ).

First, one must be aware that empirical correlations have traditionally not conveyed the broad spread of the data across the range of fits, so that the designers have little to no idea of the actual confidence intervals for the correlations. The distribution of the data is likely *non-Gaussian*, and so confidence interval estimation requires alternate statistical analyses other than mean and standard deviation.

We know that empirical correlations also have weaknesses in estimating over smaller time steps (here, sub-hourly), estimating over regions that are spatially diverse, and estimating for non-horizontal (particularly vertical) surfaces.

Translating or transforming measured irradiance values from a horizontal surface introduces errors for the best of models, with RMSE (as Root Mean Squared Error) on the order of 50 W/m^2 for vertical surfaces. Where surface areas are small and irradiance values are high, such as in a residential solar hot water or PV installation, such errors may not be a significant problem in system design and project finance. However when the surface area is high such as in building systems (e.g., solar gains on a building wall), and there is a degree of regular cloudiness associated with the majority of the mid-latitude continental regions, one should be cautious applying these correlations for light processing.

Over the past 50–80 years, the energy consumption of a building was a low priority to building occupants. A higher priority to measurement is given to systems that produce *power*, like photovoltaics. So what will this mean for PV integrated into the built environment in the future? Get measuring.

As we shall note in the next chapter considering a building as a collection of solar energy conversion system *patterns*, the radiative external boundary conditions for a building are *not measured* under current practices. Yet, the majority of energy exchange occurs through the façade or perimeter (55–80%).[26] Instead, the empirical correlations from the clearness indices are used to estimate the components of irradiation on a horizontal surface in *hourly* or *daily* increments, and then calculations using anisotropic sky models are applied to

[26] US DoE Office of Building Technology, State and Community Programs: BTS Core Databook (November 30, 2001).

again estimate the component contributions that should occur on the vertical surfaces of a building.

So what is the value of energy flux signals from a local pyranometer versus a remote airport station? *Taylor's hypothesis* states that changes of events happening in time for a fixed place (the Eulerian frame of reference) can be attributed to the passage of a phenomenon over that locale (the Langrangian frame).[27] Hence, according to Taylor's hypothesis, given wind speed and direction, the distance from a remote weather station affects the time dilation of a meteorological event like a cumulus cloud increasing the diffuse component of light. *If we are lucky*, GHI will be measured at the site of a building, or estimated by a virtual meteorological station, thus alleviating the time-space dilation occurring from Taylor's hypothesis. In a building setting, it is far more likely that GHI is not even measured, and instead data sets from Typical Meteorological Years are used as proxy for long-term energy models (30–50-year performance).

[27] G. I. Taylor. The spectrum of turbulence. *Proc Roy Soc Lond*, 164: 476–490, 1938.

In today's push for buildings with high energy efficiency however, the modeling the performance of buildings on sub-hourly time scales are called on, with control and energy management systems that need more information than the internal air temperature of a zone (room). Integrated systems of irradiance sensors in SECS can improve collector performance, and can improve the quality of the environment inside buildings, reduce energy requirements.

Quick Symbol Review:

- G_{sc} is the extraterrestrial average annual solar constant: 1361 W/m^2.
- G_0 is the extraterrestrial irradiance collected upon a horizontal surface. Units of W/m^2.
- $G = G_b + G_d$. Units of W/m^2.
- $I = \int_{t_0}^{1\,hr} G\,dt$: radiation collected over one hour on a horizontal surface. Units of MJ/m^2.
- $k_T = \frac{I}{I_0}$: hourly clearness index, using AM0 calculation in the denominator.

When radiant flux is incident upon an area of receiving surface from all directions. We term it **irradiance** (W/m^2). When radiant flux is emitted by a unit solid angle of a source or scattering surface, we call it **radiance** (W/m^2 sr) or **radiant exitance** (W/m^2).

- $k_c = \frac{I}{I_c}$: hourly clear sky index, using a clear sky model in the denominator.
- $H = \int_{t_0}^{24\,hr} G\,dt$: irradiation collected over one day on a horizontal surface. Scale of $\sim$10–15 MJ/m^2.
- $\overline{H} = \frac{1}{n_2-n_1}\sum_{i=n_1}^{n_2} H_i$: irradiation collected over one month on a horizontal surface, divided by the number of days in the month. Yields the average irradiation day for one month.

ROBOT MONKEY DOES NETWORK SOLAR RESOURCE ASSESSMENT!

From the perspective of a network of solar power stations, say a grid of PV power distributed across a region, the presence of many sites has the potential to interact in a coherent, constructive fashion (amplifying intermittent peaks and troughs in power) or more favorably in an incoherent fashion (canceling out power intermittency to yield a smooth power signal across a band of time scales of interest). We can safely assume that the movement of electrons in a power grid is much much faster than the evolution of clouds across a region, hence Taylor's hypothesis will work for assessing solar power's influence on the grid as well. From statistics, the presence of N sites across a region of network interest has the ability to reduce the net variance of power as a function of $\frac{1}{\sqrt{N}}$. The scale of reduction in net variance of power has been examined and reviewed recently and is related to the distance between sites, which tends to be a large reduction for sites within the range of meso- and microscale meteorological effects, meaning for time scales shorter than hourly intervals.[28] In

[28] T. Hoff and R. Perez. *Solar Resource Assessment and Forecasting*, chapter "Solar Resource Variability." Elsevier, 2013; M. Lave, J. Stein, and J. Kleissl. *Solar Resource Assessment and Forecasting*, chapter "Quantifying and Simulating Solar Power Plant Variability Using Irradiance Data." Elsevier, 2013.

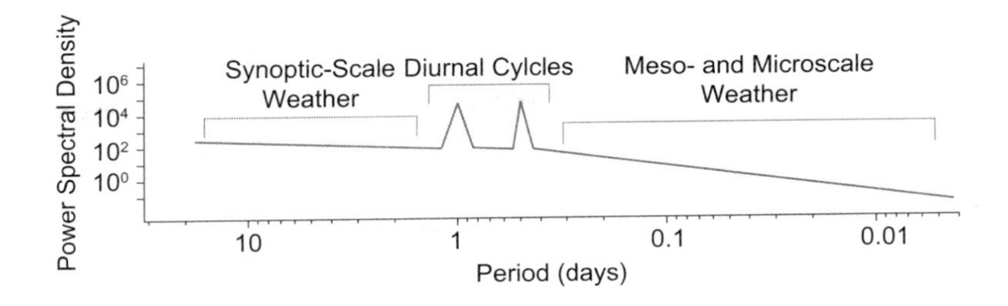

Figure 8.7: The power spectral density cartoon of various time scales.

Figure 8.7 we display a log-log plot of solar event variance (termed a power spectral density) versus period of the events happening in time (in units of days).

In any one site, the solar resource has expected values of *statistical variance* associated with the seasonal (or *synoptic*) fingerprint for the surrounding region. Specifically, if we were to collect global irradiance data for a single site from a pyranometer over three or more years, we would have a large set of data to deal with. For clients living in the mid-latitude regions of North America, we are able to divide those 3 years of data into four bins for each of four seasons (four "fingerprints"). The processed data can be used to further calculate the *power spectral density* of the data (a form of Fourier math transformation that puts the data on the x-axis into scales of periodicity or frequency). Because the data are the Fourier Transform of independent events, we may integrate under intervals of interest to arrive at expected values of *variance* for say 6–12 h events or 2–4 day events—just by integration! This is a form of uncertainty analysis that has yet to be developed extensively for economic or technological analyses.

By extension, if we were to calculate the cospectra (Cross Spectral Density, CSD) among networked sites that are separated by a physical distance, we could derive important new information as well regarding the covariance (see Figure 8.8).[29] In similar fashion, the integral of the cospectrum for a time interval of interest will result in the expected value of the covariance.[30]

Estimated variance and covariance are expected values for the spread of irradiance data within a specified period. The positive square root of the estimated variance is the standard deviation. Again following the validation of Taylor's hypothesis, a period in time is equivalent to a physical distance between

[29] Matthew Lave and Jan Kleissl. Solar variability of four site across the state of Colorado. *Renewable Energy*, 35: 2867–2873, 2010.

[30] J. Rayl, G. S. Young, and J. R. S. Brownson. Irradiance co-spectrum analysis: Tools for decision support and technological planning. *Solar Energy*, 2013. doi:10.1016/j.solener.2013.02.029.

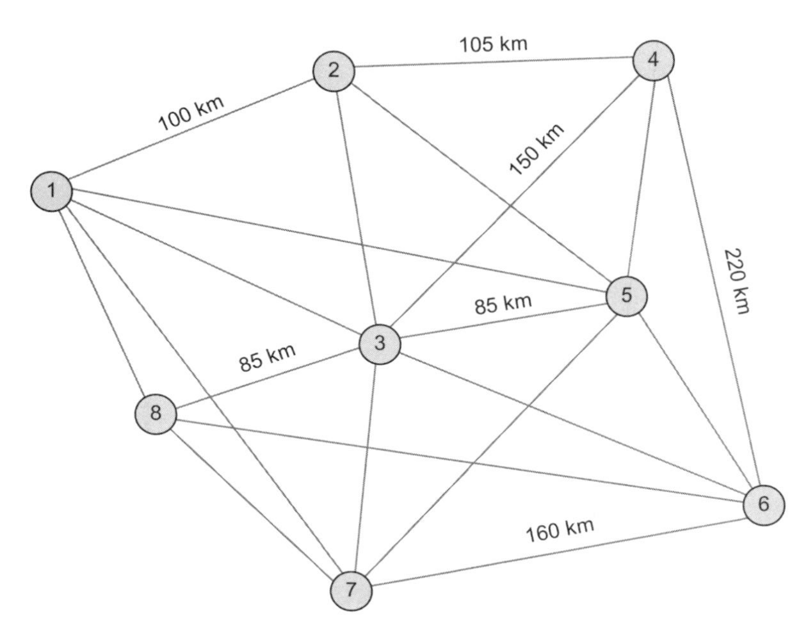

Figure 8.8: Cartoon of networked sites across a region, connected together by the electrical power grid.

sites. Hence, we can use either scales of time/period or scales of distance when planning market, policy, or engineering decisions for solar energy conversion systems. Further discussions on *uncertainty* can be found at NIST (the National Institute for Standards and Technology).

All this discussion of estimated variance and uncertainty has left open the question of the estimated probability distribution function of solar energy for different time scales. What is the distribution for solar data as it deviates from the ideal clear day? Researchers have suggested that the distribution is not Gaussian in nature (a Normal Distribution). The reason why we ask is that a probability density function would also allow us to estimate uncertainty in a method other than by the statistical analysis of a series of observations, for an individual site or for a network of sites. Furthermore, we will soon show that the dispersion of possible outcomes around an expected value not only describes the uncertainty of the scenario, but also provides the client and stakeholders with a measure of *risk* for the time horizon of interest. As we shall observe, risk is important to developers, investors, and systems operators.

Congratulations!
Robot Monkey grants you one glowing banana for getting through the mindstretch

PROBLEMS

1. Compose a short description for calculating hourly components of irradiance on a tilted surface, beginning with horizontal pyranometer measurements and hourly clearness index only.

2. Hourly solar irradiation on horizontal surface from 11h00–12h00 on January 8 in Fort Collins, CO ($\phi = 40°$) is 402 kJ/m^2. What is the k_T for that hour? From k_T, estimate the fraction of the hour's irradiation that is diffuse.

3. Daily solar irradiation on horizontal surface on January 8 in Fort Collins, CO ($\phi = 40°$) is 4480 kJ/m^2. What is the K_T for that day? From K_T, estimate the fraction of the day's irradiation that is diffuse.

4. Using TMY data found in SAM or a similar TMY database, calculate and plot the monthly average daily measured irradiation ($\overline{H}$) as a function of months for collectors in for San Diego, CA; Madison, WI; and Philadelphia, PA. Note: this is not AM0 data.

5. From the values of $\overline{H}$ in the previous problem, calculate and plot the daily clearness indices $\overline{K}_T$ for the "average days" of each month as well (use Table 8.1). Now, in 4–5 sentences, describe how the three locales are similar and how they are different based on these data that you have created.

6. On February 23 in Minneapolis, MN, the hourly irradiation has been measured from 10h00 to 11h00. $I = 1.57\,\text{MJ/m}^2$. Knowing this measured value and from the calculated value of I_0, estimate the sum total irradiation for a surface that is oriented due South ($\gamma = 0$) with a tilt of $\beta = 55°$ (40°annual optimum + 15°).[31] Do this problem using the middle of the hour (10h30) when an instantaneous value is called for. You will be required to use a component model for your estimation of the contributions for beam, diffuse, and ground reflected irradiation, so work through this problem:
   - using the Liu and Jordan isotropic model.

7. Given the same conditions for February 23 in Minneapolis, MN, do this problem:
   - using the HDKR anisotropic model.

RECOMMENDED ADDITIONAL READING

- Statistical analysis of a solar radiation dataset—characteristics and requirements for a P50, P90, and P99 evaluation.[32]
- Solar Radiation and Daylight Models.[33]
- REST2: High-performance solar radiation model for cloudless—sky irradiance, illuminance, and photosynthetically active radiation—Validation with a benchmark dataset.[34]

[31] M. Lave and J. Kleissl. Optimum fixed orientations and benefits of tracking for capturing solar radiation in the continental United States. *Renewable Energy*, 36: 1145–1152, 2011.

[32] F. Vignola, A McMahan, and C. Grover. *Solar Resource Assessment and Forecasting*, chapter "Statistical analysis of a solar radiation dataset—characteristics and requirements for a P50, P90, and P99 evaluation." Elsevier, 2013.

[33] T. Muneer. *Solar Radiation and Daylight Models.* Elsevier Butterworth-Heinemann, Jordan Hill, Oxford, 2nd edition, 2004.

[34] Christian A. Gueymard. REST2: High-performance solar radiation model for cloudless—sky irradiance, illuminance, and photosynthetically active radiation—validation with a benchmark dataset. *Solar Energy*, 82: 272–285, 2008.

- Simple Model of the Atmospheric Radiative Transfer of Sunshine, version 2 (SMARTS2): Algorithms description and performance assessment.[35]
- Solar Energy at Urban Scale.[36]
- National Solar Radiation Database 1991–2010 Update: User's Manual.[37]

[35] Christian Gueymard. Simple Model of the Atmospheric Radiative Transfer of Sunshine, version 2 (SMARTS2): Algorithms description and performance assessment. Report FSEC-PF-270-95, Florida Solar Energy Center, Cocoa, FL, USA, December 1995.

[36] Bella Espinar and Philippe Blanc. *Solar Energy at Urban Scale*, Chapter 4: Satellite Images Applied to Surface Solar Radiation Estimation, pages 57–98. ISTE Ltd. and John Wiley & Sons, 2012;Théo Pirard. *Solar Energy at Urban Scale*, Chapter 1: The Odyssey of Remote Sensing from Space: Half a Century of Satellites for Earth Observations, pages 1–12. ISTE Ltd. and John Wiley & Sons, 2012.

[37] Stephen Wilcox. National solar radiation database 1991–2010 update: User's manual. Technical Report NREL/TP-5500-54824, National Renewable Energy Laboratory, Golden, CO, USA, August 2012. Contract No. DE-AC36-08GO28308.

SOLAR ENERGY ECONOMICS

09

Some argue that the consumer can purchase warmth or work or mobility at less cost by means of coal or oil or nuclear energy than by means of sunshine or wind or biomass. The argument concludes that this fact, in and of itself, relegates renewable energy resources to a small place in the national energy budget. The argument would be valid if energy prices were set in perfectly competitive markets. They are not. The costs of energy production have been underwritten unevenly among energy resources by the Federal Government.[1]

Report of the DOE, Battelle Pacific Northwest National Laboratory(1981)

The Economics of solar technologies addresses some factors as to why we make *decisions* to make more use of the Sun.[2] From our learning, we also have an intuitive feel that *energy* is somewhere between a product and a good in demand by society, and it must be supplied by non-trivial mechanisms, at some cost for the exchange of goods. What is interesting for solar energy, is that our raw "product" is the *photon*, and we apply technologies and skilled effort to first explore for energy, then to design and deploy technologies to convert available photons into a diversity of goods that society is interested in purchasing. In fact, our creative task as the next generation of entrepreneurial solar designers is to strengthen and expand a business model of *sustainable*

[1] Nancy Pfund and Ben Healey. "What would Jefferson do? the historical role of federal subsidies in shaping America energy future." Technical report, DBL Investors, 2011.

[2] We make use of the Sun throughout our lives, but in solar design we offer compelling arguments to the client that may increase their *marginal demand* for the Sun while decreasing their demand for fuels.

Just because you *perceive* the solar resource to be weak in your region, does not mean that it cannot be successful as a technology in that society. The solar resource is *ubiquitous*, and we make use of it whether we decide to or not.

energy exploration and *environmental technology deployment*. One suite of tools for the emerging business model of energy exploration is through markets, as we will explore in this chapter. We have already found that knowledge of the solar resource is critical to aid in the SECS design process. In order to make marginally (or incrementally) more use of the Sun, we first had to develop the skills to understand and *measure* the variable phenomenological behavior of solar irradiance, as well as the important role of locale (time and space are coupled). But as of yet, we have not really addressed the role of the client as a decision maker, and the financial drivers or constraints that move clients and stakeholders to make decisions.

The Sun always rises. Solar energy is one of the most conservative investments available for an energy project. It may call for a longer time horizon, but in today's society solar energy (almost) always pays back.

As we stated in the beginning of this text, Solar Energy Conversion Systems call for an *integrative design process*. Integrative design requires all parties to be at the table from the beginning, to create *alignment* in the goals of the project. That design team must work together in iterative fashion to bring a project to maturity. In the discovery phase of project design, the integrative team brings in stakeholders following the concept of the *Four Es*: **E**verybody **E**ngaging **E**verything **E**arly.[3] It is up to the solar design team to leverage their knowledge of economics and social behavior along with technological systems behavior and meteorological/climatological phenomena in order to find compelling arguments for SECS adoption by clients in the given locale. Again, the client makes the final call in a design project, rational or irrational. Part of early engagement with the clients is to provide information and education that will help to achieve a stronger acceptable design concept, and to arrive at a common understanding of the barriers to adoption. Keep in mind that there are some barriers to design based on the perceptions and desires of your clients, and other financial and policy constraints that can emerge as even more significant obstacles to the adoption of a new solar energy conversion system.

[3] J. Boecker, S. Horst, T. Keiter, A. Lau, M. Sheffer, B. Toevs, and B. Reid. *The Integrative Design Guide to Green Building: Redefining the Practice of Sustainability*. John Wiley & Sons Ltd, 2009.

Goal of Solar Design:

- Maximize the solar utility,
- for the client,
- in a given locale.

We re-emphasize that solar design is linked with the surrounding environment (ecosystems services, meteorology) and with the social constraints of the client. The dynamic environment-society relationship in sustainable energy drives solutions to be case-dependent. One systems solution for a locale/client relation will not necessarily be applicable for the next client in a different locale.

Coincidentally, a good argument will tend to maximize the solar utility for the clients or stakeholders *and* ensure that your team has a productive business plan with an ethically sound foundation.

So what are the reasons for solar markets to emerge in society, and why have solar markets and entrepreneurial ventures in solar energy appeared in the past 100 years, only to be dissolved later? We know that the demand for energy has been increasing in response to the increasing population of the planet. At a minimum each person has basic needs that are supplied by forms of energy. We also know that demand for energy in the past has tended to increase with industrialization of a society. Finally, we know that the increasing demand (both the rate of change and physical distribution) is non-uniform across the planet. There are giant economic engines developing in Asia, South America, and Africa, which until now have demanded very little energy conversion per capita. We also know that the scientific evidence for human-caused climate change will likely result in policies that constrain the use of fossil fuel combustion.

The energy demand in countries like the USA may appear to be steady or decreasing from a national accounting, but this is largely due to outsourcing our manufacturing industry and energy demand to other countries.

We can identify two major *forms* of energy that are in demand within all societies, and which are often purchased as commodities. Across the planet we observe large demands for energy as *thermal heat* and *electrical power*,[4] each two basic forms of energy. Light, as *electromagnetic radiation* (or radiant energy), is another form of energy. We also know that light can be converted into chemical energy (biofuels, and with great time and pressure fossil fuels), and the visible band of light is extremely useful to function in society. The basis of this text is to explore conversion of light into useful electrical power and thermal heat.

Forms of Energy: radiant, thermal, electric, chemical, mechanical, etc. See the US DoE information site: Energy Explained.

[4] The terms **Heat** and **Power** have been adopted by several industries to have a specialized trade meaning. *Power* is electrical energy (as opposed to a rate of energy use), and *heat* is thermal energy (as opposed to the transfer of energy). Thus, in the energy industry we hear about *Combined Heat and Power (CHP)* for energy conversion systems that provide two useful forms in one system.

So let's think again about "collecting" light for use in society and the environment. Photons can be harvested via a solar energy conversion device. To be clear, photons are ephemeral, they are not really collected like fuel in a tank, but rather converted from one form to another with proper technologies. This is why we used the analogy of the photon as the work current in a *pump* in an earlier chapter. Once absorbed, the photon is *lost*, converted into a new form of energy that may be useful to people, plants... life. This raises some questions about how we estimate the *effective size of the solar resource* in a given locale,

and the ability for a client to access that resource in an economically feasible manner, given a real demand for a substitute to purchase fuels for electricity, heat, light, or even food.

THE VALUE OF LIGHT is dependent upon:

- the demand for light as a good or service,
- the costs of alternatives.

Think about farming and the connection to the Sun. Farmers have no use for daylight savings because much of their work depends on solar utility, and not the available evening hours to go jet skiing.

Consider that our societies are increasingly becoming urbanized—as we see in Figure 9.1. The USA alone has 10–11 emerging *megaregions* of urban development. These are amazing regions with enormous potential to develop solar technologies in, potentially relieving concentrated energy demand on the energy and fuel infrastructure. There is additional potential for adopted strategies to spread rapidly in a megaregion, due to the proximity of adopters to each other and the low visual and environmental local impact of most urban solar strategies. However, urbanized society has a consequence in terms of solar perception. Most people in urban society make use of the Sun without acknowledging its value. The more time people spend indoors, the less likely they are to be aware

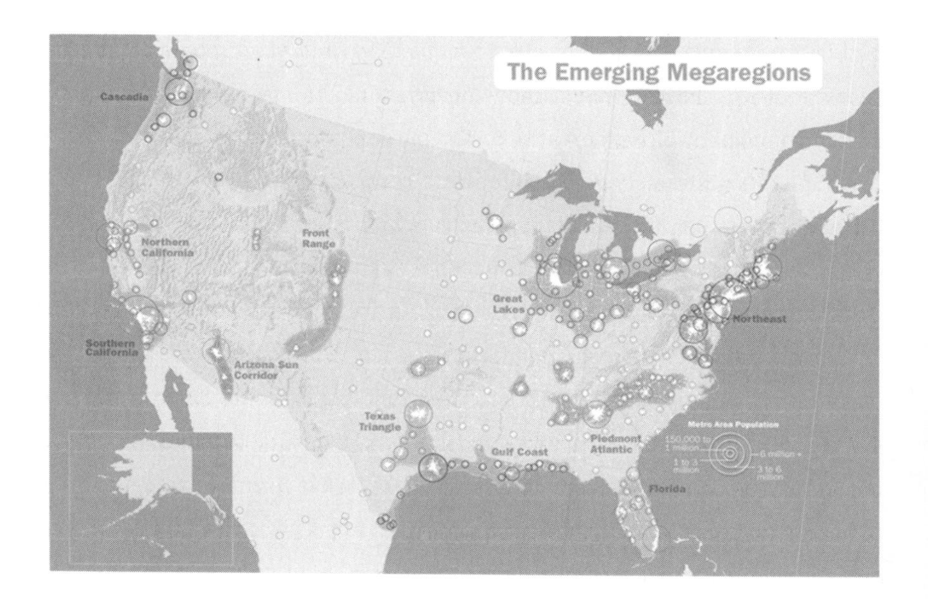

Figure 9.1: The emerging megaregions in the USA. This map, created by the Regional Plan Association, illustrates 11 metropolitan areas that are growing into megaregions.Image by America 2050, Regional Plan Association.

of little things like weather and day length. And yet, your solar design team has several advantages to succeed should you take advantage of economic criteria and social science along with solar science and engineering! By looking at the map of megaregions in Figure 9.1, you are also seeing where people can take advantage of the solar resource in great density, and even develop a decentralized, smart grid strategy for solar electricity. It is up to your design team to build an understanding of solar energy with your clients and to convey the many options that will take advantage of the resource.

FLOWS AND STOCKS

From a systems dynamics perspective, there are *stocks* of mass and energy, and there are *flows* of mass and energy within our great reservoir of the Sun-Earth-Moon system. For that matter, from a financial perspective, there are stocks of money and flows of money as well. A *stock* is essentially a storage tank of mass or energy that has accumulated through flows in the past "filling" it up. Stocks can and do change with time, and so are considered variable. A *flow* is a rate of exchange that can diminish stocks through outflows or build up stocks through inflows, or flows can be converted as they pass.

Photons from the Sun are flows of radiant energy that we can convert into other forms of energy. Photons are already converted into biomass by photosynthesis, creating a stock of biomass for food. The swamp goo converted from flows of sunshine 300 million years ago into coal is also a stock of chemical energy. Elements like indium or tellurium (critical minerals for PV) or copper found in mineral ores within the Earth's crust are also stocks, emergent from plate tectonics and magmatic inflows over hundreds of millions to billions of years (that's some slow flow). Ore stocks are depleted by erosion and mining as outflows.

In comparing energy sources, renewable resources are *stock-abundant* but *flow-limited*, while non-renewable resources like geofuels are *stock-limited* while being *flow-abundant*. So a *reserve* can first be thought of as a type of economically accessible stock within a larger stock—the physically limited reservoir in the Earth's crust.[5]

Critical minerals (and the elements derived from the minerals) perform an essential function for a specific technology or industry, with a high risk of supply chain disruption. Few substitutes may exist for that function, making the element of high risk to the industry if the supply chain is disrupted. Technological, geological, or social constraints increase the likelihood of supply chain disruption.

[5] H. E. Daly and J. Farley. *Ecological Economics: Principles And Applications*. Island Press, 2nd edition, 2011

In order to work with light in terms of its value, its reserve, and the price elasticity of demand for light, we will now pose the hypothesis. Given a significantly large solar flow (irradiance) with an extremely long-lived and large stock (e.g., fusion from the Sun, and thus irradiance will remain for billions of years), light can be evaluated using the stock-flow language of mineral commodities.

THE VALUE, RESERVE, AND ELASTICITY OF LIGHT

Light from the Sun, when converted to useful forms, is a *commodity*: it can be marketed as a good or a service to satisfy the needs and wants of individuals and society. As with *any* commodity, the value of *light* varies with the *demand for light technologies as a good or service* and the *costs of alternatives*. The *value* of an unconverted photon is a variable quantity. The technology to convert a photon to a useful form is also a commodity with some variable value.

The "quantity" of accessible light for SECS is also a variable, but it depends on more than just the annual irradiation in a given locale. It also depends on the economic viability of the technologies to convert a photon into a useful form of energy for the client. The mechanism that we can use to estimate this quantity can be drawn from mineral economics. Mineral economics is a field studying the business and economics of traditional resource extraction and the uses of those resources as commodities. A study of mineral economics demonstrates that light can be valued in similar fashion to a metal ore (say, zinc from a sphalerite deposit) or a geofuel resource such as coal. The quantity of a mineral *reserve* in the ground (in terms of thickness and depth) can expand and shrink with the change in demand, with changes in technology that reduces costs of production, and with the presence of government incentives.[6] Similarly, the quantity of an available light reserve flowing from the Sun has expanded and contracted with the

An **ore** is an unrefined rock composed of minerals, which contains a raw metal that is valued (in this case, zinc), but which must be processed to access that metal.

[6] **Reserve** is a very specific term in the USGS. See the list below.

same pressures over history. Thus, we will expand the concept of a *reserve* to include quantification of light.

> Reserves data are dynamic. They may be reduced as ore is mined and/or the extraction feasibility diminishes, or more commonly, they may continue to increase as additional deposits (known or recently discovered) are developed, or currently exploited deposits are more thoroughly explored and/or new technology or economic variables improve their economic feasibility. Reserves may be considered a working inventory of mining companies' supply of an economically extractable mineral commodity. As such, magnitude of that inventory is necessarily limited by many considerations, including cost of drilling, taxes, price of the mineral commodity being mined, and the demand for it. Reserves will be developed to the point of business needs and geologic limitations of economic ore grade and tonnage.[7]

[7] U.S. Geological Survey. Mineral commodity summaries 2012. Technical report, U.S. Geological Survey, Jan 2012. URL http://minerals.usgs.gov/minerals/pubs/mcs/index.html. 198 p.

THE SOLAR RESERVE will expand or contract with:

- Need to avoid fuel costs (which increases solar demand).
- Advances in technologies that lower materials costs and installation costs.
- Presence of government incentives to quantify externalities.

Below, we have also adapted the terminology from the US Geological Survey[6] to suggest a new working terminology for the solar economic commodity community. Given an assumption that the light resource can be evaluated in a very similar manner as a mineral commodity, we describe the established approach to estimate the viability of the solar resource, specifically in a locale that has yet to be developed. In the realm of mineral commodities, the US Geological Survey has provided standardized concepts of identifiable *resources*, *reserves*, and *reserve base*.[6] In the following Figure 9.2, we draw strongly from the pre-existing work of the USGS Mineral Commodities Summaries, and attempt

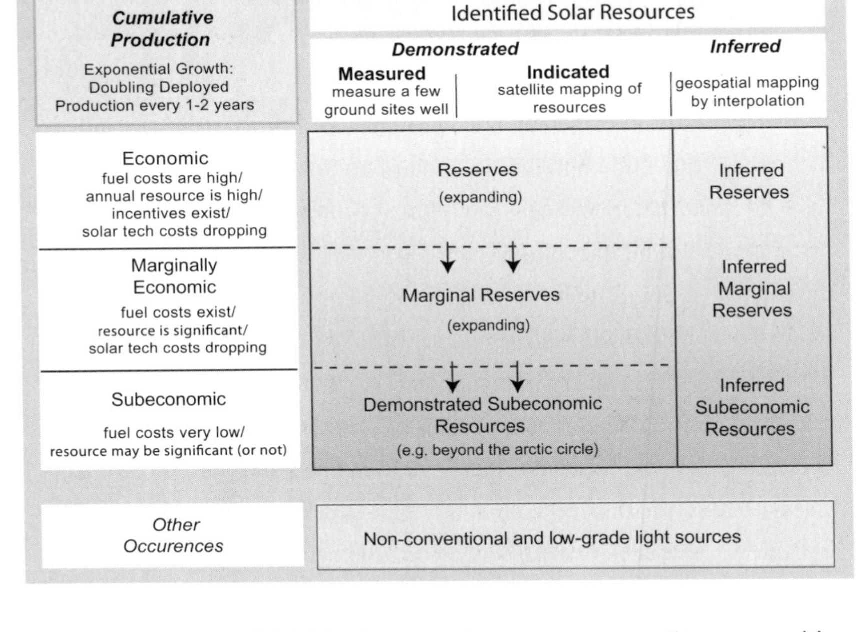

Figure 9.2: Hypothetical economic assessment typology for a solar resource, modified and adapted from the standards established for mineral reserves and resources by the USGS.

to describe the potential bridge between the present commodity communities of mineral and solar resources.

Terminology for solar resource classification paraphrased and adapted from the U.S. Geological Survey Circular 831, *Principles of a Resource/Reserve Classification for Minerals.* 1980.

Resource: A material or energy source that occurs natively in or on the Earth, with a form, concentration, and quantity such that economic collection and/or conversion of that commodity is currently or potentially feasible.

Identified Resources: Resources where one knows the location, grade, quality, and quantity from specific meteorological evidence, or where the resource has been estimated. Identified solar resources encompass *economic*, *marginally economic*, and *sub-economic* components. As a reflection of the degrees of meteorological confidence, the economic divisions can furthermore be subdivided into *measured*, *indicated*, and *inferred*.

- *Demonstrated*: A term describing the sum of measured and indicated.
- *Measured*: The quantity of which is calculated from measurements via pyranometer and pyrheliometer; the grade and (or) quality are calculated from data collected by detailed irradiance sampling in time and space. Sites included for an ensemble of measurements are spaced

so closely and the meteorological character is so well defined that variance, uncertainty, and risk for the content of the resource are well established.

- *Indicated*: The quantity and grade and (or) quality are calculated from using similar information compared to those used for measured resources, such as satellite measurements. However, the sites for inspection, sampling, and measurement are farther apart in distance or are otherwise less adequately spaced in time. Although the degree of assurance is lower than that for measured resources, the information is reliable enough to assume continuity between the various points of observation.
- *Inferred*: These are *estimates*, which are founded on the assumption of regional continuity beyond measured and/or indicated resources, for which meteorological evidence exists. Typical tools are GIS software and geospatial statistical analysis. Inferred resources may or may not be supported by samples or measurements.

Reserve Base: That part of an *identified resource* meeting minimum physical criteria related to a specified solar energy conversion technology practice currently employed, including those for photovoltaics, concentrating solar power, and lower grade solar thermal. The *reserve base* is composed of the in-place *demonstrated resource* (*measured* plus *indicated*) from which *reserves* are estimated.

Reserves: That portion of the larger *reserve base* that can be economically converted *at the time of determination*.

Marginal Reserves: That portion of the reserve base that borders on being economically producible at the time of determination.

Economic: A term implying that a profitable collection or conversion has been established under defined investment assumptions. The profitability has been analytically demonstrated, or is assumed to be credible with reasonable certainty.

Subeconomic Resources: The portion of the *identified resources* not meeting the specified economic criteria of *reserves* and *marginal reserves*.

Cumulative Production: The portion of solar energy converted from the past is not a part of the resource. Converted resources move into the classification of cumulative production.

In comparison to mineral commodities or other traditional energy commodities, the accessible solar reserve is available when economically feasible; expanding or contracting in response to pressure to use solar in avoiding fuel costs, technological advancements in materials costs and installation costs, and from government incentives (or lack of incentives). So how are commodities like fuels and solar energy tied to economics?

In economics, the measured response (in the market) of how the quantity of a product in demand is changed by the incremental change in the price of that product is termed *price elasticity of demand*.[8] The demand is considered *elastic* if a small change (like a decrease) in price (P) leads to people demanding a lot more of the product (Q). The demand is considered to be *inelastic* if a large change (again, a decrease) in price (P) does not lead to people demanding more of the product (Q). The elasticity of demand for solar power will depend on a few general rules, and we will try to contain our examples to solar scenarios for a client or group of stakeholders.

[8] N. Gregory Mankiw. *Principles of Economics*. Thomson South-Western, 3rd edition, 2004.

What? No eggs and butter or meat and potatoes. How un-provincial, how borderline cosmopolitan of you Robot Monkey.

THE ELASTICITY OF DEMAND: Demand is said to be:

- *elastic* if small $\Delta P \rightarrow$ large ΔQ.
- *inelastic* if large $\Delta P \rightarrow$ small ΔQ.

AVAILABILITY OF CLOSE SUBSTITUTES: First, one evaluates the *availability of close substitutes* for the particular SECS of interest. If the desired useful energy form or technology has many *available* close substitutes, then it will be easier for clients/stakeholders to switch among goods for the same desired feature, and the demand will tend to be *elastic*.[9] For example, visual lighting from an electric lamp can be substituted by lighting from a light pipe or window. A fraction of the electricity from the grid (typically from coal or nuclear fuels) can be substituted by power from PV system (electricity is a direct substitute

[9] With technological changes, the availability of alternatives in energy systems is often based on a lack of access rather than an absolute availability of decentralized, alternative electricity.

for electricity, making it more elastic when it is available). In contrast biofuels from solar energy might be considered to be less of a close substitute to gasoline (dominant transportation fuel in the USA) and may not be readily available at the pump in your region.

Additionally, the operating costs of electricity are often so low in the USA that people often choose electric lamps (incandescent light bulbs for example) over natural lighting solutions. You have already seen in the last vignette that without the available substitute of electricity, daylighting is the choice. In contrast, *electricity* (you know, the energy form that you never really think about) is a high-quality power source without a close substitute, and the demand for electricity (where available) is less elastic than the demand for visual lighting. There can be large changes in electricity prices, yet the consumer does not *perceive* multiple *available* alternatives. Notice that last comment, "where available". That should strike you as important. The elasticity of demand fits within a *market* where the goods and services can be bought and sold. Hence the elasticity also depends on the manner in which we define the boundaries of our market. Again, electricity is a broad category (as an entire form of energy), for which there are no good substitutes (note that we are not distinguishing *how* we get the electricity here), which makes it inelastic. However, *PV electricity* is a very narrow category and can have a very elastic demand *exactly because* there are multiple other conversion resources available that are effectively perfect substitutes for the electricity from PV modules (e.g., batteries, fuel cells, or grid power from coal-fired power plants, nuclear power plants, hydropower, wind power, etc.).

If electricity prices change from \$60.00/ MWh to \$100.00/ MWh, the demand for electricity still stays pretty high, because on the short-term, there is nothing a homeowner can do to replace it.

NECESSITY/LUXURY: The elasticity is also affected by the client's *need* for the desired feature. Is the energy form a *necessity* or is it a *luxury*? Necessities build upon our *basic needs* like water, shelter, food, sanitation and waste removal, education, and health care. Necessities tend to have inelastic demands (like the description of electricity in our Western society), while luxuries tend to have very elastic demands. In the USA, PV systems are frequently perceived as a luxury (while they are a necessity to rural communities with no power grid). When the price of photovoltaic panels decreases in New Jersey, or the apparent price of installed panels decreases because of state and federal incentives, the quantity of PV system demand skyrockets. If installed PV panel prices drop from $3/$W_p$ to $1.50/$W_p$, the demand for PV increases a lot. Again, the distinction between a good being a necessity or a luxury is defined by the preferences of the client or stakeholders (the buyers), not by an absolute scaling.

> NECESSITIES have inelastic demands. They are defined by the preference of your clients, and will vary depending on the physical climate (latitude) of the locale.
>
> - Clean water.
> - Clear air.
> - Shelter.
> - Food.
> - Sanitation.
> - Education.
> - Health Care.
> - Thermal regulation.
> - Electricity.

TIME HORIZON: Finally, we must consider the effect of time and the *time horizon* on the elasticity of demand for a solar good or service. Economic observation tells us that goods tend to have a higher elasticity of demand when considered over longer time horizons. Another way to think of a time horizon is with respect to the *period of evaluation* in project finance. For example, when the price of electricity from grid power rises in a given month, the quantity of PV systems demanded alternative to grid power does not increase very much at all. However, as the price of electricity rises over the time scale of years, people buy lower energy demand technologies and install decentralized power systems like PV. SECS like photovoltaics, windows, and solar hot water systems tend to have long lifetimes (on the order of decades) that must be considered in a project finance analysis. The decision to adopt a SECS on the part of the client will be evaluated in terms of both short- and long-term time horizons.

VIGNETTE: Millions of families across the world still live in the dark (during the day!), but that is changing. Recall the amazing viral project of appropriate technology, where one liter PET plastic soda bottles are used as light conduits (light pipes) for indoor lighting, particularly in warm climates such as in Brazil, the Philippines, India, and Africa. Material and labor costs are quite low, and the bottle is re-purposed from a waste stream. Visible light from the Sun is directly substantiated for the light from an electric or fuel source. So we can agree that the Sun is doing work (energy useful) for us even with lighting. Further, we can compare the value of a bright light source during the day for interior lighting with the avoided variable costs of purchasing electricity (if you have access to power) for a light bulb, and the fixed costs of the electrical equipment and light bulbs themselves. http://aliteroflight.org/.

ENERGY CONSTRAINT AND RESPONSE

Let us revisit the concept of the *Energy Constraint and Response* from earlier. As we stated in the introduction of the text, there is a long historical context for us to assess the adoption of and departure from solar energy design in society. We can now connect our observations with some economic perspective as well that give things a more nuanced feel. Fuels such as wood in early Greece and Rome were likely perceived as a *necessity* to add heat to homes during Winter months, as well as a necessary resource for cooking. One might observe that in Rome during the first century B.C.E. with the advent of local *hypocausts*, wood fuels had transitioned from being a need toward being a luxury for village members. So, when the price of wood increased due to over-harvesting and subsequent long-range shipping costs, people in Rome increased adoption of available, close substitutes for space heating: solar thermal space design.[10] We might suspect a high elasticity of demand for a change in the price of wood, given a known substitution of solar thermal space design. We also note that a hypocaust would be a luxury to a community, and reassessments of the increasing cost of wood over longer time horizons would introduce increased adoption of solar design.

Hypocausts were some of the earliest strategies for central heating via fuel combustion, but required constant maintenance and large quantities of fuel.

[10] Ken Butti and John Perlin. *A Golden Thread: 2500 Years of Solar Architecture and Technology.* Cheshire Books, 1980.

The hypothesis of Energy Constraint and Response: when fuels are effectively *accessible, unconstrained and hence inexpensive*, while being perceived as a *necessity*, light-induced energy conversion is not seen as an alternative. Solar energy is then deemed/perceived *diffuse* and *insufficient* for performing technical work. Correspondingly, the *reserve* of light is estimated to be small. However, for periods in history when fuels have become *constrained, inaccessible*, societal innovation has turned to solar technology solutions. During periods of fuel constraints, research into the use of solar energy, the solar resource has been counter-interpreted as *ubiquitous* and vast.[11] At these times, the *reserve* of light is estimated to be large. In turn the general public awareness surrounding solar energy increases, such that the devices and services for applying solar energy to heating of fluids and solids and power conversion are also in demand.

[11] Ken Butti and John Perlin. *A Golden Thread: 2500 Years of Solar Architecture and Technology*. Cheshire Books, 1980.

Again, we describe *fuel constraints* in several forms:

- by nature of being *physically inaccessible* over the given time horizon,
- being of *limited access* or rationed due to exceptionally high demand for fuel or constraining policy and laws,
- being a luxury item of high cost and hence *economically inaccessible*,
- or being accessible but only at *high risk*.

Consider the societal perspective that solar energy is a *diffuse* or *insufficient* technology for producing electricity or even hot water. In complement, we note a general absence of solar design from building orientation and daylighting in the last 50 years of architecture. This perspective is in keeping with the dissolution of solar markets when the alternative of fuels is unconstrained. Examination of economics and social behavior leads one to hypothesize that the perception of solar energy as diffuse is as much a result of the available geofuels in the USA as it is the perceived necessity of those fuels and the non-substitutability of geofuels for the demands of the modern public.

Yet at present and on the near horizon, our individual lives and our societies are being affected by the challenge to access and use the available fuels for combustion. The modern constraints on fuels emerge from increased global demand for forms of energy provided by geofuels, while access to fuel reserves comes from increasingly higher risk regions. Fuels are also beginning to be susceptible to price increases due to policy constraints tied to health and safety, and environmental regulation tied to sustaining our supportive biome systems. Biome disruption has included changes to ground water reserves, changes to air quality, and ecosystem disruptions from fuel spills, and aggressive land use. As a society living in the greater biomes affected by climate change, we are further limiting fossil fuel combustion linked to particulates and greenhouse gases. Even with a shift to natural gas use, global combustion for heat and power is going to be constrained. We might expect that policy and laws that reduce fuel combustion will increase the price of fuels in our future. This perspective of constrained fuel access is in keeping with the expansion of successful solar markets. However, a pure market response is not the only mechanism to follow in addressing alternative energy demand. Governments have a role to play in energy adoption and increasing awareness of energy conversion alternatives for society.

Geofuels:

- coal [*sun-derived],
- petroleum [*],
- natural gas [*],
- tar sands and oil shales [*],
- gas hydrates [*],
- fissile material for nuclear power [not from sunshine].

SO MANY GOODS: CO-PRODUCTS AND CO-GENERATION

Photons from the Sun make up a working current or a flow of energy that can be applied to numerous uses in parallel. Consider that we can use sunshine for daylighting, and at the same time warm an interior space via solar thermal space design–this is desirable solar design. Now consider a poorly designed Sun space that gets too hot in the summer time, which is undesirable.

From a historical context, when conversion of a resource (like a fuel) leads to multiple useful (valued), accessible products, the products are called *co-products*. In terms of multiple useful forms of energy, we tend to call this co-production: "co-generation". However, when conversion of a resource also yields products

The principal energy source for the most of us is the Sun. Some worms and thermophilic microorganisms living near hydrothermal vents at the ocean bottom are different, and draw energy from thermal gradients.

Solar power as **co-generation**? You bet!

In fact, solar power can be used for *poly-generation* of light, heating, power, and even cooling!

that are either undesired, harmful, or just of no value to the client, those items are called *by-products*. We close this short section by asking you to consider how the flow of solar energy can be thought of as a diverse resource, capable of parallel conversions of energy—some of which will be considered useful, while others will be of no value and perhaps even detrimental.

GOVERNMENT AND MARKETS

All energy development in our society has expanded due to government incentives. It is a fallacy and a historical failing to claim that our current "traditional" energy resources succeeded without incentives. In the USA and abroad, Government has played a role in establishing every source of energy as a fuel or technology through incentive programs. Land grant developments of the 19th century were indirect incentives to the early timber industry, while even earlier in the 1700s Pennsylvania governments allowed taxation exemption for anthracite coal production, leading to more states promoting the production and consumption of coal via incentives (including sponsoring of land surveys which significantly lowered the costs of exploration).[12]

By the 20th century, an entire taxonomy of incentives had developed for what is now the traditional energy industry: in federal tax policy (exemptions, allowances, deductions, credits), federal regulation and mandates, research and development (technology and surveys), market influence by government, government services (roads, ports, and deep channel waterways for fuel shipping), and federal financial grants (e.g., subsidizing oil tanker construction and operation costs).[12]

[12] Nancy Pfund and Ben Healey. What would Jefferson do? the historical role of federal subsidies in shaping America's energy future. Technical report, DBL Investors, 2011.

Energy Tax Policy from 1916 to 1970: Promoting Oil and Gas:

For more than half a century, federal energy tax policy focused almost exclusively on increasing domestic oil and gas reserves and production. There were no tax incentives promoting renewable energy or energy efficiency. During that period, two major tax preferences were established for oil and gas. These two provisions speed up the capital cost recovery for investments in oil and gas exploration and production. First, the expensing of intangible drilling costs (IDCs) and dry hole costs was introduced in 1916. This provision allows IDCs to be fully deducted in the first year rather than being capitalized and depreciated over time. Second, the excess of percentage over cost depletion deferral was introduced in 1926. The percentage depletion provision allows a deduction of a fixed percentage of gross receipts rather than a deduction based on the actual value of the resources extracted. Through the mid-1980s, these tax preferences given to oil and gas remained the largest energy tax provisions in terms of estimated revenue loss.[13]

[13] Molly F. Sherlock. Energy tax policy: Historical perspectives on and current status of energy tax expenditures. Technical Report R41227, Congressional Research Service, May 2 2011. URL www.crs.gov.

As we will see in the next section, markets have weak spots. Those failings are often addressed through government incentives or disincentives.

MANAGING THE GRID

As the predominant market emerging for solar energy is linked to grid-tied electricity production, it is important to discuss the different physical and market-based networks of power found in the USA. The *interconnection* is the term for a network of interconnected power grids within a region. There are two major interconnections in the USA the Eastern Interconnection and the Western Interconnection, while Texas also has its own interconnection (three separate power grids in the lower 48 states). Each interconnection in the USA has one or more forms of independent system operator managing the grid for that region.

Each has distinct structures of governance and protocols for managing power congestion—but an ISO and an RTO operate in a similar manner. We will refer to the **RTO** for simplification.

The United States interconnections are formed of Independent System Operators (ISO) or Regional Transmission Operators (RTO).[14] While both occur in the USA, to keep things simple we will use RTO only in this description. The RTO is a not-for-profit organization that acts on behalf of a group of electric utilities in a region to independently manage assets of joint power transmission. RTOs are independent in that they do not own any assets in power and do not take a preferred position with respect to the wholesale electricity market. There are several regions where the RTO also creates centralized spot markets in electric power and ancillary services, while also offering financial contracts for

[14] S. Blumsack. Measuring the benefits and costs of regional electric grid integration. *Energy Law Journal*, 28: 147–184, 2007.

Locational Marginal Prices (LMP) convey the social cost to move energy to a specific locale.

hedging/managing exposure to congestion risk (where transmission congestion is managed financially rather than physically). The centralized dispatch of the RTO provides *locational marginal prices* (LMPs) on the wholesale market, reflecting the social cost of transmitting electric power to a specific location within the managed system. Different LMPs in multiple locales signal the costs of transportation and transmission congestions to the market. Locational marginal prices vary across the time of day and for different locations. For example, heatwave in central Philadelphia, PA, would drive more people to turn on air conditioning, demanding more electricity and leading to increased transmission congestion, and as such the LMP for electricity will increase for the afternoon. The same heatwave in Pittsburgh, PA, would not have nearly the congestion, and the market would reflect a correspondingly lower LMP.

As seen in Figure 9.3, RTOs can manage large physical areas—typically spanning distances on the order of 1000 km. Within the Western Interconnect, a large and well-known system operator is CAISO. Within the Eastern Interconnect, the largest (spatially) RTOs are the Midwest ISO (MISO) and PJM. Recall that physical scales of this order are linked to times scales through *synoptic meteorology*, on the order of days to seasons. If we also refer back to Figure 9.1, we see that emerging megaregions can be associated with system operator regions.

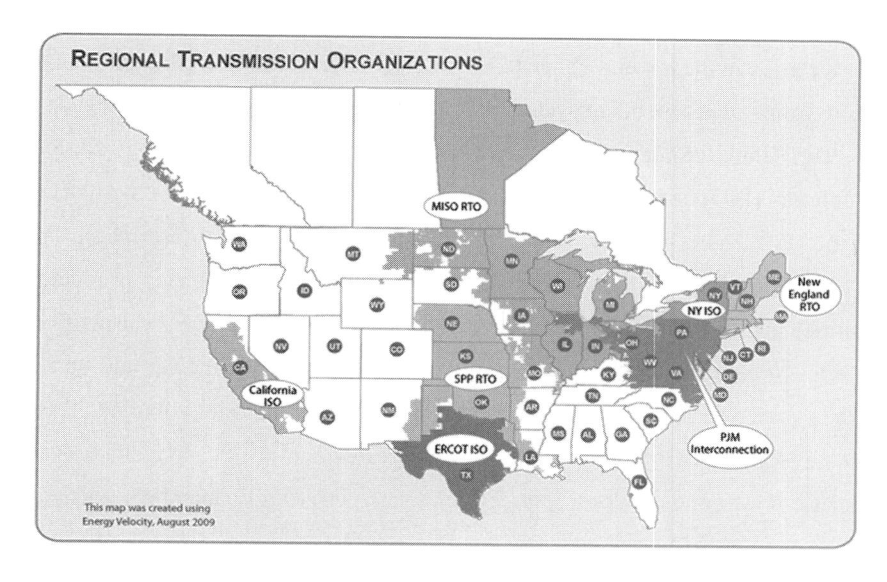

Figure 9.3: A map of RTO/ISOs in the USA. Accessed from FERC.gov on March 11, 2013.

EXTERNALITIES

Markets can fail to allocate all resources efficiently when emergent phenomena occur as a consequence of a market transaction, but are *external to the decision* of whether to buy or sell something. The emergent phenomena that are not considered in the costs of the transaction are termed *externalities: as a bystander, you didn't pay for it, you may or may not want it, but you "get it" anyway*. Given externalities are "external to the decision" whether to buy or sell, we may seek other tools beyond the market such as *taxes* or *regulation* to internalize them.

Positive externalities will have a beneficial impact on the well-being of bystanders. Open education strategies developed through MITx or Penn State's Open Educational Resource are great examples of positive externalities, as the courseware developed for a much smaller cohort of students is available to anyone with internet access. Even more generally, education has been said to improve the overall population of a country by leading to a better informed people and a better government. However, positive externalities lead markets to supply a smaller quantity of a product than is desirable. We can similarly hypothesize that by educating more individuals in solar energy design, the social value of projects having significantly higher solar utility for clients or stakeholders in numerous locales is higher than the private value of learning to design and install your own solar hot water system.

In response to a market failure, policy makers can deliver incentives to participants to internalize the externality. As seen in Figure 9.4, the social value for education of solar energy design teams is greater than the private value and

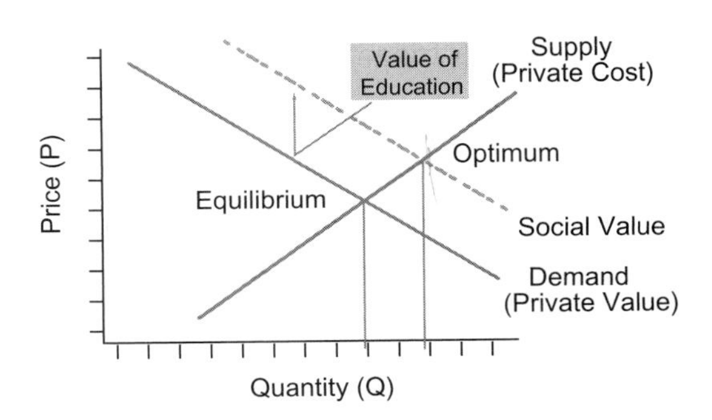

Figure 9.4: In the presence of a **positive externality**, the social value exceeds the private value. Government can correct the market failure by subsidizing the positive externality.

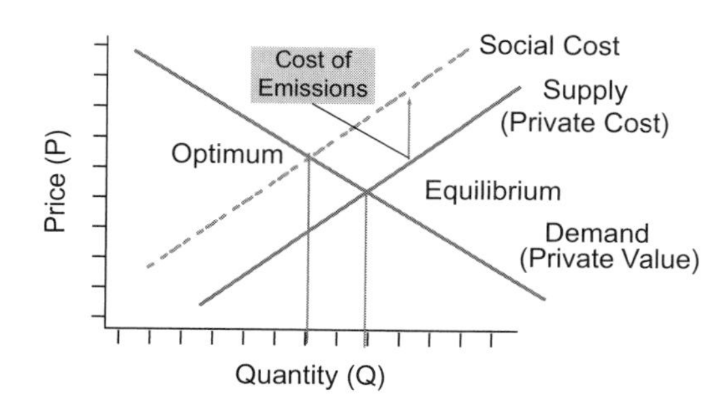

Figure 9.5: In the presence of a **negative externality**, the social cost exceeds the private cost. Policies can correct the market failure by taxing the emissions of power generation.

sits above the demand curve. Hence, a socially optimal quantity is greater than the quantity determined by the private market. Incentives have been tools used by government to shape energy development for hundreds of years.

Negative externalities will have a detrimental impact on the well-being of bystanders. For example, the CO_2 emissions from large-scale coal combustion are a negative externality because it creates a significant impact on the human-caused global warming, in turn causing adverse effects on other goods like global fresh water supplies. As a result of the negative externality from CO_2 emissions (again, not a part of the costs of operation), power producers tend to maintain coal fired power plants longer than their advised lifetime. As seen in Figure 9.5, the social cost for emissions exceeds the private cost.

In response to a market failure, policy makers can deliver incentives to burn less coal via a tax on coal combustion per ton of CO_2 emitted.[15] The effect would be to shift the supply curve upward relative to the size of the taxation, and the process would be *internalizing the externality*, because now both sellers and buyers would have a market incentive to account for the effects of their actions.[16]

[15] This type of tax to correct negative externalities is called a **Pigovian tax**, after economist Arthur Pigou (1877–1959).

[16] N. Gregory Mankiw. *Principles of Economics*. Thomson South-Western, 3rd edition, 2004.

UTILITY, RISK, AND RETURN

Goal of solar energy design: to increase or maximize the **solar utility** for a **client or stakeholders** in a given **locale**.

We have described *solar utility* earlier, when describing the context and philosophy of design, and as a core factor of the goals of solar design. Breaking the term apart, *utility* refers to the preferences of a client within a set of goods and services. Utility can also refer to a measure of derived happiness from that set of goods and services. Hence, *solar utility* is used in this text to refer to a set of goods and services originating from the solar resource, in contrast to a non-solar good or service. In economics, the framework of utility is further constrained to the consumption of goods and services at some cost for those goods. And as goods have a cost, the utility may be viewed as a function of wealth.

Furthermore, we could view solar utility in terms of a financial investment, one which we expect to return an increased amount of wealth some time in the future. The rate at which a client receives a return on their initial investment (rate of return) is also viewed as a rate of accumulating wealth. From a pure economic perspective, *utility* may be viewed as a function of *consumption*, and consumption is in turn a function of *wealth*, and wealth is a function of the *rate of return*.[17]

Utility describes the client's preference or measure of derived happiness from a set of goods and services. **Solar utility** is aligned with the set of goods and services that originate from the solar resource, rather than a non-solar good or service.

Rate of Return: rate of wealth accumulation for the client for a given investment.

[17] J. C. Francis and D. Kim. *Modern Portfolio Theory: Foundations, Analysis, and New Developments*. John Wiley & Sons, 2013.

In our position as an integrative design team, we hope to offer a client the opportunity to invest in a solar project (an economic asset) with the highest expected rate of return. This means we need to provide the client with valuable information to make that decision to invest, through project analyses and solar conversion simulations for the locale at hand. Also, given an installed SECS such as a solar farm (i.e., utility-scale solar power), the team involved in system management will look to operate the system and participate in electricity markets so as to increase profits as well, given better knowledge of solar forecasts and simulations of system performance according to those irradiance projections. However, simulations function as idealized scenarios and only estimate expected performance, whereas the actual world of meteorology and markets will have measurable *uncertainty* in the actual system returns on investment and dynamic daily performance. With uncertainty comes risk.

$$\text{Risk} = p(\text{event}) \times (\text{expected loss given the event}). \tag{9.1}$$

Risk is all about uncertainty of events in the future, and is often framed as the probability of an uncertain event occurring in the future multiplied by the expected loss should the event occur. If we know the *probability distribution* of all possible outcomes, then we also know the *expected value* of the outcome, surrounded by the *dispersion of outcomes* around that expected value. We show a simplified example of a normal probability distribution function in Figure 9.6,

Risk: is the dispersion of possible outcomes about the expected value, or the probability of a negative event times the expected loss.

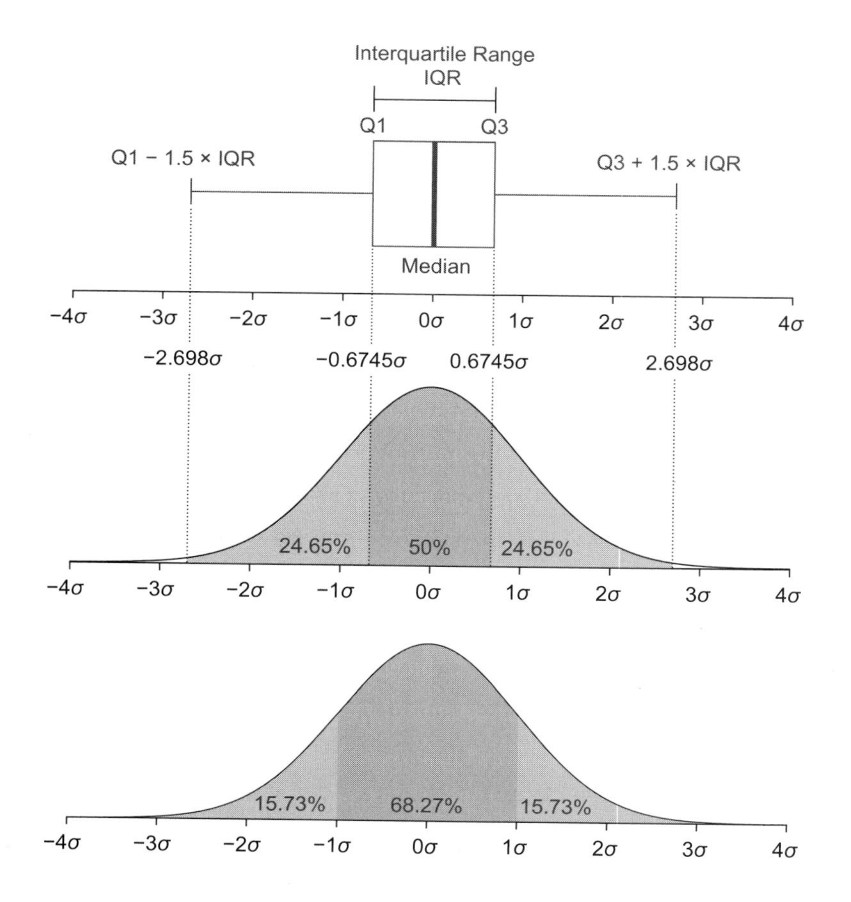

Figure 9.6: A simple probability distribution function for a normal (Gaussian) distribution, where the *mean* (here, the median) represents the expected value, and the *standard deviation* σ is proportional to the variance σ^2. Adapted from User: Jhguch/Wikimedia Commons/CC-BY-SA-2.5. March 2011.

where the expected value is the mean (μ), and the dispersion can be represented by the statistical variance (σ^2; a function of the standard deviation). Hence, in a modern sense, risk can also be viewed as the dispersion of possible outcomes around an expected value.

If we know what to expect, and the dispersion of possible outcomes is clustered tightly about that expectation, then we can adapt or make changes for the future appropriately. The greater the dispersion of outcomes, the higher the risk. So "riskier" scenarios in solar project development, or in SECS operation and management, will have a larger dispersion of outcomes around the expected value.[18] The financial risk for a solar investment could be assessed in terms of the unexpected variability for the financial returns (e.g., the internal rate or return,

[18] J. C. Francis and D. Kim. *Modern Portfolio Theory: Foundations, Analysis, and New Developments.* John Wiley & Sons, 2013.

or the payback period), brought about by a number of factors including weather, policy changes, and incentives.

When solar power becomes a significant contribution to power generation in a region, then the expected values for power generation become important, as they have for wind generation in recent years. As an example, let's consider a large installed solar farm (e.g., 20–50 MW of electric capacity). On short time horizons (hours/days/seconds) utility companies and RTOs will be concerned with grid stability and expected values of power generation in day-ahead (and to some extent hour-ahead) markets. Recall that RTOs tend to manage a region for transmission congestion financially rather than physically. We have already described how clouds are the primary disruptive meteorological feature to expected solar conversion, particularly for photovoltaic power production. Unexpected drops or surges in power production are going to cost the utility, and in turn may affect grid stability. So in this example, we would like to be aware of the dispersion of possible cloudy events about an expected value (summarized through a probability):

$$\text{Risk} = p(\textit{cloudy event}) \times (\textit{expected loss}). \tag{9.2}$$

With a probability distribution, one would want to know the amplitude and tails on the distribution, as the greater spread of possible outcomes indicates that we have low confidence that the actual value of power production will be the same as the expected value, suggesting a riskier market scenario to participate in.

In energy project finance utilities and RTOs can explore solar site networking across a region—distributed generation—can minimize intermittency in electrical power generation. Statistically, the presence of N connected sites across a region of interest will reduce the net variance (σ^2) of power as a function of $\frac{1}{\sqrt{N}}$. As such, distributed generation comprises a portfolio of physical projects that collectively can reduce risk in grid instability, while maintaining energy and market returns for a given time interval. In this case, energy project management can also follow the general concept of diversification. Locations across a region can be thought of as diverse assets. One measure of risk reduction would be to evaluate the covariance from multiple sites within the time horizon of interest, which we

We discussed **spatio-temporal uncertainty** at the end of Chapter 5 and in Chapter 8: **network solar resource assessment**.

[19] J. Rayl, G. S. Young, and J. R. S. Brownson. Irradiance co-spectrum analysis: Tools for decision support and technological planning. *Solar Energy*, 2013. http://dx.doi.org/DOI:10.1016/j.solener.2013.02.029.

Portfolio management: a process common to economic investment, but which may be applied to renewable power assets. The process involves security analysis of probability distributions for returns from investments, portfolio analysis of optimal possibilities, and portfolio selection from the menu of possibilities.

[20] Harry Markowitz. Portfolio selection. *J. Finance*, 7 (1): 77–91, Mar. 1952; and W. H. Wagner and S. C. Lau. The effect of diversification on risk. *Financial Analysts Journal*, 27 (6): 48–53, Nov.-Dec. 1971.

[21] J. C. Francis and D. Kim. *Modern Portfolio Theory: Foundations, Analysis, and New Developments*. John Wiley & Sons, 2013.

described in Chapter 8 through power spectral density and cospectra analyses.[19] Variance and covariance analyses introduce a methodology for connecting meteorological data (irradiation) and power conversion with expected values of uncertainty and risk, as well as developing collective strategies for portfolio management from solar electric technologies.

The concept of *portfolio management* through diversification is familiar in finance and investment. A portfolio is a mixture of projects or investments held by an individual, a group of stakeholders, or an institution, and key parts of managing a portfolio are *risk* and *return*. Portfolio management is a strategy to reduce the risk for the client by investing in a variety of assets, rather than putting all assets in one basket. A diversified portfolio with assets having imperfect correlation among the whole can experience less risk than the weighted average of any one asset.[20] Based on the logical implications of portfolio analysis financial analysts now work with what is termed the Capital Asset Pricing Model (CAPM), a theoretically appropriate rate of return for an asset, considered as a new addition to an already diversified portfolio, given that asset's systematic or market risk (represented by the quantity β in finance).[21] The strategy of portfolio management and the CAPM each have potential for broader scientific extension to energy project finance, hedging against weather and electricity market uncertainty, systems sustainability, and ecosystems services.

The financial *return* (because there are other values garnered in sustainability beyond financial returns) for a SECS can be framed in numerous ways. Here we introduce the return in terms of the *net present worth* (NPW), also called the net present value of a project, which is the sum of annual cash flows evaluated with respect to the time value of money, often accounting for the longer time horizon of the project life.

Most SECS projects are long-term and require a long view of return, evaluated over 30–50 years in fact. As we will see in the subsequent chapter on project finance, the *Internal Rate of Return* (IRR), or the minimum discount rate $\bar{r}$ that yields a Net Present Value of zero for the investment at the end of the period of evaluation (time t), with the time value of money factored into the calculation. When the cost of capital r is smaller than the IRR rate $\bar{r}$, the investment has a profitable return. For a portfolio, the rate of return can be the weighted average of the rates of return on the suite of assets under evaluation for the client or stakeholders.

$$NPV = \sum_{t=0}^{n} \frac{C_t}{(1+\bar{r})^t} = 0. \quad (9.3)$$

In summary, from a financial perspective many clients will be concerned with the risks of developing and managing a SECS in practice. The integrative design and management team can draw from modern portfolio theory and the capital asset pricing model to develop strategies to reduce risk and increase rates of return for the stakeholders.

PROBLEMS

1. List the weaknesses for a pure market decision (called externalities) and consider the effect of an installed and distributed PV power plant considered over periods of evaluation greater than 10 years.

2. What mechanisms can governments use to account for the social cost exceeding the private cost of a good/service?

3. What mechanisms can governments use to account for the social value exceeding the private value of a good/service?

4. What is an RTO (or ISO) and what is the role of the organization with respect to the power grid?

5. In 2012, the price of central Appalachian coal was ~$65 per short ton (907 kg), and could be used as a resource to supply fuel for a power generation plant (but you have to keep purchasing coal). In a scenario where the time-averaged cost to run a solar power from photovoltaics (PV) over the first 5 years of operation was $200 per equivalent electric power production (but

you only have to pay for it once). Given that coal power plants are already constructed, which technology would you expect an economist advise you to purchase? (Source: http://www.eia.gov/coal/news_markets/.)

6. Using available resources, investigate the history of energy incentives for the USA. In a comparison and critique, argue the way in which solar energy development as an industry is just another form of energy exploration.

RECOMMENDED ADDITIONAL READING

- Principles of Economics.[22]
- Options, Futures, and Other Derivatives by John C. Hull.[23]
- Modern Portfolio Theory: Foundations, Analysis, and New Developments.[24]
- Ecological Economics: Principles And Applications.[25]
- What Would Jefferson Do? The Historical Role of Federal Subsidies in Shaping America's Energy Future.[26]
- Red, White, & Green: The True Colors of America's Clean Tech Jobs.[27]
- Energy Tax Policy: Historical Perspectives on and Current Status of Energy Tax Expenditures.[28]

[22] N. Gregory Mankiw. *Principles of Economics.* Thomson South-Western, 3rd edition, 2004.

[23] J. C. Hull. *Options, future and other derivatives.* Pearson Education, Inc, 2009.

[24] J. C. Francis and D. Kim. *Modern Portfolio Theory: Foundations, Analysis, and New Developments.* John Wiley & Sons, 2013.

[25] H. E. Daly and J. Farley. *Ecological Economics: Principles And Applications.* Island Press, 2nd edition, 2011.

[26] Nancy Pfund and Ben Healey. What would Jefferson do? The historical role of federal subsidies in shaping America's energy future. Technical report, DBL Investors, 2011.

[27] Nancy Pfund and Michael Lazar. Red, white & green: The true colors of America's clean tech jobs. Technical report, DBL Investors, 2012.

[28] Molly F. Sherlock. Energy tax policy: Historical perspectives on and current status of energy tax expenditures. Technical Report R41227, Congressional Research Service, May 2 2011. URL www.crs.gov.

SOLAR PROJECT FINANCE

10

ENERGY ECONOMICS suggests that most people make decisions to buy a system based on the analysis of costs. Some people would even like to know how the price for fuels compares with the price for a projected SECS. The cost calculations for project design used in engineering are more appropriately specified as *project finance*. From another perspective, this section is a study of *Life Cycle Cost Analysis* (LCCA) in energy engineering, addressing whether an investment in a specific SECS project will pay off in the long run.The following methods are simplified approaches to evaluate life cycle costs, using assumptions of interest rates for central banks, mortgage rates on loans from depository banks, and inflation rates of fuels and services. Our analysis incorporates the time value of money and the unit cost for assembling and maintaining a SECS, but does not fully explore the sensitivity of these parameters as a detailed analysis. We introduce these topics to expose the broader design team to the language and principles of cost analysis in project finance. Additionally, these concepts are also wrapped into the open software SAM (the System Advisor Model) from the USA DoE National Renewable Energy Laboratory (NREL), within which one can explore a much more detailed study for future projects.[1]

[1] P. Gilman and A. Dobos. System Advisor Model, SAM 2011.12.2: General description. NREL Report No. TP-6A20-53437, National Renewable Energy Laboratory, Golden, CO, 2012. 18 pp; and System Advisor Model Version 2012.5.11 (SAM 2012.5.11). URL https://sam.nrel.gov/content/downloads. Accessed November 2, 2012.

TIME VALUE OF MONEY

Fundamental to finance over long spans of time is the *time value of money*. The *present value* (PV) is the worth of an asset, money, or cash flows in today's dollars when rate of return is specified. The *future value* (FV) of money is the worth of an asset or cash flow, being evaluated in today's dollars. It should make sense that *fuel costs* (FC) or *fuel savings* (FS) have values that we can follow as annual (or other periodic) cash flows.

While we do not have the ability to predict the future, we can make an assumption that the future costs or savings can be discounted to the present value using a *market discount rate* (d). The market discount rate is tied to the interest rates that central banks charge to depository institutions (credit unions, savings and commercial banks, and savings and loan associations).[2] Here, we make a simplified assumption that the discount rate for the market is of a similar scale to the interest rate of the relevant central bank. However, as seen in Table 10.1 the effects of the global recession in the early 2000s have resulted in very low central bank rates for the USA, Japan, and Europe. We must be cognizant that these rates will likely rise over the course of a 20–40-year evaluation period of life cycle cost analysis.

[2] Fdic law, regulations, related acts, September 15 2012. URL http://www.fdic.gov/regulations/laws/rules/1000-400.html.

$$PV = \frac{FV}{(1+d)^n}, \tag{10.1}$$

where d is a market discount rate, and PV is evaluated for "now" (time = 0) and the FV is evaluated for period n (here n is in years).

We apply finance rate concepts when we explore the use of loans or mortgages on a system, and we need to encapsulate the time value of money paid on a loan as well as the depreciation of taxes paid on the system (in the USA, the Modified Accelerated Cost Recovery System (MACRS) is used for tax depreciation). We also apply rates to describe the escalation of fuel costs or utility electricity rates, to describe the inflation of operating costs such as maintenance and insurance, or to describe the change in production-based government incentives over the period of evaluation.

In contrast, *future costs for goods* like fuels or electric power (labeled FC for "fuel costs" here), *future costs for services* like systems maintenance and

Country/Zone	Central Bank	Currency	Key Interest Rate
Australia	Reserve Bank of Australia	Australian dollar (AUD)	3.25% (03.10.2012)
Brazil	Banco Central do Brasil	Brazilian real (BRL)	7.25% (10.10.2012)
Canada	Bank of Canada	Canadian dollar (CAD)	1.00% (20.07.2010)
China	People's Bank of China	PRC Yuan (CNY)	6.00% (05.07.2012)
Eurozone	(Multiple)	Euro (EUR)	0.75% (05.07.2012)
India	Reserve Bank of India	Indian rupee (INR)	8.00% (17.04.2012)
Japan	Bank of Japan	Japanese yen (JPY)	0.0–0.1% (16.11.2008)
Mexico	Bank of Mexico	Mexican peso (MXN)	4.50% (17.07.2010)
Turkey	Central Bank of Turkey	Turkish Lira (TRL)	5.75% (04.08.2011)
United Kingdom	Bank of England	Pound sterling (GBP)	0.50% (05.03.2009)
United States	Federal Reserve System	US dollar (USD)	0.25% (16.12.2008)

Table 10.1: A list of market interest rates, given the central bank rates of late 2012. Data accessed from http://www.banksdaily.com/central-banks/ on November 24, 2012.

insurance, and *future savings on income tax* are each expected to inflate over the years, although each may have different rates of inflation (i). The *future value* (FV) of *costs* (C) or *savings* ($S = -C$) can be evaluated with a simple uniform rate of inflation (i) for period n in the future, represented in the generic Eq. (10.2).

$$FV = C(1+i)^{n-1}. \tag{10.2}$$

In many cases of SECS design and deployment, the fixed costs of installation are too large for a direct payment from the client. Much like the costs for a car or a house, or college education, the costs call for a combined investment of a

Present value and present worth are used equivalently here.

If, for some special case, the inflation rate were equal in scale to the discount rate ($i=d$), then the *PWF* would be proportional to the period of evaluation only: $PWF = \frac{n}{n+1}$.

cash *deposit* and a *loan* or mortgage. As such, we will show how the concept of the present value ties together for loan payments or for future costs on goods and services. The *present worth in year n* for costs (*C*) on loan or mortgage payments is shown in Eq. (10.3).

$$PW_n = \frac{C(1+i)^{n-1}}{(1+d)^n}. \tag{10.3}$$

If we separate the costs (*C*) from Eq. (10.3), we are left with the Present Worth Factor, which has identifiers *n* (period in the future), *i* uniform inflation rate, and *d* the uniform discount or loan interest rate: $PWF(n,i,d)$.

$$PWF(n,i,d) = \sum_{j=1}^{n} \frac{(1+i)^{j-1}}{(1+d)^j}. \tag{10.4}$$

Hence, the present worth can be evaluated for each year, as an annualized calculation for a given year *n*, with systems costs *C*, an inflation rate of fuels of *i*, and a market discount rate of cash *d*. The *total present worth* can also be evaluated for the total arc of the period of evaluation, as presented in Eq. (10.5).

$$TPW = C[PWF(n,i,d)]. \tag{10.5}$$

SOLAR SAVINGS

In 1977 Beckman, Klein, and Duffie (University of Wisconsin–Madison) introduced the concept of *solar savings* (*SS*) in project finance analysis.[3] In fact, much of what we will address decades later as solar energy project finance comes directly from that earlier work. Solar technologies tend to substitute for fuels and their associated costs, leading to *avoided fuel costs* (*FC*) or *fuel savings* (*FS*). The solar savings are calculated as the sum of the *FS minus* the fixed and variable costs for the intended solar energy conversion system (C_S) (see Eq. 10.6). The *SS* are furthermore evaluated over a *time horizon* or *period of evaluation* that includes the life of any loans tied to the system, and a significant portion of the lifetime of SECS (on the order of 15–30 years).

[3] W. A. Beckman, S. A. Klein, and J. A. Duffie. *Solar Heating Design by the f-Chart Method.* Wiley-Interscience, 1977; and John A. Duffie and William A. Beckman. *Solar Engineering of Thermal Processes.* John Wiley & Sons, Inc., 3rd edition, 2006.

- *SS*: Solar Savings.
- *FS*: Fuel Savings.
- *FC*: Fuel Costs.

Initially, the solar savings approach was applied to evaluate solar hot water systems. A SECS like a hot water system might have some annual costs for maintenance and insurance of the collector. We call these costs *incremental*, because there are already established costs associated with installing a non-solar energy conversion system and we are trying to do a financial comparison of the marginal difference in cash flow for a SECS that would have some additional costs associated with the fuel savings and potential incentives.

$$\begin{aligned} SS = FS &- \text{incremental mortgage/loan payment} \\ &- \text{incremental maintenance/insurance} \\ &- \text{incremental parasitic energy costs} \\ &- \text{incremental property taxes} \\ &+ \textbf{tax credit incentives} \\ &+ \textbf{production credit incentives}. \end{aligned} \tag{10.6}$$

For SECS, incremental operational costs include fees for annual maintenance or repairs, additional insurance costs, and any additional property tax costs. The incremental costs to power equipment like pumps or tracking systems are *parasitic energy costs*. Parasitic energy costs as so-named because they will sap out a small portion of energy and thus add a small cost to total solar savings. For example, by paying an electric bill for a water pump to circulate solar hot water from a rooftop to a basement tank we experience a small monthly cost.[4] Additional parasitic energy costs will arise in large-scale solar projects, where energy losses to the environment are also considered parasitic, as they do not contribute to the usable, or billable portion of the solar load (L).

Incremental incentives will vary across regions, states, and countries. This is another reason that we advocate the goal of solar energy design to include the *locale*. It's not just about the weather, it is also the climate of policy and government incentives for energy in your area of interest! We have separated the direct financial incentives similar to the open project design software SAM (System Advisor Model) from the NREL.[5] Tax credit incentives will deliver money as credit to interest paid on a mortgage and on the first-year payment of the system (ITC: an Investment Tax Credit). Production credit incentives are those offered as a form of *payment incentive*

[4] Of course, we now have technologies that produce geyser pumps, which require no electricity for solar hot water. For an example, see the Sunnovations company.

[5] P. Gilman and A. Dobos. System Advisor Model, SAM 2011.12.2: General description. NREL Report No. TP-6A20-53437, National Renewable Energy Laboratory, Golden, CO, 2012. 18 pp; and System Advisor Model Version 2012.5.11 (SAM 2012.5.11). URL https://sam.nrel.gov/content/downloads. Accessed November 2, 2012.

for production of carbon-free energy. At this time the energy valued for production is solar electric power, and in several states in the USA for every megawatt-hour (energy unit: MWh) of electricity produced a *Solar Renewable Energy Certificate* (SREC) can be earned and then sold on the market. SRECs are tradeable, non-tangible energy commodities in the USA tied to government incentive carve-outs specific to the solar electric industry. Renewable Energy Certificates are proof of 1 MWh electricity generated by renewable energy, and are a part of a state's Renewable (or Alternative) Energy Portfolio Standard.

We have since observed that the price of electricity is really a fuel cost, and photovoltaic systems can be evaluated in the same manner in project design, as well as in most other solar projects. When evaluated using fuels, *energy is money*. We purchase a supply of energy to meet our time-dependent energy demands. It is up to the solar design team to find effective means to reduce and replace fuel demands, and so we build a common language around the general flows of energy in a system. Again, given the urban trend for society, we will focus our project finance efforts on the modern era of solar systems hybridized with our local heat and power network. There are scenarios where decentralized solar technologies will grow, disconnected from the utility networks of the grid or natural gas pipelines, but even these will typically be supported by local fuel use or alternative energy storage strategies (electric storage and thermal storage).

LIFE CYCLE COST ANALYSIS

There are several economic figures of merit that we can look at in assessing a new project.[6] Each is related to the time horizon and the value of the project in *present worth* (today's dollars). First, when we invest in an energy system we do so with the money valued today, and we know from the *time value of money* that future costs or savings will be discounted over any time horizon in terms of *Present Worth* (*PW*).

Life Cycle Cost Analysis (LCCA): As stated in the section on elasticity of demand, the *time horizon* is an important criteria to consider in evaluating and comparing options for your client.

Here, time horizon and evaluation period are synonymous.

The *Net Present Worth* (*NPW*) of a system over the period of evaluation is called the Life Cycle Savings (LCS). The LCS is the difference between the life cycle costs of a conventional fuel-only energy system and the life cycle costs

[6] John A. Duffie and William A. Beckman. *Solar Engineering of Thermal Processes.* John Wiley & Sons, Inc., 3rd edition, 2006.

of a SECS with auxiliary fuel costs. We can evaluate the LCS using discrete or continuous (calculus) mathematical methods. In the discrete methods we first calculate the *annualized life cycle costs*, or average yearly cash flow summed from the contributing costs and savings (in complement we can evaluate the *annualized life cycle savings*). We then apply the time value of money to place the annualized cash flows in terms of present worth (*PW*). Finally, we sum the annualized cash flows in present worth, to arrive at the LCS or *NPW*. Using a continuous method of analysis, we can obtain the net present worth of each cost and savings contribution for the full period of evaluation, and then directly sum the contributing factors to arrive at the LCS.

So how long is a "life cycle"? Is it 10 years, 20 years, 50 years? Recall that the time horizon is a factor in the elasticity of demand. Very short time horizons (1–2 years) tend to have an inelastic demand. In our project assessment, we typically make sure that the evaluation period encompasses the period of the mortgage, and then decide upon a reasonable extension beyond that which depends on the time horizon of interest to the client.

COSTS AND SAVINGS

In project finance, we have *investments* and *operating costs*. The investment relates to the *fixed costs* of materials and installation to deliver a SECS to the client or stakeholders. The client only has to make this payment once, and the system will last for decades. In the case of systems like photovoltaics or solar hot water, you don't buy half of a module, so the size of your system increases in lumps, rather than in smooth increments. *Lumpy costs* are common to energy systems deployment, especially large-scale power systems for utilities. In comparison to a nuclear power plant the investment in incremental *PV* modules is fairly fine-grained and smooth. But from the perspective of a residential homeowner, the incremental investment in solar technologies appears to have a high fixed cost, and to be a lumpy cost.

For a reasonable complementary design tool to explore this section, please download and open a case study in SAM, the System Advisor Model from the USA DoE National Renewable Energy Laboratory.

$$C_S = C_{dir}(\$/\text{unit}) \cdot n(\text{units}) + C_{indir}(\$/\text{unit}) \cdot n(\text{units}). \tag{10.7}$$

As seen in Eq. 10.7 SECS *investments* (total costs: C_S) are further divided into *direct capital costs* (C_{dir}) and *indirect capital costs* (C_{indir}), both of which can have a scale dependency. From an economics perspective, we would frame the scaled dependency of the size of the project in terms of the *unit cost*, that is the fixed and variable costs per characteristic unit of performance or area. Area dependency is one type of unit in residential solar hot water systems, and costs per unit area (C_{area}, in \$/m^2) are proportional to the area of the aperture for the hot water collector system (A_c, in m^2). However, panels themselves have discrete sizes (area per unit) and so we can take unit cost one step further, by multiplying the number of units times the performance per unit, delivering and estimate of the net performance for the intended systems under common testing conditions.

UNIT COST SCALING FOR COLLECTORS:

- Cost per unit: \$/unit (module)
- Cost per unit area: \$/m^2
- Cost per unit power: \$/W$_p$ or \$/W$_{dc}$ or \$/W$_{ac}$ or \$/MW$_e$
- Cost per unit energy: \$/MWh

In solar project design, the costs for materials are a substantial part of *direct capital costs*. Direct costs also include labor for installation, and installer margin and overhead. Each of these can be broken down into *unit costs*, where the scaling unit is \$ per kW_{dc} (in terms of direct current under AM1.5 testing conditions) for all but the inverters, where the scaling unit is \$ per kW_{ac} (in terms of alternating current under AM1.5 testing conditions).

For a photovoltaic system, direct costs will include the *PV* modules, inverters, and the remaining balance of systems equipment costs (BoS-equipment). In this case BoS-equipment is used specifically to describe the remaining structural and electrical parts such as mounting components, racking components, and wiring. In a broader evaluation of *PV*, BoS is used to describe all costs outside of the direct capital costs for the modules themselves. Extensive studies are progressing to reduce the costs of the modules as well as reducing the remaining BoS costs to the investment.[7]

The unit cost is derived from knowledge of the performance per unit for each *PV* module or inverter (kW_{dc} or kW_{ac}, respectively) tested under standardized conditions, such as Air Mass 1.5 ($1000\,W/m^2$). By multiplying the number of units times the performance per unit, we achieve the net performance for the intended systems. But one also needs the cost per unit of performance (e.g., $\$/kW_{dc}$), as our goal is to include the *unit cost* in our project systems costs.

Indirect capital costs include permitting for project development, taxes, as well as costs for environmental studies, engineering design of the system, and costs for grid interconnection, and land costs. Again, indirect costs can be scaled in terms of *unit cost*, and specifically land costs will be proportional to the acres of land being developed.

Operating costs are *variable costs* of monthly or annual operations. These costs include unit costs for maintenance, insurance, fuel, and electricity that are not being served by the SECS. They are also costs separate from any loan or mortgage payments and additional property taxes linked to project financing.

$$\begin{aligned}\text{Annual cost} = {} & FC + \text{loan/mort.} \\ & + \text{maintenance/insurance} \\ & + \text{parasitic energy cost} \\ & + \text{property taxes} \\ & - \textbf{income tax savings} \\ & - \textbf{SRECs or other payment incentives}. \end{aligned} \tag{10.8}$$

Interest and property taxes can be deducted in proportion to the effective tax rate from a client's gross annual income. This *income tax savings* is further divided between non-income producing SECS (e.g., residential, local ownership, not a for-profit firm) and income-producing SECS (e.g., utility power, PPA power production, industry, for-profit use). Non-income producing SECS can roll the interest payment on a solar loan or mortgage into the income tax savings.

[7] L. Bony, S. Doig, C. Hart, E. Maurer, and S. Newman. Achieving low-cost solar *PV*: Industry workshop recommendations for near-term balance of system cost reductions. Technical report, Rocky Mountain Institute, Snowmass, CO, September 2010.

Each of these costs (or savings) can be represented in columns across a standard spreadsheet, beginning at Year 0 (when the system is installed). Then each additional year each of the columns can be evaluated as a new row. This is the discrete mathematical method of calculating the Life Cycle Savings for a SECS. The summation of annual fuel savings and costs across the columns for each year will provide the annualized solar savings, and by summing the annualized solar savings each year (already in terms of present worth), we arrive at the *cumulative solar savings*: equivalent to the *Life Cycle Savings*. To repeat: the LCS of a SECS in comparison with a conventional Fuel Energy Conversion System (or access to electricity) is expressed as the difference between *avoided fuel costs* and *incremental increases in expenses* incurred due to the additional investment for the SECS.

A template for tabulated analysis can be found in the accompanying website for the text.

DISCRETE ANALYSIS

When tabulating results for a discrete analysis over the years of evaluation, each of these contributions to *SS* can be input as a separate column of data in a spreadsheet. The data will increment from the present ($n = 0$) to the end of the time horizon in annual steps. From simple analysis of contributions, there will be both positive cash flows (from fuel savings and incentives) as well as negative cash flows (from variable costs of operation) in each year, n. The sum of these cash flows in a given row on the spreadsheet must then be discounted according to the Present Worth Factor ($PWF(n, 0, d)$) for year n and the corresponding market discount rate d to determine the annualized costs or savings. We sum the present worth of each year to arrive at the *cumulative solar savings*. Finally, the *cumulative solar savings* in terms of present worth is defined as the *life cycle savings* (LCS).

We can see this tabulation in Table 10.3, and the accompanying loan tabulation is listed in Table 10.2. In this example, the client intends to make an investment in a SECS at \$11,000 in fixed costs. The *investment* includes a down payment (DP) of −\$2750, and a loan of \$8250, extended over 10 years at 4% annualized interest. We represent the *Present Worth Factor* of the loan payments as $PWF(10, 0, 0.04)$.

Year	Loan Payment	Interest Paid	Principle Paid	Balance	Periodic Payment *PWF* (10, 0, 0.04)
0				\$8250.00	8.111
1	−\$1017.15	\$330.00	−\$687.15	\$7562.85	
2	−\$1017.15	\$302.51	−\$714.64	\$6848.21	
3	−\$1017.15	\$273.93	−\$743.22	\$6104.99	
4	−\$1017.15	\$244.20	−\$772.95	\$5332.04	
5	−\$1017.15	\$213.28	−\$803.87	\$4528.17	
6	−\$1017.15	\$181.13	−\$836.02	\$3692.15	
7	−\$1017.15	\$147.69	−\$869.46	\$2822.68	
8	−\$1017.15	\$112.91	−\$904.24	\$1918.44	
9	−\$1017.15	\$76.74	−\$940.41	\$978.03	
10	−\$1017.15	\$39.12	−\$978.03	\$0.00	

Table 10.2: A synthetic table of a 10-year loan to demonstrate annualized payments with interest and balance reductions. The loan amount is \$8250, with Periodic Payment *PWF*(10, 0, 0.04) of 8.111.

By dividing the loan total by the $PWF(n,i,d)$, seen in Eq. (10.8). It should be noted that for loan or mortgage payments, we have no inflation ($i=0$) and the interest on the loan is equivalent to the discount rate ($d=0.04$). You may find that placing the interest rate for a loan in the position of the discount rate is confusing. But consider that these are costs to the client, not earnings. You are estimating payments into the future for the client, which requires a discount rate. The client is not actually collecting interest on their own payments, and the client's obligation to pay the loan is not inflating each year (rate $i=0$).

$$\begin{aligned} PWF(10,0,0.04) &= \frac{1}{d-i}\left[1-\left(\frac{1+i}{1+d}\right)^n\right] \\ &= \frac{1}{0.04}\left[1-\left(\frac{1}{1.04}\right)^1 0\right] = 8.111, \end{aligned} \tag{10.9}$$

$$\text{Loan Payment} = \frac{-\$8250}{8.111} = -\$1017.15, \tag{10.10}$$

$$\text{Interest Paid}_1 = \$8250.00 \cdot 0.04 = \$330.00 \text{ at Year 1.} \tag{10.11}$$

We see that the annual loan payment is a negative value, because it is a cost to the client. Furthermore, there is a 4% interest to be collected each year on the Balance. At the end of one year (from present to Year 1), the balance is \$8250, and upon receiving the first annual payment, the interest collected by the bank or credit union is \$330. This leaves a remainder of \$7562.85 on the balance for the second period of one year (from Year 1 to Year 2).

$$\text{Interest Paid}_2 = \$7562.85 \cdot 0.04 = \$302.51 \text{ at Year 1.} \tag{10.12}$$

And so the second annual payment has a slightly smaller interest amount collected because the balance has been reduced slightly. As we observe in Table 10.2, the initial years of payment on a loan are small and then they grow quickly near the end of the loan period, with a parabolic curve shape.

Now, if we look at Table 10.3, we see that the down payment (DP) of −\$2750 on the described project occurred in the present *at Year 0*. We also see

Table 10.3: A synthetic table of *FS* and C_S to demonstrate annualized cash flow and cumulative solar savings (*SS*). The down payment is −\$2750, with a 10-year loan of −\$8250 at 4% annualized interest (*PWF*(10, 0, 0.04)).

Year	*FS* (\$) (+1.05 per Year)	Loan Costs (\$) (fixed)	Operating Costs (\$) (+1.04 per Year)	Property Taxes (\$) (+1.04 per Year)	Income Tax Savings (\$)	*SS* (\$)	*PW* of *SS* (\$) (Use *PWF* for Each Year)	Cumulative *SS* (\$)
0						−\$2750	−\$2750	
1	\$1658.57	−\$1017.15	−\$120.00	−\$300.00	\$90.00	\$311.42	\$288.35	−\$2461.65
2	\$1741.50	−\$1017.15	−\$124.80	−\$312.00	\$192.60	\$480.15	\$411.65	−\$2050.00
3	\$1828.58	−\$1017.15	−\$129.79	−\$324.48	\$188.10	\$545.25	\$432.84	−\$1617.16
4	\$1920.00	−\$1017.15	−\$134.98	−\$337.46	\$183.42	\$613.83	\$451.18	−\$1165.98
5	\$2016.00	−\$1017.15	−\$140.38	−\$350.96	\$178.55	\$686.06	\$466.92	−\$699.06
6	\$2116.80	−\$1017.15	−\$146.00	−\$365.00	\$173.48	\$762.14	\$480.28	−\$218.78
7	\$2222.64	−\$1017.15	−\$151.84	−\$379.60	\$168.22	\$842.28	\$491.46	\$272.68
8	\$2333.78	−\$1017.15	−\$157.91	−\$394.78	\$162.74	\$926.67	\$500.65	\$773.34
9	\$2450.47	−\$1017.15	−\$164.23	−\$410.57	\$157.04	\$1015.56	\$508.03	\$1281.37
10	\$2572.99	−\$1017.15	−\$170.80	−\$426.99	\$151.12	\$1109.17	\$513.76	\$1795.13
10						\$4400.00	\$1887.08	\$3682.21

that the DP is a *cost* (a negative cash flow for the client) to the annualized Solar Savings (*SS*), and that it has not been discounted. As we only discount *Future Values*, and the DP occurred in the present ($n=0$), there is no devaluation of the present worth.

In Year 1 we collect the costs and savings of the various columns, representing the variable costs/savings for the year. There is an initial (Year 1) *Fuel Savings* (*FS*) of +$1658.57 based on the performance of the SECS, the solar resource in the given locale, and the cost of fuel/electricity at the given locale. These are the *avoided fuel costs* which will offer the largest annual returns for our example. Each year following Year 1 is estimated to have a fuel inflation rate of $i=0.05$, which means we multiply the preceding row by 1.05. The loan costs are fixed annual payments with no inflation, as we described earlier. The costs of operation (maintenance, insurance, and parasitic costs) were estimated at −$120 for the first year, with a general inflation rate on goods and services of $i=0.04$. Notice how this rate can be different than the fuel inflation rate. The incremental increase in property taxes of −$300 has been estimated to also have an inflation rate of 4% in this simplified example.

In the US there are savings allotted which have a return to the client based on reduced taxable income or a savings on income tax linked to the costs of installing a SECS. In Table 10.3, we apply a income tax savings of 0.30 to both the interest paid on the loan each year and to the property taxes paid on the system. This results in a positive cash flow (savings) of +$90 in Year 1.

One can review the detailed incentives for renewable energy at the Federal and State level within the *DSIRE* website: http://www.dsireusa.org/. For example, in 2005, the *Energy Policy Act* established a tax credit of 30% for systems purchased and installed as residential photovoltaic and solar water heating (at the time, up to $2000). In October 2008, the *Energy Improvement and Extension Act* extended the tax credits until December 31, 2016, while the *American Recovery and Reinvestment Act* of 2009 removed the maximum credit amount for technologies placed in service after December 31, 2008. Hence, all qualified residential SECS installations received a Residential Renewable Energy Tax Credit of 30%, applied in the year following the installation (Year 1).

DSIRE is the "Database of State Incentives for Renewables and Efficiency," maintained by North Carolina State University, in partnership with the Interstate Renewables Energy Council, the National Renewable Energy Laboratory, and the USA Department of Energy.

We observe this credit in the cash flows for Year 1 when using the open energy simulation software SAM from NREL.[8]

[8] P. Gilman and A. Dobos. System Advisor Model, SAM 2011.12.2: General description. NREL Report No. TP-6A20-53437, National Renewable Energy Laboratory, Golden, CO, 2012. 18 pp; and System Advisor Model Version 2012.5.11 (SAM 2012.5.11). URL https://sam.nrel.gov/content/downloads. Accessed November 2, 2012.

Again, we sum the columns of the table to the column of solar savings (*SS*) for each year, and then place the annualized *SS* in terms of present worth. In this case, the *market discount rate* that we are estimating is going to be assumed to be 8%. This may be high for real markets, given the relatively recent global financial recession of the Earth 2000s, but it will suffice for this example. Present Worth Factor for the Solar Savings is $PWF(n, 0, 0.08)$, applied to each year n. We have already calculated the costs/savings given various rates of inflation (i) in each year n, which take care of the numerator of the present worth relation. Hence, in order to bring the Future Values of the costs and savings into Present Value, we apply Eq. (10.13) to the *SS*.

$$PW_n(d) = \left(\frac{1}{d}\right)^n = \left(\frac{1}{0.08}\right)^n. \tag{10.13}$$

The results for this example have been tabulated in Table 10.4.

Now, imagine if we increased the market discount rate. An increase in a hypothetical discount rate would mean that the present value of the solar savings would also decrease, and as such the *cumulative solar savings* (which is synonymous with the LCS) would decrease. If the market discount rate were increased sufficiently high then, the LCS for the period of evaluation would eventually reach *zero*. This critical rate of return for the evaluation period is termed the *internal rate of return*, or IRR. Comparative analysis of projects would suggest that projects falling below the IRR have a net present worth that is greater than a non-solar alternative, while projects above the IRR have a net present worth that is less than the fuel alternative. If the market discount rate is a separate cell in a spreadsheet, linked to the present worth factors and the resulting cumulative solar savings at present worth, then the design team can vary the market discount rate by trial and error to find the IRR for the period of evaluation.

In some texts, the IRR is equated to the ROI, or return on investment—we do not advocate this approach.

Year	Present Worth for Savings $(n, 0, 0.08)$
1	0.926
2	0.857
3	0.794
4	0.735
5	0.681
6	0.630
7	0.583
8	0.540
9	0.500
10	0.463

Table 10.4: A table of present worth factors (*PWF*) to be applied to the annualized Solar Savings.

GOTTA GOTTA PAYBACK

The concept of a *payback time* is used extensively in business. Yet we shall observe that there are several points on the time horizon at which we achieve different stages of *payback*. In Figure 10.1, we see five critical points that all indicate a change in the cash flow scenario for the evaluation period n_e. Given the tendency of solar project finance analysis to occur over periods longer than a decade, we assume that monetary discounting is considered. One common use of the payback time is the period for which the discounted *cumulative fuel savings* is equivalent to the initial *investment* (loan + down payment). However, we also show that the payback where the cumulative solar savings reaches zero (point C, in Figure 10.1) will occur within a similar period of time.

Feel free to play James Brown at this point on your local media source.

In the example presented, the *cumulative fuel savings* have been tabulated as well. We can review the process in Table 10.5, and observe that the fuel savings will pay back the investment as of Year 8 (\$11,115 > \$11,000 investment). In comparison, we observe in Table 10.3 that the *cumulative solar savings* for this synthetic case will cross from negative to positive cash flow as of Year 7.

Absorbed solar energy from G_t is termed *S*.

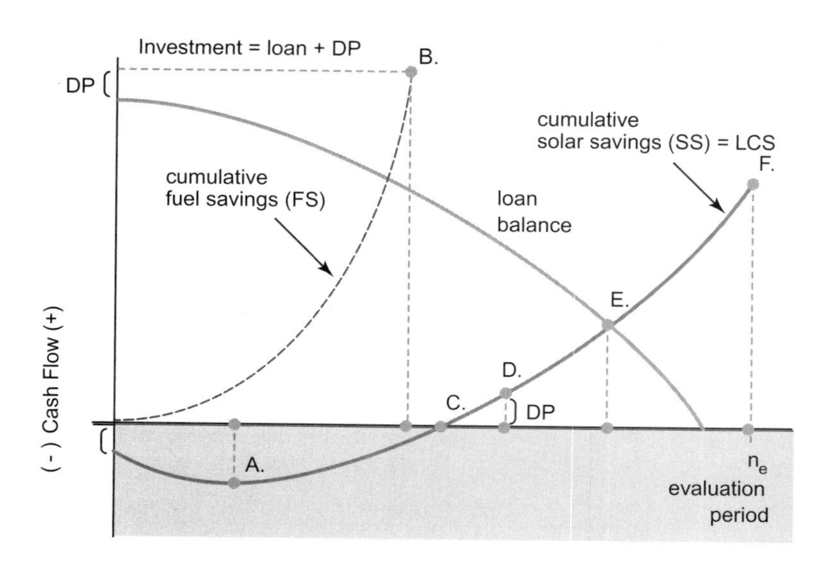

Figure 10.1: For a given evaluation period n_e, there are numerous critical points in time that communicate concepts of "payback" for the client. The following shows discounted payback and savings. Adapted from Duffie and Beckman (2006).

Table 10.5: A table of cumulative *FS* for the example system.

Year	Fuel Savings (*FS*)	*PWF* for Savings $(n, 0, 0.08)$	Discounted *FS*	Cumulative *FS*
0	\$0	1	\$0	\$0
1	\$1658.57	0.926	\$1535.71	\$1535.71
2	\$1741.50	0.857	\$1493.06	\$3028.77
3	\$1828.58	0.794	\$1451.58	\$4480.35
4	\$1920.00	0.735	\$1411.26	\$5891.61
5	\$2016.00	0.681	\$1372.06	\$7263.67
6	\$2116.80	0.630	\$1333.95	\$8597.62
7	\$2222.64	0.583	\$1296.89	\$9894.51
8	\$2333.78	0.540	\$1260.87	\$11155.37
9	\$2450.47	0.500	\$1225.84	\$12381.22
10	\$2572.99	0.463	\$1191.79	\$13573.01

GAINS, LOADS LOSSES

The two crucial concepts in SECS project analysis are *gains* and *loads*. *Solar gains* (or just *gains*) are quantities of *useful energy* that vary over the day, provided from a SECS. We acknowledge that solar energy has potential *usefulness* for energy conversion only if it is absorbed. We term the absorbed irradiance in a SECS as S, where $G_t \rightarrow S$ according to the performance of the SECS.

Keep in mind from our earlier chapters that *useful* energy (*work*) from a system can have a broad definition, as the needs of the client might go beyond *heat* and *power*. Traditionally, "gains" and "losses" were conceived to be purely thermal in nature, tied to design of the built environment. However, there is no reason for solar-derived electricity not to be considered a gain as well. Imagine that every energy system has an energetic threshold, a level above which energy flows to the client to be useful in a prescribed form of energy (thermal, electrical, radiant, etc.). As seen in the cartoon for solar energy gains in Figure 10.2, below that threshold in the energy system, all energy quantities are termed *losses*, because they do not serve the prescribed form of energy in demand by the client. In solar energy conversion systems that threshold has been termed the *critical radiation level* ($G_{t,c}$), where the absorbed irradiance by the system is equal to the losses of light and thermal power.[9] We will describe the critical radiation level in more detail in a subsequent chapter on optocaloric conversion systems. Every energy conversion system has a critical level of energy input flow, below which the system sputters to a stop.

[9] John A. Duffie and William A. Beckman. *Solar Engineering of Thermal Processes*. John Wiley & Sons, Inc., 3rd edition, 2006.

When we use the converted energy for a characteristic application like electricity for making coffee or cooling air, or hot water for a shower, the quantities of use are called *loads* (L). Process loads represent energy needs that

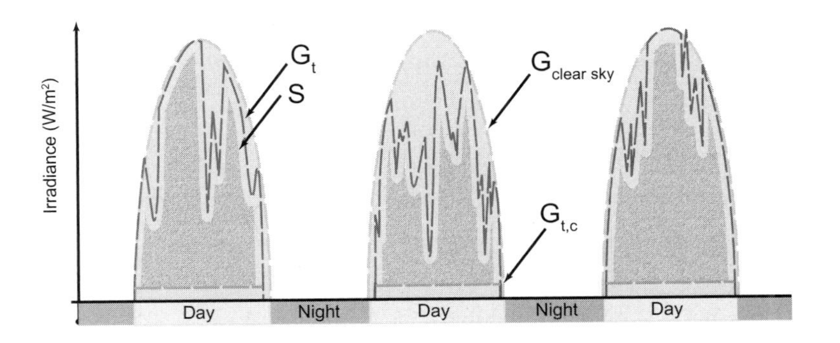

Figure 10.2: Solar Gains: encompass the absorbed irradiance S and critical radiation levels for useful work $G_{t,c}$ and marked in green. Also shown are global tilted irradiance G_t and estimated clear sky irradiance $G_{clear\ sky}$. Adapted from Duffie and Beckman (2006).

It may seem unusual to the reader to classify traditionally purchased energy as "auxiliary." This is a tradition of the solar field—perhaps in the near future it won't seem unusual any more.

vary throughout the days and months of a year, met by a combined energy systems strategy. Loads are composed of two parts: useful energy for work and energetic losses. Load estimation must account for the sensible heat requirements of the client, the thermal losses that occur from the distribution system (e.g., hot fluids losing energy when passing through piping), and the additional losses that occur during storage (e.g., thermal losses that occur within storage tanks). Keep in mind that losses in solar energy conversion systems are both *thermal* and *optical*, as some of the light incident upon a system is not absorbed due to transmission and reflection losses.

$$L = Q_{work} + Q_{(thermal\ loss)} + (Q_{(optical\ loss)}), \tag{10.14}$$

where Q refers to traditional heat exchange, or energy transfer from thermodynamics. We placed the optical losses in parentheses in Eq. (10.14) because the optical losses will also be considered in the evaluation of absorbed irradiance (S) for a cover-absorber system in subsequent chapters.

Total loads can be evaluated as rates of energy transfer ($\dot{L}$) or as integrated quantities (L), while the portion of a load that is supplied by solar energy is $\dot{L}_S$ and the remaining load that is supplied by an auxiliary supply of purchased fuel (natural gas, fuel oil, electricity derived from geofuel) is $\dot{L}_A$.

$$\dot{L} = \dot{L}_S + \dot{L}_A, \tag{10.15}$$

where loads are represented as time-dependent quantities (units of W).

SOLAR FRACTION

We have discussed Loads (L) and Costs (C), and now we move on to the metric of the annual solar fraction (F), and the monthly solar fraction (f). One of the parameters that we left out in the hybridization of energy systems was the total fraction of energy savings in an interval of evaluation that would be attributed to the installed SECS. In solar hot water cases, we can imagine that a system which meets all energy needs in the month of January for a residence in Buffalo, New York, but is then over-sized for the hot water *loads* in the NY summers. In such a case, there would be no need for auxiliary fuel to heat the hot water tank all year long, but the over-sized system would not be cost effective. It would be a large investment used to simply dump heat back out into the roof area during the summer months. In contrast, we might imagine a system that meets 100% ($f=1$) of the June or July hot water loads, but meets 30% of the hot water loads in January. Perhaps the operating costs of auxiliary fuel each year in exchange for lower investment costs are in fact of greater annual utility to the client than a super-sized SECS (or no SECS at all).

$$F = \frac{L_S}{L} = \frac{L - L_A}{L} = \frac{\sum f_i L_i}{\sum L_i}, \tag{10.16}$$

$$f_{sav,i} = \frac{C_i - C_{A,i}}{C_i}. \tag{10.17}$$

The solar fraction can be used as the proportionality factor in project finance to estimate an optimal size for a client in a given locale. This is in part because a large solar fraction will entail more *units* of modules and balance of systems, increasing the total systems investment according to the *unit cost*. There are two extremes to be aware of: a solar fraction of *zero*, where the client opts for no installation of a new SECS; and a solar fraction of *one*, where the SECS covers all energy loads for the entire year. $F=0$ will entail the client to the highest fuel/energy costs (FC) of any system alternative solar system (trivial case). $F=1$ will have the highest solar costs (C_S) and the lowest annual energy costs (second trivial case). Between the two extremes we hope to find a maximum return on investment, and a return that will also be net positive in cumulative solar savings.

CONTINUOUS ANALYSIS

From another perspective, one could also use basic calculus to assess the economic performance of a system, or to find a financial optimum for a SECS in a given locale. Instead of summing across fields of annualized cash flow (discrete discounting per year), and subsequently converting those values into present worth, we will simply determine the present worth for each column given the end of evaluation time horizon (continuous discounting for the entire period). Then, the sum of the present worth fields will deliver the end of evaluation net present value. As such, the LCS for a SECS relative to a conventional system will be represented mathematically as the difference between the avoided fuel costs for the system and the increased costs attributed to the marginal additional investment required for the SECS. This LCS is provided in Eq. (10.18).

$$\mathrm{LCS} = P_1 \cdot C_{F1} \cdot F \cdot L - P_2 \cdot C_s, \tag{10.18}$$

where P_1 and P_2 are bundled parameters to represent the continuous, time-dependent factors that affect the costs associated with traditional fuels being used in the current system ($C_{F1} \cdot F \cdot L$), and factors that reflect the costs for the SECS (C_s).

P_1: The ratio of fuel cost savings for the life cycle, relative to first-year fuel savings.

P_2: The ratio of additional expenditures from the SECS investment to the initial investment, considered within the period of the life cycle. A multi-parameter factor.

P_1 represents a simple present worth factor calculation for the inflation rate of fuels (i_F), given either an effective federal income tax rate t_e for commercial systems ($X=1$) or no available income tax rate for non-commercial systems ($X=0$). Again, the period of evaluation is n_e, with a discount rate of d.

$$P_1 = (1 - X \cdot t_e) \cdot PWF(n_e, i_F, d). \tag{10.19}$$

P_2, the continuous, time-dependent factor affecting the costs for the SECS, is actually divided into seven parts (the same basic parts that make up the columns for the discrete analyses). The parts include the down payment D, the LCC for the mortgage/loan and associated interest, any income tax deductions from that interest, operation and maintenance plus insurance and parasitic costs, the net property tax costs, a straight line depreciation tax deduction, and the present worth for the SECS at year n_e.[10]

[10] John A. Duffie and William A. Beckman. *Solar Engineering of Thermal Processes.* John Wiley & Sons, Inc., 3rd edition, 2006; and Soteris A. Kalogirou. *Solar Energy Engineering: Processes and Systems.* Academic Press, 2011.

$$P_2 = P_{2,1} + P_{2,1} + P_{2,3} - P_{2,3} + P_{2,4} + P_{2,5} - P_{2,6} - P_{2,7}, \quad (10.20)$$

1. $P_{2,1} = D$,
2. $P_{2,2} = (1-D)\frac{PWF(n_{min},0,d)}{n_L,0,d_m}$,
3. $P_{2,3} = (1-D) \cdot t_e \left[PWF(n_{min}, d_m, d)\left(d_m - \frac{1}{PWF(n_L,0,d_m)}\right) + \frac{PWF(n_{min},0,d)}{n_L,0,d_m}\right]$,
4. $P_{2,4} = (1 - X \cdot t_e) \cdot M_1 \cdot PWF(n_e, i, d)$,
5. $P_{2,5} = t_p \cdot (1 - t_e) \cdot V_1 \cdot PWF(n_e, i, d)$,
6. $P_{2,6} = \frac{X \cdot t_e}{n_d} \cdot PWF(n_{min}, 0, d)$,
7. $P_{2,6} = \frac{R}{(1+d)^{n_e}}$,

where

- n_e: The years of economic analysis.
- n_{min}: The years in which deductions for depreciation contribute to the analysis.
- D: Ratio of down payment to the initial investment.
- M_1: The ratio of costs associated with maintenance, insurance, and operation in the first year relative to the initial investment.
- V_1: The ratio of the SECS value in the first year relative to the initial investment.
- t_p: The property tax for the SECS, based upon an assessed value.
- R: The ratio of the resale value estimated for the end of SECS life relative to the initial investment.

From the basis of continuous analysis, one may then find the optimal size of a SECS in accord with the maximum LCS relative to a proposed collector size. The optimum can be found from the partial derivative of the LCS with respect to the partial derivative of the system size, or from the marginal change in the annual solar fraction relative to the marginal change in the system size. Furthermore, an analyst could use the continuous P_1, P_2 method to assess the sensitivity of that maximum with respect to changes in other system parameters.[11] Sensitivity analyses in financial and performance assessment are available within the System Advisor Model software from the US DoE National Renewable Energy Laboratory.[12]

[11] John A. Duffie and William A. Beckman. *Solar Engineering of Thermal Processes.* John Wiley & Sons, Inc., 3rd edition, 2006.

[12] System Advisor Model Version 2012.5.11 (SAM 2012.5.11). URL https://sam.nrel.gov/content/downloads. Accessed November 2, 2012.

Concept Review:

- *Solar Fraction* monthly or annual evaluation of solar Gains relative to total Demand (Loads + Losses).
- *Gains* Absorbed solar energy and Useful solar energy to do work required.
- *Losses* Energy demanded from the system, but rejected as being not useful to do work required.
- *Loads* Useful energy demanded from the system, used to do work required.
- *Life Cycle* period of n_e years of evaluation for a given energy system under study. Typical life cycles encompass the period of years for a mortgage or loan (n_L).
- *Solar Savings (SS)* Sum of cash flows in a given year: includes fuel savings and income tax savings (positive cash flow) with Costs for mortgage payments, maintenance, insurance, parasitic energy requirements, and extra property taxes (negative cash flow).
- *Life Cycle Savings (LCS)* the cumulative solar savings (*SS*), valued in todays dollars (total *PW*), for the period of n_e years.
- *Present Worth (PW)* value of cash flows in today's dollars.
- *SECS* solar energy conversion system.

PROBLEMS

1. What is the meaning of the annual solar fraction (F), and how can it be used to optimize the solar utility for a client in a given locale?

2. The initial cost of a solar energy system is $11,500. Given the initial investment is paid with a 20% down payment and the balance is borrowed at 5.4% interest for 8 years, calculate what the annual payments and interest charges would be. The market discount rate is assumed to be 6%. Beyond this initial tabulation, estimate the present worth of the annual payments.

3. Calculate the fuel costs for a non-solar energy system. Evaluate over a period of 8 years, given the total annual Load has been estimated to be 184 GJ and the unit cost of fuel is $13.50/GJ. The market discount rate is assumed to be 6%, while the inflation rate of fuel is assumed to be 3% per year.

4. Tabulate (1) the annual cash flows, (2) the life cycle savings, and (3) the IRR for a SECS with an initial investment cost of $9000 for a 16-year analysis, using a 20% down payment and the remainder under loan at an interest rate of 5.75%. The fuel savings *FS* in year 1 is $1430. The market discount rate is assumed to be 6%, the fuel inflation rate is assumed to be 6%, while costs associated with both maintenance, insurance, and parasitic losses will inflate at a rate of 1% annually. Finally, the resale value at the end of the period of evaluation is assumed to be 40% of the initial cost.

RECOMMENDED ADDITIONAL READING

- Solar Engineering of Thermal Processes.[13]
- Fundamentals of Solar Energy Conversion.[14]
- Achieving Low-Cost Solar *PV*: Industry Workshop Recommendations for Near-Term Balance of System Cost Reductions.[15]

[13] John A. Duffie and William A. Beckman. *Solar Engineering of Thermal Processes*. John Wiley & Sons, Inc., 3rd edition, 2006.

[14] Edward E. Anderson. *Fundamentals of Solar Energy Conversion*. Addison-Wesley Series in Mechanics and Thermodynamics. Addison-Wesley, 1983.

[15] L. Bony, S. Doig, C. Hart, E. Maurer, and S. Newman. Achieving low-cost solar *PV*: Industry workshop recommendations for near-term balance of system cost reductions. Technical report, Rocky Mountain Institute, Snowmass, CO, September 2010.

THE SUN AS COMMONS

11

Economics evolved in a time when labor and capital were the most common limiting factors to production. Therefore, most economic production functions keep track only of these two factors (and sometimes technology). As the economy grows relative to the ecosystem, however, and the limiting factors shift to clean water, clean air, dump space, and acceptable forms of energy and raw materials, the traditional focus on only capital and labor becomes increasingly unhelpful.

Donella Meadows, Thinking in Systems: A Primer (2011)

...neither the state nor the market is uniformly successful in enabling individuals to sustain long-term, productive use of natural resource systems. Further, communities of individuals have relied on institutions resembling neither the state nor the market to govern some resource systems with reasonable degrees of success over long periods of time.

Elinor Ostrom, Governing the Commons: The Evolution of Institutions for Collective Action (1990)

An ill-informed observer might suspect that "sustainability" has little meaning in the business world. Check the Dow Jones Sustainability Index or the Goldman Sachs SUSTAIN focus for the changing culture in the business world. Sustainable practices tend to outperform the industry average in the market.

WE HAVE ESTABLISHED that the Sun has value to individuals and to society. In the consumer society, one might assume that a limiting factor in solar energy proliferation is then financial in nature—as profits to the supplier or as return on investment to the stakeholders. However, for solar energy conversion, the full value for the client in their locale may extend far beyond a specific fiscal reasoning. There may be additional constraining factors within the broader society of the client, in terms of simple access to high-quality energy forms like cooking heat, or clean indoor air and access to clean water, even providing light in a dark indoor space, or light at night to read and study. Clients can also hold strong convictions for energy independence, either on the international scale or on the local level. Clients and communities can choose to adopt solar for ethical reasons, for health and safety, and for improved ecosystems services overall tied to sustainability. Then again at the broader level of the firm, sustainable practices have been found to outperform the industry averages during financial slowdowns too.[1]

And yet the Sun is a bit different than oil or coal as a resource for energy, isn't it? For one thing, light is not actually scarce on the global scale. Using 2005 numbers for energy demand, more solar energy falls upon half of the Earth in an hour (4.3×10^{20} J, the other half of Earth is experiencing night) than the entire world population consumed in that year (4.1×10^{20} J).[2] As of 2012, the world energy consumption climbed to 5.7×10^{20} J, and by 2020 demand is expected to rise to 6.5×10^{20} J, so by 2020 we will need to adjust this bold statement to *an hour and a half*.[3] There are diurnal phenomena (days and night), but for the next few billion years we have a very high certainty that the Sun and its photons will still "be there" for all life to access. In fact, you can put solar down as one of the most conservative energy strategies available.

We sometimes design our buildings without knowledge of the Sun, and in doing so we "artificially" make light scarce by enclosing spaces without daylight access, or orienting building that will respond well to solar gains in inappropriate ways. As we have noted before, a revolution in solar home design (to allow daylight into the building space) has included My Shelter Foundation's "Isang Litrong Liwanag" project that has spread from South America to the Philippines to Africa, Central America, Southeast Asia, and India. Also in Germany, every worker must have

[1] D. Mahler, J. Barker, L. Belsand, and O. Schulz. "Green Winners": The performance of sustainability-focused companies during the financial crisis. Technical report, A. T. Kearny, 2009. URL http://www.atkearney.com/paper/-/asset_publisher/dVxv4Hz2h8bS/content/green-winners/10192. Accessed March 2, 2013.

[2] Nathan S. Lewis, George Crabtree, Arthur J. Nozick, Michael R. Wasielewski, Paul Alivasatos, Harrient Kung, Jeffrey Tsao, Elaine Chandler, Wladek Walukiewicz, Mark Spitler, Randy Ellingson, Ralph Overend, Jeffrey Mazer, Mary Gress, and James Horwitz. Research needs for solar energy utilization: Report on the basic energy sciences workshop on solar energy utilization. Technical report, US Department of Energy, April 18–21, 2005. URL http://science.energy.gov/~/media/bes/pdf/reports/files/seu_rpt.pdf.

[3] Data collected from the US Dept. of Energy's Energy Information Administration (see EIA Analysis).

access to some daylight in their work space by law. This has affected the manner in which large-scale commercial buildings are designed—seen in the *Commerzbank Tower* in Frankfurt, Germany. The tower has a central atrium permitting central daylight access and nine levels of sky-gardens, where plants are cultivated.

From another perspective, the energy budget received from the Sun is affected by conditions of latitude, meteorology, topography, and local surroundings from built structures or trees. All of these influential factors are locally derived, or dependent upon the client's *locale*. Hence solar irradiation can be seen as both globally distributed and locally specific. Light is also transient—a flow of energy, and is not stored as a stock. We do not collect, package, and ship photons as energy from Wyoming to Idaho. We use them for work right where they encounter our collection systems.

Consider that we can transport photons as information in a fiber optic cable, but we do not move photons as energy in this manner. Perhaps light as information is valued higher than light as energy?

Potential: A driving force to flow.

Stock: Stored potential, or a resource system that replenishes itself.

Flow: Change in mass, energy, or information with respect to time.

A central concept in the text thus far has been that by improving our knowledge of the Sun and its relation to society and the surrounding ecosystems services. We also hold a strong sustainability ethic for our approach to developing solar as appropriate and environmental technologies, linked with society and the environment. We continue to demonstrate that solar energy collector design requires a systems approach, calling upon a transdisciplinary team of contributors to design and deploy projects. We have also advocated the *integrative design* process to break through barriers, creating an effective communication plan and design strategy to deploy a SECS that has high solar utility for the client or stakeholders in a given locale. Compelling cases can be developed for adoption by the client or stakeholders by being more knowledgeable about the solar resource and its applications. Now we turn to thinking of the Sun and the mineral resources used to make SECSs—either as a *commons* or a *public good*; a resource like water and air that is shared by all.

A **commons** is a natural resource that is accessible to all members in a society.

GOODS AND VALUES

The Sun is both an energy source and a *commons* for society, which is why we find solar energy conversion so interesting. Certain technologies and supporting materials that are scarce in society (PV, SHW panels, batteries, etc.), but the photons arriving to Earth from the Sun are a common resource that almost all life forms use directly every day. Keep in mind that the Sun's light (mainly shortwave irradiation) warms the surface and atmosphere of the planet to a sufficient temperature for H_2O to exist in liquid form, to evaporate water (driving the water cycle), and for endothermic biochemical reactions such as photosynthesis to proceed. Our bodies even make essential vitamin D from our own cholesterol using solar UV light. In essence, the work current or flow of energy from the Sun creates one of our most important baselines for energy conversion on the planet. We cannot take away the Sun from society and the environment, and the light from the Sun is not scarce.

Vitamin D is the sunshine vitamin! Vitamin D enriched foods are actually exposed to UV lamps to convert 7-dehydrocholesterol to cholecalciferol (vitamin D_3).

The *value* of an unconverted photon is a variable quantity, and there are no "mineral rights" on the flow of light.

Consider that *markets* allocate scarce resources via the collective forces of supply and demand. Consider that *governments* can ration resources as well by deciding who can use the resource, when, and to what degree through permitting and penalties. What happens if a resource is common to all individuals, but is regionally distributed, and how does society allocate and manage a common resource that is not necessarily scarce? Existing strategies suggest that managing a common resource can be performed in ways alternative to an otherwise unsatisfying binary selection from extremes of rational markets or a central governing authority.[4] We would like to examine those strategies as they apply to the Sun and Solar Energy Conversion Systems.

Again, markets are based upon an assumption of people coming together with shared information for the exchange of goods and services, behaving as rational producers and consumers of those goods. As we have demonstrated in our earlier study of economics, markets can and do fail in identifiable areas related to weather, climate, and energy systems. Under real conditions people are not rational, they often try to obscure or hide information, and they are exposed to real environmental perturbations or crises of weather and climate—the latter two both create information asymmetries. A common failure is a gap in addressing the market for energy according to the externalities associated with energy supply and demand. Recall that in the presence of a *positive externality* the social value exceeds the private value, and policies can correct the market failure by subsidizing the positive externality. In the presence of a *negative externality* the social cost exceeds the private cost, and policies can correct the market failure by taxing the emissions of power generation.

[4] Elinor Ostrom. *Governing the Commons: The Evolution of Institutions for Collective Action.* Cambridge University Press, 1990.

As we have noted, centralized control strategies aligned with the state (government policies) can also be applied to manage resources. However, real governing bodies will also display failures in allocating resources. Resources need to be accurately monitored, reliably sanctioned, and allocated, when there are real limitations on the availability of accurate information, limited bandwidth in the governing institution to monitor the pool of resources appropriately, questionable reliability or permitting or sanctioning, and one must address the question of supporting the costs tied to administration of a centralized control and management.[5] Hence, there are also limitations in managing goods and services from the state. So how might we consider valuing and managing goods and services linked to solar energy conversion within a broader framework?

The State is a compelling option. Many ecologists of the past have argued that environmental problems can only be solved through action by the government. In her seminal work, *Governing the Commons*, Ostrom entitles this option *"Leviathan as the 'only' way,"* following the Hobbes text. Ostrom demonstrates how these pure government strategies also contain significant failures for allocating resources efficiently.

[5] Elinor Ostrom. *Governing the Commons: The Evolution of Institutions for Collective Action*. Cambridge University Press, 1990.

In Figure 11.1 four broad type goods are present, described by their relative states of *excludability* and *rivalry*. Although the terms may be new, you are probably familiar with examples of goods that can have restricted access if not paid for, and goods that only permit one person per unit to consume them. Goods and service with restricted access are called *excludable* goods. In contrast, good is non-excludable when it cannot be guaranteed restricted access from a lack of payment for access. For example, the air in the sky is non-excludable, as is water from a public drinking fountain.

Consider the scenario where the consumption of a good by one person prevents simultaneous consumption of the same good by another. These goods are called *rivalrous*, or *subtractable*. A good is rivalrous/subtractable when one

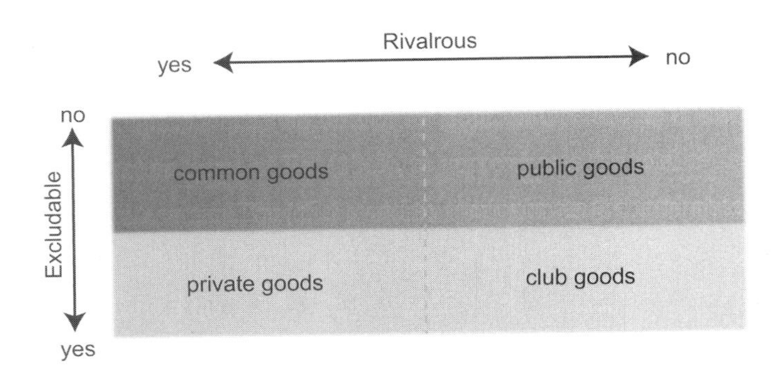

Figure 11.1: Typology of goods according to degree of excludability and rivalry.

Can you imaging trying to wear someone's pants while they are in them? Don't answer that question. A personal pair of pants is rivalrous.

person's use of that good can decrease access to the good for others intending to use it. Clothing is both highly excludable and rivalrous. In contrast, a public weather forecast available on the internet or by radio does not subtract from the availability of that forecast to others.

We can identify scenarios in which a person's use of the Sun's light (or the technologies applied to converting light to useful work) is found to be *rivalrous* or not. If we have a tall tree (which uses light for photosynthesis) that grows to significantly block our neighbor's PV system, we create a *rivalry* for the solar good. An electronic device like a photovoltaic module could be initially perceived as a private good (excludable and rivalrous), but it might also be purchased and managed by a community to be shared at a library. Under such a condition, the electronics would be considered a *common-pool resource*.

From our reading in Chapter 2, we observed that rivalry can prompt policy development and legal structures enabling access to the solar gains from daylighting, photovoltaics, solar heat gains in the winter that would reduce consumption of fuels.[6] For the purpose of examining the solar resource, and creative strategies for ecopreneurship and common-pool resource management, we will explore opportunities in all four quadrants of Figure 11.1.

[6] Sara C. Bronin. Solar rights. *Boston University Law Review*, 89(4):1217, October 2009. URL http://www.bu.edu/law/central/jd/organizations/journals/bulr/documents/BRONIN.pdf.

PRIVATE GOODS

First, let us examine the seemingly obvious social scenario for which a good is both rivalrous and excludable. This is called a *private good*, found in the lower left quadrant of Figure 11.1. Privates goods entail fairly straightforward project finance assessment, as well as personal loads to finance the large initial investment. The private good scenario is also likely to impact only clients with significant available liquid capital, in a scenario where the SECS is considered a luxury good. In essence a small fraction of the larger societal whole.

In the case of energy systems, while goods for energy conversion might be subtractable (e.g., a PV panel), the electricity provided is shared with the power grid. The natural gas and electricity networks in many nations have been connected as resource systems. They are both *stocks* and have representative *flows*. Both electrons and molecules of methane are flows of mass through a resource system (as a power grid or pipeline). Because of this embedded technological ecosystem in energy and power, there is a significantly higher cost to make a private good in energy truly independent from the system. At the same time, in regions that have no energy network the value in having any energy installed locally may outweigh the cost of acting independently.

We can imagine scenarios where a private homeowner wishes to install photovoltaic panels integrated into the structure of the home, to serve the goal of removing the home's electricity from the larger grid. We frame this example because this is often the case study on the mind of the beginning student. The panels in a private scenario can only be owned by the owner and in buying those panels, they cannot be used by another homeowner. However, the excess electricity cannot be acquired by the grid and re-purposed to serve other clients either. In such a case, as the electricity cannot be stored, a large battery backup will need to be installed to maintain this severance from the broader power and societal ecosystem or infrastructure. Also, due to the variable nature of the solar resource over the course of a year, the array of PV collectors will tend to be over-sized during the summer in order to meet the demand for electricity during the Winter months. In such a case, the excess energy would just be shed as waste heat, creating an unused profit stream to the client. The more likely scenario is a client where the exchange of electricity (in the case of PV) is met with the infrastructure of the grid, which is managed by the independent system operator. As such the electricity is part of a common-pool resource, as we shall explore shortly.

CLUB GOODS

Next we open up the framework where the Sun is non-rivalrous—the use from the one will not take away the ability to use the good by the many—and excludable—those who do not contribute (or pay) for the good can be excluded from having access to it. This is called a *club good*, and examples include movie theaters, copyrighted cloud access media, and private parks. From a solar perspective, one might envision a solar park, or a *solar garden* that is

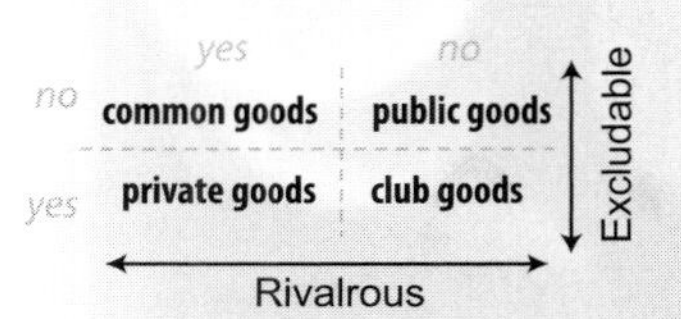

owned and accessed by a small community. A contractual distributed heating system which delivers district heat among the paying community would also be a club good.

COMMON-POOL RESOURCE:

A **common-pool resource** is an environmental or human-made system of resources large enough such that it would be a very expensive enterprise to exclude stakeholders from obtaining beneficial outcomes in its use.

Resource system: An environmental **stock** (could be a human-established institution). Remember that humans are an integral part of our environment.

Resource units: The rate of **flow** of a useful resource from a resource system.

A *Common-Pool Resource* (CPR) is a *non-excludable good* that is *rivalrous* (subtractable). Common-pool goods are some of the most interesting related to society and sustainability, and can be thoroughly explored for the present and future of the solar field. A CPR could be part of the surrounding environment or a human institution, and has a general requirement in size. The requirement for a CPR is that the *resource system* (e.g., the stock) be sufficiently large such that the process of excluding potential stakeholders from obtaining benefits in its use is essentially prohibitively costly (but not impossible). Recall the use of *stocks* and *flows* in describing a resource system. The stock refers to the size of the resource system, while the flow refers to the transfer of *resource units* from the system to the client. Stock and flows are interdependent.

In 2009, Elinor Ostrom was the first woman awarded the Nobel Prize in Economics. Dr. Ostrom was recognized for "her analysis of economic governance, especially the commons." Her efforts in *common-pool resources* helped describe how humans interact with ecosystems to properly maintain long-term sustainable resource yields. She was able to describe pathways of collective action arrangements by which CPRs are managed to the benefit of the local society. In effect, she described the ways in which humans interact with their supporting ecosystems in a manner that will maintain the long-term yields of a sustainable resource.[7]

Dr. Ostrom was a faculty at Indiana University, and passed away in June, 2012. For more information, on Elinor Ostrom see "The Sveriges Riksbank Prize in Economic Sciences in Memory of Alfred Nobel 2009." 17 June 2012 http://www.nobelprize.org/

Resource systems could include groundwater aquifers, fields for grazing, parking garages, cloud computing banks, the electrical power grids or interconnections, and in our case, the Sun. Resource systems (as stocks) and the flow of the resource in use shape a common-pool resource. A resource system has conditions for producing an upper limit of *flow* without disturbing or harming the constitution of the *stock*. In fact, this is how we define a *renewable*

[7] Elinor Ostrom. *Governing the Commons: The Evolution of Institutions for Collective Action*. Cambridge University Press, 1990.

resource: a resource where the average flow rate of conversion or withdrawal from the stock does not exceed the rate of resource replenishment.[8]

[8] Elinor Ostrom. *Governing the Commons: The Evolution of Institutions for Collective Action*. Cambridge University Press, 1990.

In contrast, the mineral resources within the Earth's crust that we use to manufacture our energy technologies *are* subtractable. It is also difficult and costly to exclude individuals from a sizable mineral resource, typically because it includes owning the overlying property (with the exception of the USA, which separates mineral rights from property rights). The terawatt scale and exponential growth of global energy demand, tied to the minerals associated with that demand, suggest that there are inherent crowding effects or unsustainable demand of the mineral stocks. Hence, mineral resources can also be explored as common pool goods for society. In order to assess the flows of mineral resources from the crustal stocks to various phases of application, we use the process of Life Cycle Assessment.

A **renewable resource** has a rate of withdrawal from the stock that does not exceed the rate of resource replenishment.

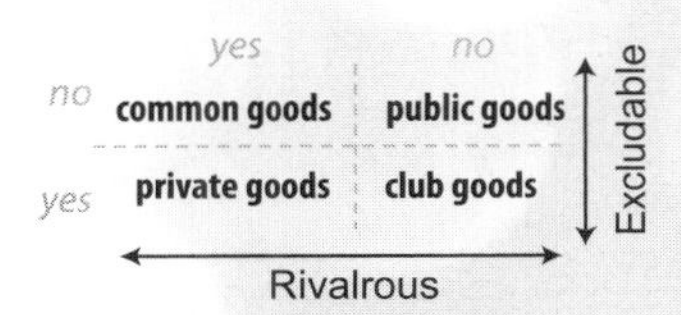

PUBLIC GOODS

We can also envision many scenarios where the use of the Sun is not rivalrous at all, just due to distance and geographical spread. The Sun is also non-excludable, in that we cannot restrict access to light. For the scenario of a *non-excludable good* that is *non-rivalrous*, we have something called a *Public Good*, shown in the upper right quadrant of Figure 11.1. The air around us, and in many open scenarios, the light from the Sun that we have access to is a Public Good.

Although we do not "renew" a photon at Earth's surface, the resource system or stock is the Sun, not the light incident upon the Earth. The Sun does replenish photons at a rate far exceeding our ability to convert it for SECSs, and so light is a renewable energy resource.

We are already familiar with the resource units from the Sun, they are photons, measured in units of irradiance (W/m^2). The Sun itself is our seemingly boundless resource system. Photons are created by both thermal and stimulated emission of plasma from hydrogen and helium within the photosphere, chromosphere, and corona, the visible layers of the Sun. The collection of photons for SECSs on Earth does not disturb the resource system of the Sun, and photons are replenished at a rate far exceeding our ability to collect them. Hence, solar energy in the form of light is a renewable energy resource, and a common-pool resource, and an energy resource that is not scarce.

On a global or regional scale, light from the Sun is therefore a public good. A solar power plant in Spain does not subtract from my own ability in the USA or France to use the Sun. It is fairly difficult to prevent someone from using the solar resource (without geo-engineering machines of doom).[9] We do not pay for access to the Sun on a regional scale, and others do not have the ability to restrict our general access to the Sun. This means that the light from the Sun as a resource is a *non-excludable good*.

[9] I recommend a viewing of *Bladerunner* and *The Matrix* for examples of blocking out the Sun by geo-engineering.

MANAGING THE COMMONS

In the late 1960s, Garrett Hardin posed the hypothetical case of the *Tragedy of the Commons*, where social and private incentives diverge and the resource is over-consumed.[10] While this text has been extremely influential in suggesting a disastrous result from an unmanaged common resource, the basis of the argument was ultimately flawed by (1) assuming a common resource pool has a uniform pattern globally, when essentially each challenge of the commons is bound to a local place and time horizon; (2) assuming the appropriators of the resource units (the flow) are prisoners of the scenario, uninterested in communicating or negotiating collectively; and (3) assuming the common resource is *unmanaged*.

[10] Garrett Hardin. The tragedy of the commons. *Science*, 162:1243–1248, 1968.

In solar energy, the common resource is only "common" to the locale of the client and stakeholders, within the time horizon of their use of solar energy for social activities. For photovoltaic and CSP deployment, the local power grid is another managed commons. Consider that the most informed participants (or appropriators of resource units) in a resource system will be the installer/design teams in the given locale, as they will know both the solar resource and the permitting and code structures for that locale. Hence, all solar design challenges are dependent upon the *locale* and the *stakeholders*. So neighborhoods, municipalities, and states tend to develop and manage the deployment of solar energy technologies for heat and power faster than federal governments.

Solar design and installation is integrative and *local*. Improving the process to manage SECS integration into the technological ecosystem of the locale is a **collective action** challenge.

"The key fact of life for coappropriators is that they are tied together in a lattice of interdependence so long as they continue to share a single CPR…

When appropriators act independently in relationship to a CPR generating scarce resource units, the total net benefits they obtain usually will be less than could have been achieved if they had coordinated their strategies in some way. At a minimum, the returns they receive from their appropriation efforts will be lower when decisions are made independently than they would have been otherwise. At worst, they can destroy the CPR itself."

–Elinor Ostrom. *Governing the Commons: The Evolution of Institutions for Collective Action*. Cambridge University Press, 1990 (p. 38).

Research has found that the vast majority of our resource systems are indeed managed through collective actions of local communities. We observe the same for managing the solar resource in local community governments or state governments via permitting and code regulations, and Solar Access and Solar Rights laws. Solar energy resource systems are also increasingly managed through local sharing of information among the appropriators of solar energy (be they designers, installers, or clients). The appropriators of solar energy find themselves interdependent among each other and tied to the power grid as a technological ecosystem, as well as the physical resource of the Sun. More SECSs installed for electricity generation means more information to share, and more incentive for structures to manage the increase in solar integration into society.

The **power grid** is a **common-pool resource**.

Solar energy can be framed as a CPR in an urban or suburban environment facing challenges of congestion as well, because the access to solar energy by tall structures in a congested region is ultimately subtractable from the whole. Your client in Philadelphia may wish that their two-story building be equipped with rooftop solar panels, yet the eight-story building to the Southwest will subtract many MWh per year of solar electricity by blocking access to the solar resource in the afternoons.

The necessity to be aware of these scenarios has led many regions (States) to develop contracts or collective action arrangements surrounding local sunlight access. Because of the common interest in solar energy, many states have *Solar Access Laws*, specifically to establish precedent that an individual may not use sunlight and diminish another person's ability to use it. The incentive to enact Solar Access Laws is to avoid a commons loss, where the resource system is congested at a loss to the larger community.

Solar Access Laws were an ancient strategy used in early Rome to guarantee access to solar gains in buildings, thus decreasing the amount of fuels needed for heating and lighting.

FRAMEWORK: ENTER THE CLIENT AS ARBITRATOR

In the case of managing a common-pool resource, the players can always agree to an external third-party arbitrator, who can enforce contracts in the long-term. The third-party arbitrator helps to resolve disputes within the confines of the agreed upon rules for the commons. Athletic leagues employ this strategy regularly.[11] In such a case, the participants within the commons themselves determine the scope of the contract, based on detailed information from engaging in the commons, rather than either an outside market or an outside governing body.

[11] Elinor Ostrom. *Governing the Commons: The Evolution of Institutions for Collective Action.* Cambridge University Press, 1990.

In the case of a Power Purchase Agreement (PPA), a third-party owner of PV technologies agrees by contract to sell electric power to a client at a rate that is below that of the local utility rate. Normally, there is only the client and the utility provider in the exchange of electricity as goods for cash. Although the PV array will be installed on the client's property, the PPA will own and manage the technologies as separate third-party private goods, where the managed good is the delivered electric power for the client at a known rate.

The owner of the system and the holder of the PPA contract maintains control of the physical SECS, and also earns financial returns in terms of the Solar Renewable Energy Certificates (SRECs) as dollars per megawatt-hour ($/MW h) certificate. It is interesting to note that accurate and complete information is essential to solar resource assessment, which is the cornerstone of low risk SECS project development. The third-party owners of the PV technology have detailed and accurate information regarding the dynamic solar resource and the hourly or monthly performance expected from the system, which is far more information that a standard client would have access to or desire to acquire. In essence, the *client* becomes the arbitrator between the PPA owner and the utility provider. Should the PV system greatly under-perform, the client will seek reparations enforcing the contract for fixed electricity costs.

In SECS project development, accurate and complete information regarding solar resource of the **locale** is critical to reducing risk.

FRAMEWORK: EMERGING LOCAL POLICY STRATEGIES

Creative new strategies are emerging each year to expand and explore the legal space surrounding solar access, rights, regulation, and market inefficiencies. We point out a case that emerged from Arizona, from the Solar Commons Project. The team at the SCP have established a community land *trust* to maintain SECS in the public *right of way* zones as a part of the commons. The emergence of trusts in solar energy solutions is yet another use of a third party for CPR management.

A **trust**: Common law institution enabling a third party to hold property on behalf of another party. The practice of a trust dates back to the Middle Ages and the crusades.

A trust is a common law institution allowing property to be held ("in trust") on behalf of another. The practice of a trust falls back to the Middle Ages, as a mechanism to allow property of an individual on crusade to be held by a peer, and along with the legal concept of "equity" should the property not be well maintained in absence. The practice of *right of way* is included in the Forest Charter of the *Magna Carta* (The Great Charter of the Liberties of England) of the 13th century, but was formally established as a common law tradition of the 16th century. In the 1500s, right of way allowed peasants to cross through forests and fields that had once been English commons spaces, but were converted to private lands in the course of industrialization.

Equity: The consequence or compensation apportioned to a party, enforced by a third party, should a trust not be well maintained.

Right of Way: An easement, a specific ability to use property and access it, without possessing the property.

The Solar Commons Project is one example for common pool management, where an institution out of Phoenix, AZ, has used *trusts* and *right of way* as creative legal mechanisms to build a stronger local common-pool resource management strategy. They have tied their evolving process for producing renewable solar energy (using photovoltaics) to the tenets of Ostrom's research—emphasizing transparency, accountability, and access as key principles of a successfully managed commons system. The project has based their pilot effort out of

Phoenix, selecting areas identified as the public right of way, then developing those areas with PV and holding the benefits of the PV commons in the public trust, such that the systems cannot be sold into private hands for profit ever.

> "The Solar Commons is a plan to produce renewable energy in our public right of way and capture the common wealth it generates in a community trust dedicated to local social equity. Using principles of commons management, the Solar Commons is a powerful organizing tool for civic organizations to partner with government and investors to grow a commons sector of the green economy." (http://solarcommons.org)

New alternative cases for common pool management are also emerging from a community solar perspective. The *solar garden* is a shared community array with subscribers connected to the grid. Even if individual residences or business places do not have direct access to sunlight (e.g., shaded by trees), the subscribers receive credit toward their energy bill by using "virtual net metering." The Solar Gardens Institute (SGI) out of Colorado was formed to educate communities about developing multiple energy subscriptions to offer solar arrays as useful engines for community development.

Finally, crowd-sourced photovoltaic projects are emerging, where the photovoltaic project is a solar farm, but the ownership, cooperative investments, and benefits of that farm are decentralized among a crowd-sourced pool of small investors. The Mosaic Project has had recent success with crowd-sourced PV project development.

FRAMEWORK: SYSTEMS SUSTAINABILITY ASSESSMENT AS LCA

Expanding upon the concept of goods, we observe that photons from the Sun can be used for many functions, often simultaneously. In contrast, when one transforms a mineral resource into the basis for a photovoltaic module, those elemental goods are subtracted from the whole of the ore body, as is noted in mineral commodity assessments by institutions such as the US Geological Survey.[12] Therefore, at the terawatt scale of photovoltaic deployment (or similar technology sequestering massive amounts of elemental material), the light from the Sun will remain non-rivalrous on a global scale, while the supply chain for the fundamental materials may grow more rivalrous, speaking to the need for study of the life cycle of the materials used in SECSs.

[12] US Geological Survey. Mineral commodity summaries 2012. Technical report, US Geological Survey, Jan 2012. URL http://minerals.usgs.gov/minerals/pubs/mcs/index.html. 198 p.

In SECS development where an environmental impact assessment is called for, particularly using solar energy for industrial process heating and agriculture, we need to quantify and interpret the flows and impacts of all goods. We assess the material resources that were called for "upstream" (events prior to technology deployment) and from the impact of using the SECS in place "downstream" (events following technology deployment). Upstream analysis of solar technologies suggests looking at the energy demands, raw materials, and chemicals that go into a photovoltaic panel or solar hot water system. Downstream analysis suggests looking at the impact of deploying a system within a locale over a significant time horizon. The full arc of study from mineral acquisition from ore to a disposed module is termed a *cradle-to-grave* assessment (see Figure 11.2). Other arcs can be defined by *gates* within the life cycle time horizon, and are called *cradle-to-gate* or even *gate-to-gate* studies. It should be noted however, that abbreviated or non-life cycle assessments can bring on questions of the ethics of the study, and can obscure an otherwise transparent process of research and documentation.

Upstream analysis: Referring to the earlier stages of technology processing, looking back and assessing the processes, materials, and costs used to create the current technologies being deployed—can be a part of **inventory analysis** in a **Life Cycle Assessment** (LCA).

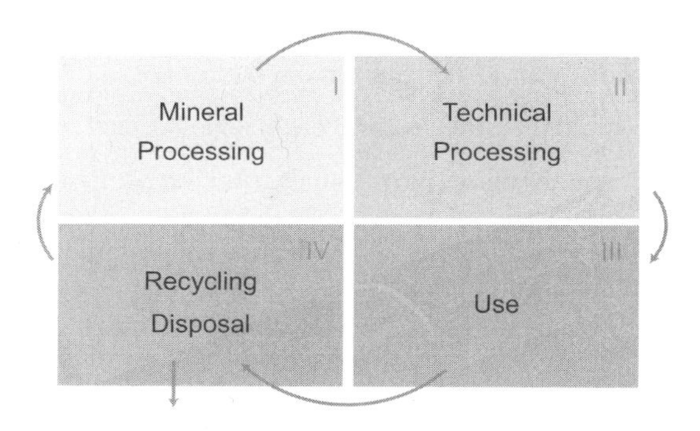

Figure 11.2: Potential life cycle stages for materials.

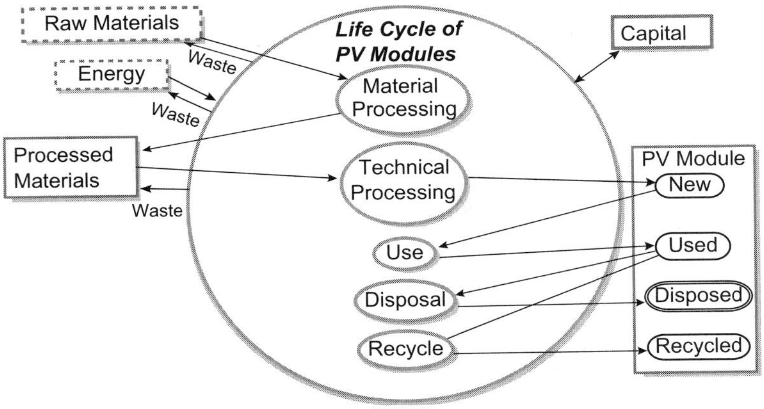

Figure 11.3: A graphical Object-Process model of the generic life cycle encompassed for a PV module. Objects like resources, energy, modules, and capital are boxed. Processes like manufacturing a module, transporting and using a module, disposing of a module, and the entire life cycle are enclosed in circular bounds. The module itself can be represented in four states of new, used, disposed, or recycled.

To understand the relevant ecosystems services or ecological impacts linked to SECS manufacture, deployment, use, and end of life, we must have methods to catalog and quantify the flows of energy, materials, biology and biodiversity, as well as cost equivalents. Solar Energy Conversion Systems are themselves made of components of raw materials from mineral (metals, ceramics), biological, and geofuel-based resources (see Figure 11.3).

A diverse integrative design team can investigate the impact of both engaging use of the solar resource through solar technologies and the impact of the materials tied to the SECS using Life Cycle Assessment (LCA). Life cycle assessment provides a process for making strategic decisions within a network of information, given scope and goals of assessment. LCA is used as a methodology for evaluating, modeling, documenting, and communicating the impact of selected materials, processing techniques, and complimentary services as they affect the environment and society over the life cycle of the relevant technologies and products. Physical properties can be measured and inventoried in the proposed or deployed systems, analyzed and interpreted relative to their environmental and energetic attributes, and linked to the constraining scope and goals from our social values and physical needs for well-being.[13] Hence, LCA is a widely applicable set of tools that supports comparisons between valued options in a sustainable environment.

Life Cycle Assessment (LCA), used to assess materials and energy flows, is distinct from **Life Cycle Cost Analysis** (LCCA) used in project finance.

LCA is distinct from the more familiar Life Cycle Cost Analysis (LCCA; discussed in our chapter on Project Finance), which applies criteria linked to

[13] Environmental management–life cycle assessment–principles and framework. ISO 14040, International Organization for Standardization: ISO, Geneva, Switzerland, 2006.

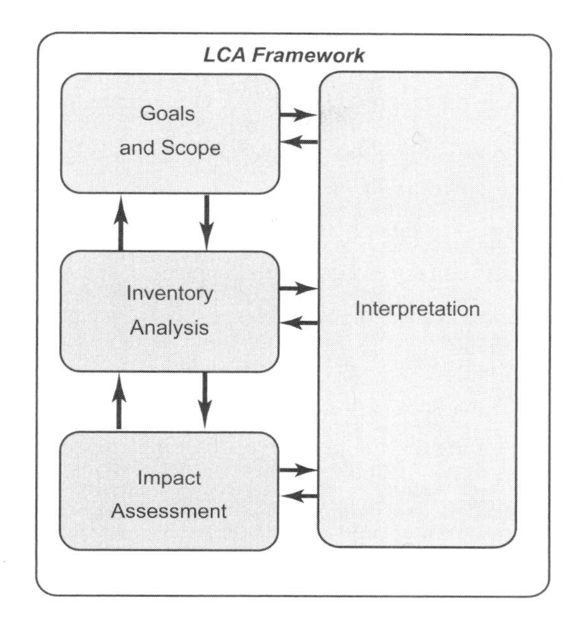

Figure 11.4: The iterative Life Cycle Assessment Framework.

highest economic utility through cost/performance comparisons and optimization. Unfortunately, LCCA has been shown to marginalize environmental and social impacts by limiting the defined system boundaries (scope) of the analysis. However, LCA can be developed in supplement with LCCA. It is very important to recognize that LCA is an *iterative process*, (see Figure 11.4), so assessments performed in the first cycle of assessment will by nature have wide ranges of confidence, while assessments that have occurred over decades of iteration will converge toward more certain values, and will incorporate the dynamic changes in technology processing that occur. In compliment, LCA reports from a decade past are unlikely to have significant value unless they have been harmonized with current data sets. It is up to the team and our peer group to stay up-to-date with progress in LCA data sets. Fortunately, the solar field has been performing LCA for decades already as a natural part of the justification for solar energy exploration.

Life Cycle Assessment is an iterative process. LCA is only as good as the data and the understanding of the flows of materials and energy within their context of society and environment.

LCA is often used in renewable energy systems studies to evaluate the impact of renewable energy technologies in terms of greenhouse gas (GhG) emission estimates. There is a common public misunderstanding of the impact of processing photovoltaic modules—(e.g., surely there must be a horrible impact such that PV will never recover the CO_2 or the energy required to produce modules from coal and petroleum). LCA can address the concern, and uses the available data to interpret the GhG impact of PV. In this case, an LCA is designed with a goal and scope to assess both the GhGs associated with the generation of electricity and all indirect emissions that can be associated with fuels and materials upstream, in a cradle-to-grave assessment. Upstream processes include mining and mineral processing, associated chemical processes, transportation of materials, and PV manufacturing plant construction. Downstream processes include decommissioning the PV plant, recycling materials, and waste streams for disposal. In such an example, studies have evaluated the question of GhG as grams of carbon dioxide equivalents emitted per kilowatt-hour of energy production (g CO_2-eq/kW h), and continue to iterate through new interpretations with new information. Relatively recent studies have demonstrated that PV GhG embodied energy can be recovered within 1.5–3 years (assuming Southern European irradiation conditions of 1700 kW h/(m^2 *y*)), while associated GhG emissions are on the order of 35–55 g CO_2-eq/kW h.[14] In comparison, coal combustion has associated GhG emissions two orders of magnitude larger, around 1000 g CO_2-eq/kW h.[15] Similar LCA studies have been performed for energy payback times of photovoltaic modules, with respect to the entire industry and per unit of PV produced.[16]

[14] V. Fthenakis, H. C. Kim, and E. Alsema. Emissions from photovoltaic life cycles. *Environ. Sci. Technol.*, 42:2168–2174, 2008. doi:10.1021/es071763q; and D. D. Hsu, P. O'Donoughue, V. Fthenakis, G. A. Heath, H. C. Kim, P. Sawyer, J.-K. Choi, and D. E. Turney. Life cycle greenhouse gas emissions of crystalline silicon photovoltaic electricity generation: Systematic review and harmonization. *J. of Industrial Ecology*, 16, 2012. http://dx.doi.org/10.1111/j.1530-9290.2011.00439.x.

[15] B. K. Sovacool. Valuing the greenhouse gas emissions from nuclear power: A critical survey. *Energy Policy*, 36:2940–2953, 2008. doi:10.1016/j.enpol.2008.04.017.

[16] M. Dale, and S. M. Benson. Energy balance of the global photovoltaic (PV) industry—is the PV industry a net electricity producer? *Envir. Sci. & Tech.* 47(7): 3482–3489, 2013. http://dx.doi.org/10.1021/es3038824.

PROBLEMS

1. Consider a scenario where you are an entrepreneur in a new solar energy technology startup. Your challenge is to convince them that sustainability is a secure business investment. How might you integrate the principles of sustainability and environmental technology with your solar technology into a convincing pitch for venture capital investors?

2. Again, for an entrepreneurial scenario as a solar tech startup: discuss if you would or would not include the ethics of sustainability and a strategy for

addressing solar ecosystems services in your business plan, and why (or why not).

3. Compose a strategy to convince policy makers that there is merit in devising a progressive solar policy of your choice. The decision maker could be your town council, mayor, or an elected member of state/federal government. You will have two minutes (elevator speech) to convey your interests to the individual. Record your pitch on video and share with the class.

RECOMMENDED ADDITIONAL READING

- Governing the Commons: The Evolution of Institutions for Collective Action.[17]
- The Story of Stuff: How Our Obsession with Stuff is Trashing the Planet, Our Communities, and Our Health—and a Vision for Change.[18]
- Home Economics.[19]
- Thinking in Systems: A Primer.[20]
- A Pattern Language.[21]
- Why Eco-efficiency?[22]
- Environmental management—Life cycle assessment—Principles and framework.[23]
- Emissions from Photovoltaic Life Cycles.[24]

[17] Elinor Ostrom. *Governing the Commons: The Evolution of Institutions for Collective Action.* Cambridge University Press, 1990.

[18] Annie Leonard. *The Story of Stuff: How Our Obsession with Stuff is Trashing the Planet, Our Communities, and our Health—and a Vision for Change.* Simon & Schuster, 2010.

[19] Wendell Berry. *Home Economics.* North Point Press, 1987.

[20] Donella H Meadows. *Thinking in Systems: A Primer.* Chelsea Green Publishing, 2008.

[21] C. Alexander, S. Ishikawa, and M. Silverstein. *A Pattern Language: Towns, Buildings, Construction.* Oxford University Press, 1977.

[22] Gjalt Huppes and Masanobu Ishikawa. Why eco-efficiency? *Journal of Industrial Ecology*, 9(4):2–5, 2005.

[23] Environmental management—life cycle assessment—principles and framework. ISO 14040, International Organization for Standardization: ISO, Geneva, Switzerland, 2006.

[24] V. Fthenakis, H. C. Kim, and E. Alsema. Emissions from photovoltaic life cycles. *Environ. Sci. Technol.*, 42:2168–2174, 2008. doi:10.1021/es071763q.

12 SYSTEMS LOGIC OF DEVICES: PATTERNS

The only principle that does not inhibit progress is: anything goes.

Paul Feyerabend

We have spent a significant amount of time so far working through the context of the solar resource in space and time, both in terms of the meteorology of the atmosphere affecting the quality of the light impinging upon our collectors, and the diurnal phenomena occurring due to the moving bodies of the Sun and Earth. We have also spent significant effort establishing photoconversion within the modern basis of heat transfer and thermodynamics, but have yet to specialize our discussion into the types of materials which work together to yield a *solar energy conversion system.*

The truth is that the solar resource is much more complicated than the actual devices converting solar energy to useful forms for society. As seen in Figure 12.1, there is an entire *solar ecosystem* linking society and technology with our environment and solar resource. Following the complex systems context of the solar resource, addressing ecosystems services in addition to serving the client in their locale is equally entangled and challenging.

Solar Energy Conversion Systems (SECSs):

- Aperture
- Receiver
- Distribution mechanism
- Storage
- Control mechanism

However, when we review our solar technologies available today, most devices have minimal moving parts and operate at significantly lower temperatures than internal combustion systems. Also, topics of device design in *optoelectronics* and *optocalorics* can be significantly less complicated than most fuel-based conversion technologies. And yet there are a great diversity of possibilities to

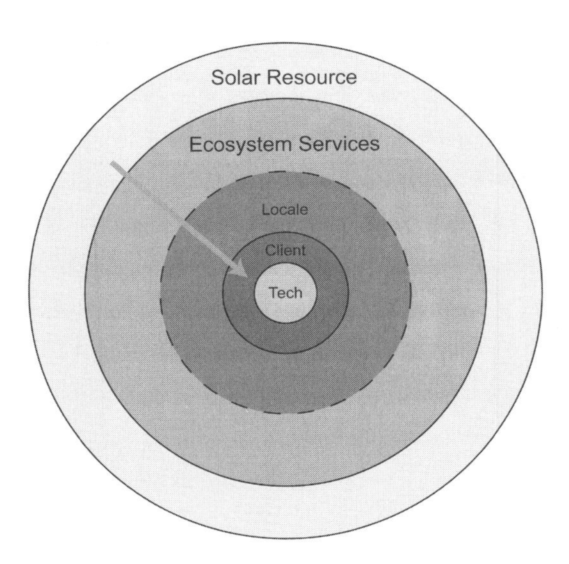

Figure 12.1: Layers of systemic connections within the **solar ecosystem** for design. Notice the multiple layers of influence between the **solar resource** and the deployed **technology**.

explore, and enormous potential for new technologies to emerge in the near future. For the scope of this text, where we are exploring creative systems integration of SECSs into a new energy ecosystem, we will describe integrative design using primarily "off-the-shelf" technologies. We would like the solar design team to be familiar with current technologies, but also ready to adapt to completely new technologies as they come to market. As new technologies emerge and prove themselves financially robust, the same design principles will allow for their incorporation into future projects.

If we recall from our introduction to light as a work current in a solar energy pumping system, we design or engage our systems to use light to induce three responses:

1. **Optoelectronic response:** the excitation of electrons inside of a material to yield increased electronic (semiconductor) or electrical (metal) conductivity,
2. **Optocaloric response:** the excitation of a material to induce thermal vibrations, or
3. **Photoelectrochemical response:** the excitation of electrons inside of a molecule to yield photosynthesis and vision (also termed: optochemical or photochemical).

Every SECS will also retain several key technical components that we can identify, and simple motifs allowing the system to function in a manner useful to society and the environment. The following components convey information regarding the manner in which light is directed through the systems: from an aperture (opening) onto to a receiving surface (an absorber), there redistributed as converted heat/light/electricity/fuel, potentially stored, and the whole flow of energy is subject to one or more control mechanisms.

- Aperture (enlarged relative to the receiver when applied for concentration).
- Receiver.
- Distribution Mechanism (internal to the system).
- Storage (however, not all systems will have storage).
- Control Mechanism.

Note that applied methods to evaluate and scale storage systems linked with converted solar energy (both thermal and electrochemical) are not emphasized, nor are detailed strategies for passive solar building design. If you are interested, there are numerous texts in the literature that expand upon the topic of thermal and electrochemical storage in great detail, including the engineering calculations to contribute to a systems design with high solar utility for the client in their given locale and demands.[1]

[1] John A. Duffie and William A. Beckman. *Solar Engineering of Thermal Processes.* John Wiley & Sons, Inc., 3rd edition, 2006; and Soteris A. Kalogirou. *Solar Energy Engineering: Processes and Systems.* Academic Press, 2011.

PATTERN LANGUAGE IN SOLAR ENERGY CONVERSION SYSTEMS

In 1977 a transformative text was published by Alexander, Ishikawa, and Silverstein addressing architecture, urban design, and community planning together as liveable spaces.[2] Alexander and coauthors detailed a *pattern language* of design solutions

[2] C. Alexander, S. Ishikawa, and M. Silverstein. *A Pattern Language: Towns, Buildings, Construction.* Oxford University Press, 1977.

to common problems in the field of architecture and urban planning, having a vocabulary, syntax, and grammar, with a web of relationships among the patterns. In this text we focus on solar energy conversion processes as they exist in the context of a given locale and environment for a client, and these functioning systems fulfill another broad *pattern language* of opportunities to explore and build upon.

Exploring the concept that *design is pattern with a purpose*, let us identify some common patterns in the realm of Solar Energy Conversion Systems. Again, sustainability science in solar energy requires integration and alignment among multiple disciplines, integration of science and society, and specifically requires solutions that reference the patterns tied to *locale* coordinated with solutions over time.[3] In the following sections, we reveal systematic patterns in SECS, and relations between different motifs. The goal is to see patterns as opportunities for maximizing solar utility for clients in their locales, and to open up our imagination to new solutions that combine or add new patterns to the whole. Remember, solar energy is experiencing a renaissance of ideas, and there are so many possibilities to integrate motifs and patterns in ways that will benefit our society transitioning into a sustainable future.

[3] Chrstian U. Becker. *Sustainability Ethics and Sustainability Research.* Dordrecht: Springer, 2012.

With experience and application, we shall start to see solar applications as low-hanging fruit everywhere around us. The following sections have been divided into *motifs* and *patterns*. The motifs are raw concepts of light interacting with surfaces, absorbing, reflecting, transmitting, and emitting. The motifs also incorporate the desired functionality of light conversion to thermal energy and electronic energy: absorber, receiver, control, distribution. The patterns described are more complex interplays among these motifs, and among other patterns. Again, we would hope that the engaged student of SECS would link this pattern knowledge within a larger integrative design team to solve our problems in society and the environment.

CAVITY COLLECTORS—THE PATTERN OF ROOMS

A room with a view is also a room with an *aperture* to transmit solar gains on the inside walls. The cavity collector is probably one of the most familiar patterns in

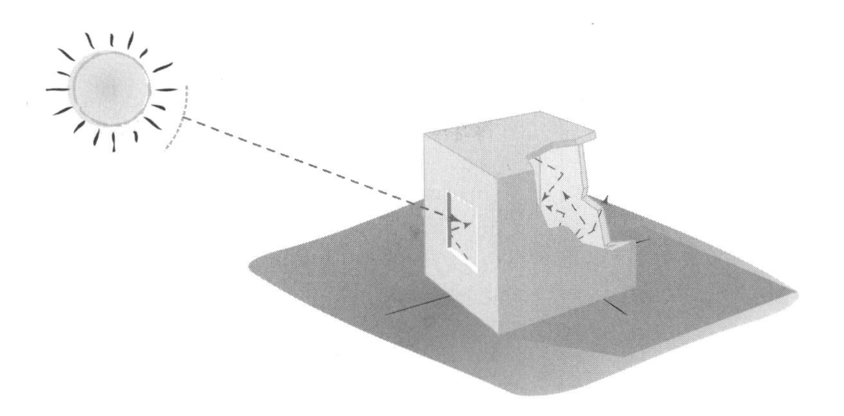

Figure 12.2: Schematic of a cavity absorber with no cover.

the language of SECSs. Everyone is familiar with a room that is exposed to the outside Sun via a window.

In a window-room pattern of the cavity collector, the window serves as a *cover* preventing warm air from escaping, while transmitting shortwave irradiance into the perimeter zone of a building.[4] The walls and floor are then *absorbing motifs* or *reflecting motifs* for the *optocaloric* energy conversion processes driven by our solar work current (see Figure 12.2).

Windows as apertures can have single/double/triple pane glazings (or none!). A sheet of glass is termed a "lite," and will have two surfaces (just like our model of the atmosphere), one that faces the exterior and one that faces the interior of the room. Thus, in a double pane insulated glass unit (IGU) a window will have four surfaces (1, 2, 3, 4 from outside to inside), upon which one can deposit specially tailored thin films to make *selective surfaces*.[5] Depending on the desired tailored outcome of the advanced window, specific to the locale and climate zone and client needs, glass companies will coat the #2 or #3 surfaces (or more) with selective surface thin films that result in a *low-E* window.

Now, we have been exposed to a more detailed level of knowledge in this text and so we can properly call a "low-E" window or glazing a *low-*ϵ or *low emittance* glazing. Considering that glass installed in windows are not likely to exceed a working temperature of 400 K (or 127 °C/260 °F), we can use this temperature as an order of magnitude estimate for the most probable wavelength of light to be emitted from a window.

[4] From a building systems perspective, a "zone" is a volume of space coincident with a room or set of rooms that have similar thermal properties (and similar air chemistry like humidity).

[5] For a review of *selective surfaces* please review the prior chapter: "Physics of Light, Heat, Work and Photoconversion."

Using Wien's Displacement Law from Eq. (12.1), we can estimate that 400 K would have a most probable wavelength at 7245 nm—well into the longwave band of light.

$$\lambda_{max} T = 2.8978 \times 10^6 \text{ nm K},$$
$$\lambda_{max} = \frac{2.8978 \times 10^6 \text{ nm K}}{T(K)}. \tag{12.1}$$

So if low emittance windows are not effective emitters of graybody radiance, why do we seek them out in modern windows? Let's recall our graybody accounting one more time from Eq. (12.2), and our knowledge of the radiosity equation:

$$1 = \tau + \epsilon + \rho, \tag{12.2}$$

$$J_{window} = \rho_w G + \epsilon_w E_{b,w}. \tag{12.3}$$

We offer our knowledge that a window is highly transparent in the shortwave band, so the majority of shortwave solar irradiance (G) will not be absorbed and only a small fraction of shortwave light will be reflected by the window as well. However, if the emittance of a window is low in the longwave band, then the absorptance is also low (by *Kirchoff's Law of Radiation*). More important, if the emittance in the longwave band is low then the *reflectance* of longwave band light will be *high*. So, a low-E window is actually a low-ϵ surface, selectively reflecting longwave irradiance. In fact, modern windows that are classified as "low-E" will have high transparency in the visible band with high reflectivity in the ultraviolet and IR sub-bands of the shortwave in addition to high reflectivity in the longwave (this is due to the material properties of a low-E thin film), making the thin film a "mid-pass" optical filter.

For engineering solutions we have tools that can estimate energy transfer through window systems and into the adjoining rooms/zones. The computer tool REFSEN was developed to evaluate the solar and thermal behavior of glazings with thin film coatings.[6]

[6] Robin Mitchell, Joe Huang, Dariush Arasteh, Charlie Huizenga, and Steve Glendenning. *RESFEN5: Program Description (LBNL-40682 Rev. BS-371).* Windows and Daylighting Group, Building Technologies Department, Environmental Energy Technologies Division, Lawrence Berkeley National Laboratory, Berkeley National Laboratory Berkeley, CA, USA, May 2005. http://windows.lbl.gov/software/resfen/50/RESFEN50UserManual.pdf. A PC Program for Calculating the Heating and Cooling Energy Use of Windows in Residential Buildings.

Now returning to the receiving walls absorbing or reflecting in a cavity collector, recall that light intensity still exhibits a non-linear decay with distance. Hence, the surfaces far from the window (aperture) can be less affected than those closer. That being said, a bright sunny day (beam irradiance) with the right window design can allow light to penetrate deep into a room. In designing for SECS, the integrative team can increase the visual utility of daylight in numerous ways. In fact the discipline of lighting design and engineering is far more extensive than can be covered in this text alone.[7] Entire degrees in higher education have been dedicated to understanding light within the visible spectrum (R/G/B only), and the comfort and energy value of placing and sensing lighting conditions for the occupants in a space. However, even at the introductory level we may find value in selecting a palette of *paints* or other materials for a room that have higher values of reflectance in addition to the color and hue desired, thus decreasing the demand for artificial lighting (while increase daylighting and avoiding electricity costs).

[7] T. Muneer. *Solar Radiation and Daylight Models*. Elsevier Butterworth-Heinemann, Jordan Hill, Oxford, 2nd edition, 2004.

Keep in mind also that the choice of paint color (other than beige and off-white) in a room is useful even with artificial lighting, due to the change in the physics of light emission in luminaires. Today's compact fluorescent and LED sources are significantly more "white" than old incandescents. Society in the age of incandescents used to impart color in a room with lights (often a yellow tone with a thin film to adjust the color), while now we do so by imparting color with the receiving and reflecting surfaces instead. Again, we emphasize the value of being aware of the directionality of light and the material reflecting properties of a surface to shortwave or longwave light. Think creatively to solve problems with light!

And finally, all apertures for a cavity collector have the option of a *control mechanism* to permit light to enter the room. This control is brought about through shading strategies via blinds, curtains, horizontal folding shutters, or vertical roller shutters (common in Europe).

METHOD FOR SOLAR THERMAL ESTIMATION: Solar gains within the space of a cavity have been evaluated in the past by the *solar load ratio* method and by the *unutilizability* method. Both are explored in detail in more advanced solar thermal engineering texts.[8] These are important methods to explore the science of a cavity collector, although the approaches are not well known to communities beyond the building science field.

[8] John A. Duffie and William A. Beckman. *Solar Engineering of Thermal Processes*. John Wiley & Sons, Inc., 3rd edition, 2006.

COMPLEMENTARY PATTERNS to the *cavity collector* are those related to the super pattern of *optocalorics*: *flat plate collectors, parasoleils, distillation, and dryers*. The complementary pattern of visual and thermal comfort for the occupant is inclusive, but beyond the scope of this text, and we strongly encourage the integrative design process to include lighting design experts tied to decisions within the built environment.

FLAT PLATE COLLECTOR—THE PATTERN OF SIMPLE SURFACES

The flat plate collector (FPC) is also one of the more familiar in the language of SECSs. A solar hot water panel that is constructed as a large box is also a flat plate collector for optocaloric utility. We will clearly define a flat, non-concentrating photovoltaic panel as an optoelectronic flat plate collector (with some additional undesirable optocaloric properties) (see Figure 12.3). From the roots in engineering, we typically refer to flat plate collectors for photovoltaic modules or solar hot water panels. But going beyond the simple box model, what else could be viewed in the pattern of the flat plate collector? A wall or a roof top could be a flat plate collector, but so could a parking lot or the canopy on a picnic table. Flat plate collectors are also ancient strategies for maintaining microclimate in urban centers, as seen in *piazzas* (stone plazas).[9]

[9] Robert D. Brown. *Design With Microclimate: The Secret to Comfortable Outdoor Spaces*. Island Press, 2010.

FPCs tend to be of two varieties: covered or uncovered. When an FPC is covered, the system behaves as a *cover-absorber ensemble*, the cover working in concert with the absorbing surface to yield a solution with higher utility. The cover-absorber system can be explored in great detail for systems design.

COMPLEMENTARY PATTERNS to the *flat plate collector* are also those related to the super pattern *optocalorics*: *tubular collectors, parasoleils, distillation, and dryers*, and to the super pattern of *optoelectronics*.

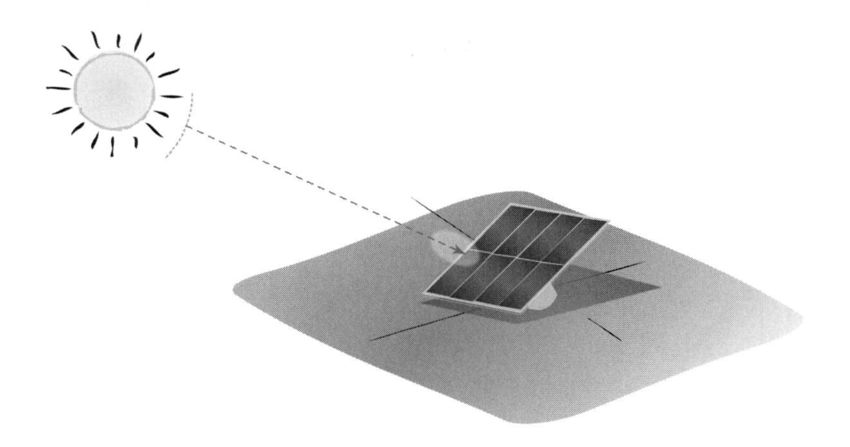

Figure 12.3: Schematic of a flat plate absorber as a non-concentrating PV module.

TUBULAR COLLECTOR—THE PATTERN OF A GEOMETRY WORKING WITH THE SUN

The tubular design (or designs that follow the arc of the Sun) has been explored in both optoelectronic and optocaloric super patterns. In both cases, the systems still use cover-absorber strategies to manage light and thermal energy in addition to environmental robustness, applied in concentric rings.

COMPLEMENTARY PATTERNS to the *tubular collector* are also those related to the super pattern *optocalorics*: *flat plate collectors, concentration,* and even to the super pattern of *optoelectronics*.

Thanks to the innovative design of the Solyndra PV system. Keep in mind that the design and the materials were good, it was the cost of the system that was not sustainable at the time.

PARASOLEILS—THE PATTERN OF SHADING

We introduce a term here in solar design, the *parasoleil*, as a solar umbrella (or parasol) to encompass the pattern of solutions that encompass structured shading of our environment to maintain cool spaces during summer seasons. So we ask ourselves, when is *shading* a strategy to increase the solar utility for the client, and under what conditions of the locale. In this case, avoided solar gains are measured as *avoided fuel costs* for air conditioning systems coupled

Shading helps to create microclimates.

[10] A. Dominguez, J. Kleissl, and J. C. Luvall. Effects of solar photovoltaic panels on roof heat transfer. *Solar Energy*, 85(9):2244–2255, 2011. doi: http://dx.doi.org/10.1016/j.solener.2011.06.010.

[11] Robert D. Brown. *Design With Microclimate: The Secret to Comfortable Outdoor Spaces*. Island Press, 2010.

There is nothing passive about the Sun, but the buildings that we design may have appropriate technology implementations, called "passive."

to our living spaces. We know of trees shading a building to maintain a cooler microclimate on a hot, bright day, but did you know that PV panels on your roof can also create a cooler microclimate and reduce cooling loads?[10] Parasoleil strategies can induce comfort and reduce energy costs for the client. Parasoleil strategies have also been common to vernacular architecture in moderate and warm climate regimes for thousands of years, as seen in the design of porticos (patios or porches).[11]

In particular, we *re*-introduce the pattern of the parasoleil for SECS design because of their *absence* from design strategies in engineered systems of the age of modern design. Energy simulation models for building simulations with shading are notoriously poor at accurately dealing with the shadow-solar gain interplay. The use of shading is a common strategy within the realm *passive solar* design, yet another broad field beyond the scope of this text. So why is this?

Much of the work on solar simulation in buildings was developed during a time when *measurement* was not an inexpensive option. Hence, strategies were developed to derive *empirical correlations* of expensive measurements into projections of the solar components onto building surfaces. Unfortunately, computational energy was also very expensive at that time (1970–1980s), and as we have learned in earlier chapters radiative transfer is both *non-linear* and rich in systems *complexity*. In an effort to minimize calculations, radiative balances on walls and awnings for shading were only incorporated into *window* energy calculations, the most direct area for energy transfer and solar gains.

In the case of developing a parasoleil pattern solution, we ask our integrative design team several questions. Given that our concept of "locale" is relevant to both space and time:

- What hours of the day do we most want to shade? This will determine the orientation and size of the parasoleil system.

- What portion of the year (in terms of declination range) are we most interested in shading?

- Given the cost of installation, should the parasoleil be fixed, tracking, or variable in position?

- What is the shared value of the heat capacity in the wall materials receiving the irradiance or blocking the irradiance (will the heat be retained to "glow" during the night?).

- What are the avoided fuel/electricity costs projected for the parasoleil solution?

There are additional refinements to consider when using living structures like trees. Should one use a deciduous tree versus an evergreen? For a deciduous tree, leaves lost during Winter will potentially allow some desirable solar gains on the surface, while evergreens will likely provide annual shading. What will be the maximum height and shape of the trees if one plants them as saplings, or how will a tree develop as a parasoleil over the years with respect to the shaded surface?

VIGNETTE:

The use of unusual strategies in Geo-engineering to create "super umbrellas" or "dark skies." In a far-flung dystopian future, members of society may choose to "hack" the atmosphere by dispersing stratospheric aerosols of sulfur. This will do nothing with respect to the challenge of ocean chemistry acidification, but has the prospect of quickly reducing the incoming shortwave irradiance. Unfortunately, geo-engineering of this kind removes the concept of locale entirely from the design framework. These strategies are untested, and potential "pattern-busters" as they would recreate the ecosystem around us. Earth scientists advise extreme caution, as the strategy reduces the solar resource for warming effects, but also reduces our ability to apply energy conversion (like adding water to a gas tank), and CO_2 emissions will still result in ocean acidification that destroys the ocean biome supporting much of our world culture.

PROBLEMS

1. Explain how a room with a window (a single thermal zone) is a cavity collector system. What are the optical and thermal functions within the system, and what could be the favorable value of paint colors or stone floors within the room?

2. Describe three flat plate collector memes for solar energy conversion.

3. Describe three parasoleil memes that could be coupled with SECSs to help control the performance of a technology such as a house.

RECOMMENDED ADDITIONAL READING

- Thinking in Systems: A Primer.[12]

- A Pattern Language.

- Design with microclimate: the secret to comfortable outdoor spaces.[13]

- Let it Shine: The 6000-Year Story of Solar Energy.[14]

- Solar Technologies for Buildings.[15]

[12] Donella H Meadows. Thinking in Systems: A Primer. Chelsea Green Publishing, 2008.

[13] Robert D. Brown. *Design With Microclimate: The Secret to Comfortable Outdoor Spaces*. Island Press, 2010.

[14] John Perlin. *Let it Shine: The 6000-Year Story of Solar Energy*. New World Library, 2013.

[15] Ursula Eicker. *Solar Technologies for Buildings*. John Wiley & Sons Ltd, 2003.

SYSTEMS LOGIC OF DEVICES: OPTOCALORICS

13

WE HAVE USED the preceding chapter to explore the qualitative differences of various patterns for Solar Energy Conversion Systems. Now we take the time to quantify the performance of systems which make use of those pattern differences. For the process of optocaloric conversion, the absorbing surface will absorb shortwave irradiance from the Sun and convert that electromagnetic light into thermal vibrations. The thermal energy is then conveyed into a distribution mechanism and finally a storage and control tank. As we see in Figure 13.1 the light-matter interactions connect energy inputs (irradiance (G) to useful solar gains (S)) with minor optical losses. The system is thermodynamically *open* (mass and energy exchange with the surroundings) and the system is at non-equilibrium. The solar thermal absorber/collector converts light to thermal energy via a change in sensible heat (a change in the temperature of a material) and potentially via a change in latent heat (a phase change at a given temperature, like from ice to water or from water to steam). Some thermal energy is lost to the surroundings in the process of "pumping up" the optocaloric system, and then the collector translates the thermal energy into a heat transfer fluid, such that it may leave the system to do work (as a heat engine). Often in a solar thermal system, the hot fluid is transferred into a tank or stock, to be stored for a period until demanded by the client.

- **Sensible heat:** Energy exchange that changes the temperature of a material.
- **Latent heat:** Energy exchange process that occurs without a change in temperature, here from a phase transition of a material (solid ↔ liquid ↔ gas).

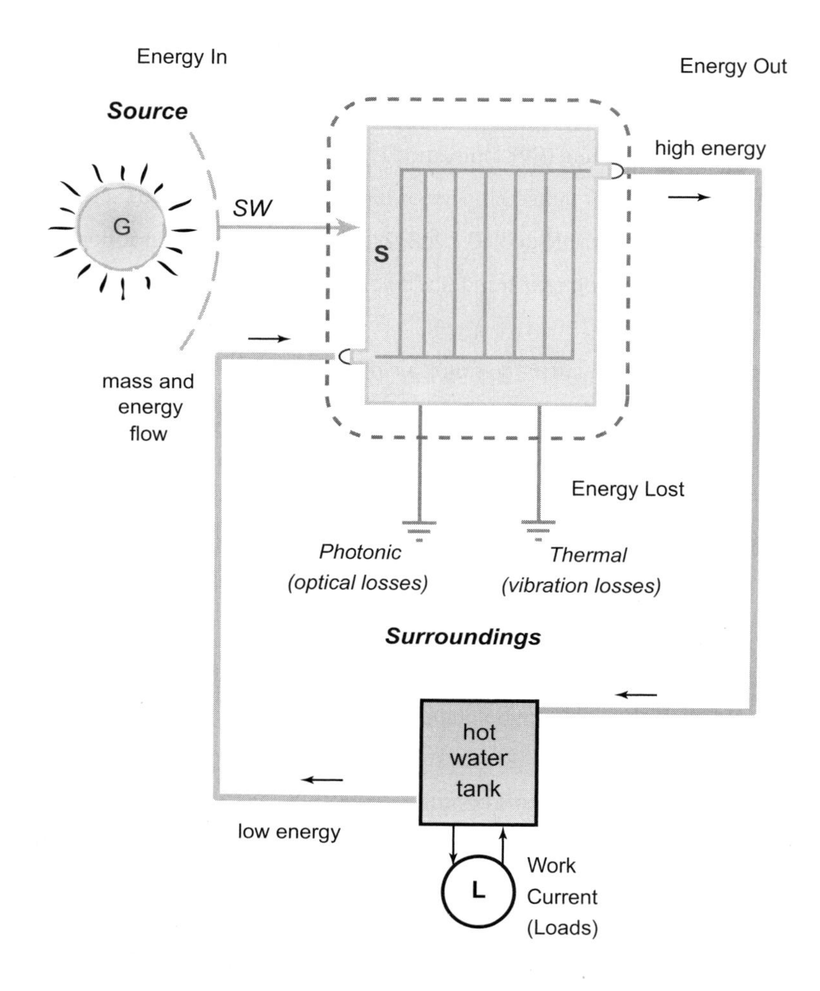

Figure 13.1: Schematic of a optocaloric device interacting as an open energy conversion device.

Q is a multiple of the area of the collector A_c with the energy exchange per unit area q.

Recall that for a flat plate collector the area of the *aperture* and the area of the *receiver* are equivalent: $A_c = A_{ap} = A_{rec}$. Energy exchange is represented by Q, and a rate of energy exchange may occasionally be represented as $\dot{Q}$.

$$Q = A_c \cdot q \tag{13.1}$$

or

$$\dot{Q} = A_c \cdot \dot{q}. \tag{13.2}$$

One should also be aware that the thermal response to radiative heat transfer will not be instantaneous in the morning (warm-up) and will also lag after sunset (cool-down)—these are transient conditions that can be simulated in the future. For the moment, we will focus on *steady state conditions*, where the optocaloric flat plate is in a continuous excited state due to the radiant "pumping" from the Sun.

Steady state is a bit like mechanically pumping up a tire with a leak in it. The tire remains pressurized, but there is air entering and leaving the system.

There are solar thermal collectors of many different varieties, flat systems (large flat boxes) with or without transparent covers that assist in retaining thermal heat called flat plate collectors (FPC), systems that are composed of sets of evacuated glass tubes housing a single absorbing fin (evacuated tube collectors, ETC)—used in concentrating and non-concentrating applications. Solar thermal collectors have been used to heat liquid fluids (water, glycol, ethanol), and to heat ambient air. In fact, special unglazed transpired collectors were developed for heating ambient air using the vertical façade of a building as a mounting surface.[1] These systems are also termed solar walls.

- **FPC:** Flat plate collectors.
- **ETC:** Evacuated tube collectors.
- **Transpired collectors:** Unglazed hot air collectors.

[1] Graham L. Morrison. *Solar Energy: The State of the Art*, ISES position papers 4: Solar Collectors, pages 145–222. James & James Ltd., London, UK, 2001.

COVER-ABSORBER: OPTICAL PERFORMANCE

As a recap, we can calculate values of reflectivity (r) based on the refractive indices (n) of two materials. The first material is often the air that light is propagating through within our sky, while the second is our semitransparent material that will reflect and refract light as a cover, protecting the absorber beneath. The losses to transmission will be due to the combined effects of absorption and light scattering (either at the surface or within Material 2). The sum of energy contributions per wavelength will still obey Eq. (4.5), in that there will be fractional contributions from absorptance (α) and reflectance (ρ). But again, both absorptance and reflectance values can be derived from the reflectivity (r) based on the refractive indices (n) of two materials.

In the case of a cover interacting with a thermal absorber, there is a composite performance condition that we can define, $f(\tau,\alpha)$.[2] This is a function of the transmittance of the semitransparent cover (τ_{cover}) and the absorptance of an opaque absorber material (α_{abs}). Using the example of systems design to trap radiant energy (as in a greenhouse), shortwave light will pass through the cover

[2] John A. Duffie and William A. Beckman. *Solar Engineering of Thermal Processes.* John Wiley & Sons, Inc., 3rd edition, 2006.

The tau-alpha term has been represented in the past simply as "(τ, α)," however this method leads to misinterpretation of the actual function. Hence, we use the $f(\tau, \alpha)$ convention for disambiguation.

and be partially absorbed and scattered (in small proportion). Then a fraction of the shortwave light will be absorbed by the material below, with a small, but measurable fraction being reflected back up to the cover. Of course the cover will also reflect a fraction of the light back down to the absorber plate a second time. And so the light may "bounce" several times before the density of photons is scattered and diminished to a negligible level. We describe this coupled interaction as the *cover-absorber* system.

The function $f(\tau,\alpha)$ is a performance factor used to estimate optical losses in the system ($0<f(\tau,\alpha)<1$, and is typically >0.7), regardless of subsequent thermal or electronic applications for the solar conversion process. In Figure 13.2 we can observe that the steps contributing to the function $f(\tau,\alpha)$ are slightly more complicated than a simple multiplication of τ_{cover} and α_{abs}, particularly in scenarios where there are more than one cover. In Eq. (13.3), we see that by multiplying irradiance on a tilted surface (G_t) we find the net *absorbed irradiance* S for the *cover-absorber* sub-system[3]:

$f(\tau,\alpha) \neq \tau \cdot \alpha$, but it can be close.

[3] Yes, irradiation calculations for I_t can also be used to estimate S.

$$S = G_t \cdot f(\tau,\alpha)\ [\mathrm{W/m^2}]. \tag{13.3}$$

The purpose of calculating $f(\tau,\alpha)$ to provide more accurate values of S from measured or estimated G_t values, when considering the multiple bounces of light propagating through a thin gap. In Eq. (13.4) we see that the zero-order contribution of $\tau \cdot \alpha$ is reduced in proportion to an infinite sum of n reflections, propagating between the absorbing surface (where $\rho_{abs}=(1-\alpha)$) and the diffuse

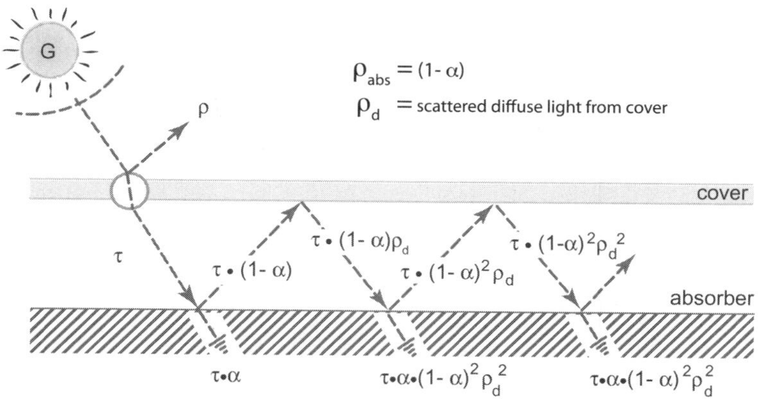

Figure 13.2: Schematic of a cover-absorber system.

reflectance of the upper cover (as ρ_d). There is an analytical solution to this discrete infinite sum, shown at the end of Eq. (13.4).

CONTRIBUTIONS TO THE COVER-ABSORBER METRIC $f(\tau,\alpha)$: $(1-\alpha)$: We use Kirchoff's Law of Radiation to represent the reflectance of the absorber layer (ρ_{abs}) in terms of the absorptance.

ρ_d: The scattered diffuse light reflected from the internal surface of the top cover. This is not ρ_{abs}, and the angle of incidence can be approximated as 60°.

τ: The transmittance of the semi-transparent *cover* sheet.

α: The absorptance of the opaque *absorber* material.

$$f(\tau,\alpha) = \tau \cdot \alpha \sum_{n=0}^{\infty}[(1-\alpha)\rho_d]^n = \frac{\tau \cdot \alpha}{[1-(1-\alpha)\rho_d]}. \tag{13.4}$$

Typically, for a single glass cover and a high-quality absorber film, $f(\tau,\alpha)$ values are reported in the range of 0.70 to 0.75 using traditional window glass. For higher quality glazings (e.g., low-iron glass), the tau-alpha values are in the range of 0.80 to 0.85.[4] As more covers of glass are added (to increase the thermal performance), the tau-alpha value decreases.

So how do we find the values for ρd, the diffuse reflectance from the semitransparent cover? We have already learned in our chapter dealing with the physics of light and selective surfaces that one needs to be familiar with Snell's Law, the angle of incidence, and the material properties of the optical interfaces to process values of reflectance and transmittance. Let us proceed through a quick review of these important topics to the cover-absorber performance metric.

For the simple case of a *single cover* only, we can apply a simplified rapid estimate of transmittance, reflectance, and absorptance, demonstrated in the following equations. This allows us to approximate the total transmittance, τ (no subscripts) in terms of Eq. (13.5), to approximate the total reflectance, ρ (no subscripts) in terms of Eq. (13.6), and to approximate the total absorptance, α in terms of Eq. (13.7):

$$\tau \cong \tau_a \tau_\rho, \tag{13.5}$$

[4] Graham L. Morrison. *Solar Energy: The State of the Art*, ISES position papers 4: Solar Collectors, pages 145–222. James & James Ltd., London, UK, 2001.

$$\rho \cong \tau_\alpha(1 - \tau_\rho) = \tau_\alpha - \tau, \tag{13.6}$$

$$\alpha \cong 1 - \tau_\alpha. \tag{13.7}$$

Most covers of glass and shortwave-transparent polymers (PMMA, polycarbonate, some fluoropolymers) as well as for the sky as a composite cover system. The contribution of τ_α will be close to 0.9, while the reflectivity in the shortwave will approach 0.1.

The *transmission losses due to absorption of light* (τ_α) are calculated knowing (1) the *angle of refraction* (θ_2), (2) the *extinction coefficient* of the material (k) in the band of interest, and (3) the thickness of the material or distance that light must propagate through (d). We see this in Eq. (13.8). The basis for Eq. (13.8) is described from the Beer-Lambert-Bouguer law of materials absorbing light:

$$\tau_\alpha = \exp\left(-\frac{kd}{\cos(\theta_2)}\right). \tag{13.8}$$

Review Figure 13.2, the effective "tilt" of the cover relative to the light reflected up from the absorber to the incident surface of the inside cover can be evaluated as $\beta=0°$.[5] We also observe in Figure 13.3 that this type of diffuse light can have an "effective angle of incidence" of 60° that we will draw from, using the research of Brandemuehl and Beckman from 1980.[6] We will need the effective angle of incidence upon the cover, the index of refraction for the cover, and the extinction coefficient for the cover to estimate the various losses of light that contribute to the cover-absorber performance.

Extinction coefficients convey a quantifiable measure of both transmittance and absorptance in a select band of irradiance. The inverse of the extinction coefficient is the thickness of material that would be required to cause absorption losses of about 36% (decreasing by a factor of $1/e$). As we noted earlier, extinction coefficients tend to be reported in two different scales of inverse distances in the solar industry.[7] For semitransparent cover materials, extinction coefficients are of the scale 4–30 m^{-1}. In contrast, photovoltaic absorbers and

[5] Feel free to turn the page upside down to check this.

[6] John A. Duffie and William A. Beckman. *Solar Engineering of Thermal Processes*. John Wiley & Sons, Inc., 3rd edition, 2006.

[7] Christiana Honsberg and Stuart Bowden. Pvcdrom, 2009. URL http://pvcdrom.pveducation.org/. Site information collected on January 27, 2009.

$$\alpha_a = \frac{4\pi k}{\lambda}.$$

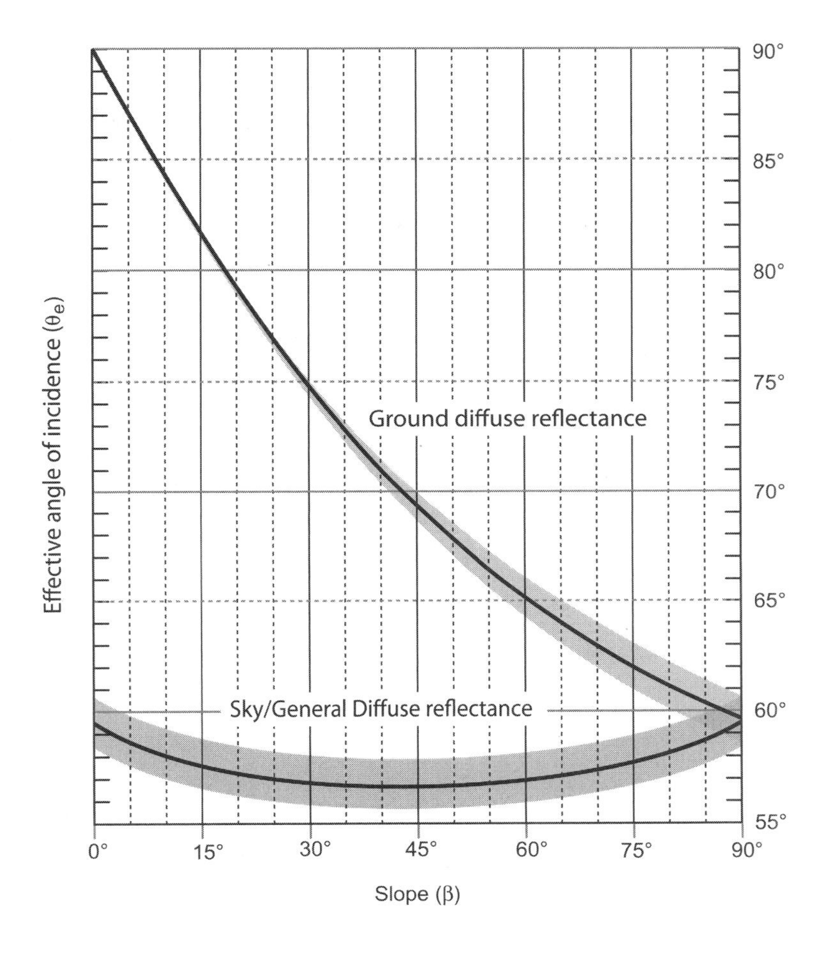

Figure 13.3: Effective angles of incidence used for isotropic diffuse sky irradiance and ground reflected irradiance.Figure adapted from Brandemuehl and Beckman (1980).

selective thermal absorbers will report extinction coefficients on the order of 10^{12}–10^{13} cm^{-1}.

Also, *transmission losses due to reflection* (τ_ρ) can be accounted for in Eq. (13.9), for N layers or covers of transparent material (distinct from n bounces of light in $f(\tau,\alpha)$). One requires the measure of *reflectivity* to calculate the transmittance with respect to losses from scattered light. In turn, we need the *angles of incidence* and *refraction* as well from Eq. (13.10) to solve for $r_\perp$ and $r_\parallel$. Recall that we explored this earlier in Chapter 4, with Eqs. (4.31) and (4.32). We list a short group of common shortwave cover materials below.

$$\tau_\rho = \frac{1}{2}\left(\frac{1-r_\parallel}{1+(2N-1)r_\parallel} + \frac{1-r_\perp}{1+(2N-1)r_\perp}\right). \tag{13.9}$$

SHORTWAVE/VISIBLE BAND INDICES OF REFRACTION FOR COMMON COVER MATERIALS:

- Air is a mixture of gases, the majority of which is N_2: $n = 1$.
- SiO_2; glass: $n = 1.53$.
- PMMA; poly(methylmethacrylate) is a transparent thermoplastic, with the trademark of Plexiglas® (for the hockey fans): $n = 1.49$.
- PTFE; polytetrafluoroethylene is a transparent fluoropolymer, PTFE is sold under the trademark Teflon®: $n = 1.37$.
- PC; polycarbonate is another transparent thermoplastic, with the commercial trademarks of Lexan™ and Makrolon®: $n = 1.60$.

The angle of incidence and the angle of refraction are related using Snell's Law. Shortwave light will propagate through the cover sheet(s), which typically have indices of refraction greater than one (n_{air}). For our goal of finding ρ_d, we will need to know the angle of refraction (θ_2):

$$n_{cover} = n_{air} \left(\frac{\sin(\theta_1)}{\sin(\theta_2)} \right), \tag{13.10}$$

Review Figure 13.4 with Eq. 13.10.

where θ_1 is the angle of incidence θ, and θ_2 is the angle of refraction required for transmittance calculations (see Eq. 13.10).

All of these optical calculations have permitted us to find the value of $f(\tau, \alpha)$, the optical performance measure for a solar energy conversion system. The tau-alpha performance value provides the solar designer with a scaling factor to communicate the diminished absorbed solar energy due to minor optical losses in the *cover-absorber* subsystem, yielding the absorbed irradiance/irradiation values of S.

Typical values of $f(\tau, \alpha)$ are ~0.70 to 0.75 for standard window glazings (single pane), and ~0.80 to 0.85 for specialized low-iron glass.

We can use the tau-alpha function for a flat plate collector (FPC) like a simple solar hot water system, or a photovoltaic system (we will see that PVs get hot in addition to generating electrons). Additionally, tau-alpha is used for utility-scale concentration systems, for the tubular receivers in a solar trough concentration system; and tau-alpha can be used to describe light in a cavity-absorber motif (windows and rooms) as well. However, an optocaloric system is both an optical system and a thermal system. Hence, we also need to incorporate thermal losses (represented as Q_{loss}) into the total assessment of systems performance.

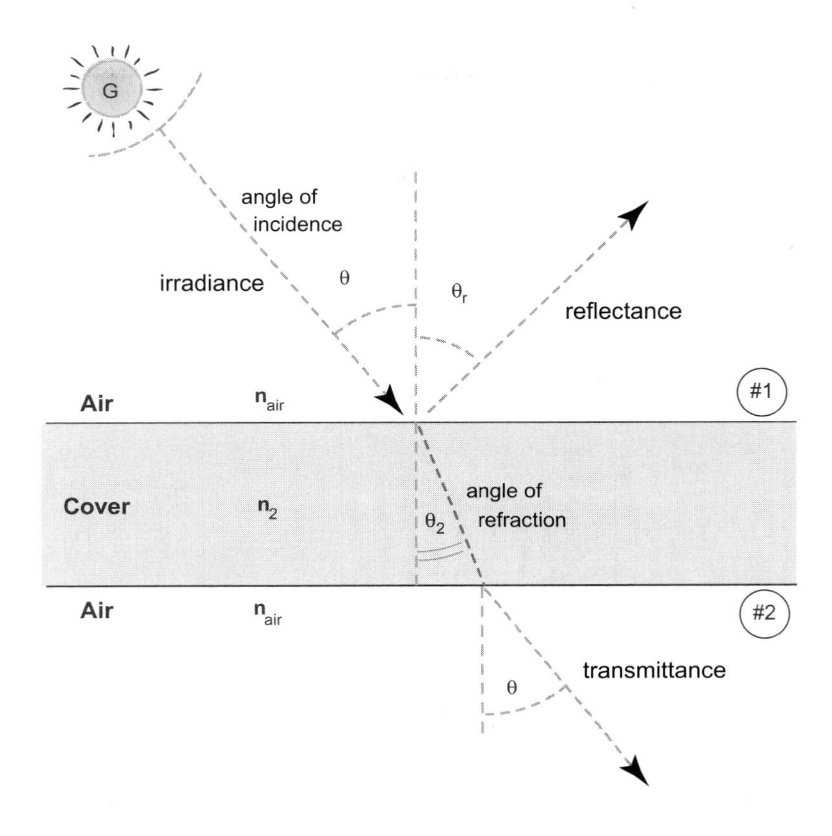

Figure 13.4: The angle of incidence upon surface #1 is equal to the angle of reflection, but different from the *angle of refraction*, according to Snell's Law. Note how the exiting angle into air from surface #2 is also equivalent to the angle of incidence, another result of Snell's Law.

As $f(\tau,\alpha)$ is dependent upon the angle of incidence, the values are expected to vary over the day, and with respect to the contributing components of light across the sky dome. Hence, the use of tau-alpha is broken up according to the components of beam, diffuse sky, and ground reflected light. In Eq. (13.11) we demonstrate this breakdown for the simple isotropic diffuse sky model.

$$S = I_b R_b \cdot f(\tau,\alpha)_b + I_d \cdot f(\tau,\alpha)_d \left(\frac{1+\cos\beta}{2}\right) + \rho_g I \cdot f(\tau,\alpha)_g \left(\frac{1-\cos\beta}{2}\right) \tag{13.11}$$

In order to simplify the use of the tau-alpha function, researchers have determined an empirical approximation for an average tau-alpha value, based

upon the calculated $f(\tau,\alpha)$ for the beam component of solar irradiance $(f(\tau,\alpha)_b)$[8]:

$$f(\tau,\alpha)_{avg} \cong 0.96 \cdot f(\tau,\alpha)_b. \tag{13.12}$$

For a known G_t, we can apply $f(\tau,\alpha)_{avg}$ to estimate S:

$$S \cong f(\tau,\alpha)_{avg} \cdot G_t. \tag{13.13}$$

ROBOT MONKEY "COVERS"
REFLECTIVITY OF POLARIZED LIGHT

Again, thanks to Augustin-Jean Fresnel we have a working model to estimate *reflectivity* of polarized components of light (perpendicular ($r_\perp$) and parallel ($r_\parallel$) polarized) from basic materials properties, seen in Eqs. (13.14) and (13.15) (where θ_1 is set to θ, the common angle of incidence used in all of our solar calculations). Finally, the average reflectivity for the material (r) is calculated in Eq. (13.16) (where G_i is shortwave solar irradiance and G_r is the solar irradiance reflected):

$$r_\perp = \frac{\sin^2(\theta_2 - \theta)}{\sin^2(\theta_2 + \theta)}, \tag{13.14}$$

$$r_\parallel = \frac{\tan^2(\theta_2 - \theta)}{\tan^2(\theta_2 + \theta)}, \tag{13.15}$$

$$r = \frac{G_r}{G_i} = \frac{1}{2}(r_\perp + r_\parallel). \tag{13.16}$$

The detailed equations which couple reflectivity (r) and transmittance (τ), reflectance (ρ), and absorptance (α) for each component of polarization are listed below. For perpendicularly polarized light of a given wavelength, λ:

[8] Soteris A. Kalogirou. *Solar Energy Engineering: Processes and Systems.* Academic Press, 2011.

$$\begin{aligned}
\tau_\perp &= \frac{\tau_\alpha(1-r_\perp)^2}{1-(r_\perp\tau_\alpha)^2},\\
\rho_\perp &= r_\perp(1+\tau_\alpha\tau_\perp),\\
\alpha_\perp &= (1-\tau_\alpha)\left(\frac{1-r_\perp}{1-r_\perp\tau_\alpha}\right),
\end{aligned}$$

while for parallel polarized light of a given wavelength, λ:

$$\begin{aligned}
\tau_\parallel &= \frac{\tau_\alpha(1-r_\parallel)^2}{1-(r_\parallel\tau_\alpha)^2},\\
\rho_\parallel &= r_\parallel(1+\tau_\alpha\tau_\parallel),\\
\alpha_\parallel &= (1-\tau_\alpha)\left(\frac{1-r_\parallel}{1-r_\parallel\tau_\alpha}\right).
\end{aligned}$$

τ_α is transmittance considering only absorption losses.

For general angles of incidence found in G_t, Eq. (4.30) should be used in combination with Eq. (4.33). The measured values are in terms of *reflectivity*, and *not reflectance*, as *reflectance* is a function of both *reflectivity* and *transmittance* for a given wavelength of light ($\rho = f(r_{\lambda,\perp}, r_{\lambda,\parallel}, \tau_\lambda)$). The two mechanisms for optical energy losses include light that is scattered or reflected away and light that is absorbed as it passes through the cover material (even in small quantities).

Congratulations!

Robot Monkey grants you one glowing banana for getting through the mindstretch

We know that the angle of incidence θ varies over the course of a day as the Sun follows the arc from East to West across the sky dome (in terms of hour angle ω). That angle also varies with respect to the time of the year (in terms of declination δ), the location of the site (latitude ϕ), and the slope/tilt and azimuth of the collector (β, γ):

$$\begin{aligned}\theta = \cos^{-1}[&\sin\phi\sin\delta\cos\beta - \cos\phi\sin\delta\sin\beta\cos\gamma \\ &+ \cos\phi\cos\delta\cos\omega\cos\beta + \sin\phi\cos\delta\sin\beta\cos\gamma\cos\omega \\ &+ \cos\delta\sin\omega\sin\beta\sin\gamma].\end{aligned} \tag{13.17}$$

Research for practical modifications of a tau-alpha value to accommodate changes throughout the year has resulted in the use of an angle of incidence modifier factor, K_θ, seen in Eq. (13.18). The calculated values offer rapid use of a fractional modifier (from 0 to 1) for deviations from a collector oriented normal to the Sun, when given calculated angles of incidence:

$$K_\theta = \frac{f(\tau,\alpha)}{f(\tau,\alpha)_n}. \tag{13.18}$$

FIN-RISER-FLUID: THERMAL PERFORMANCE

We can envision a full flat plate collector (such as a solar hot water system) as a series of metal fins that have been coated with a absorbing thin film. Metals are not good absorbers, so the fins will always be coated with an opaque material. While black carbon (even black paint) will work to absorb solar light, in modern systems the absorber material will be a metal sub-oxide or a complex layering of materials that imparts a high absorbance in the shortwave (low SW reflectance), and a low emittance in the longwave band (low-ϵ, a selective surface absorber; also high LW reflectance). Each long fin will have a metal tube (called a riser) bonded/welded to the surface, and heat transfer fluid will be mechanically pumped through the tubing to collect the absorbed and converted thermal energy. It should make sense that some of the converted thermal energy will be unintentionally lost from the system to the surroundings, despite good insulation and systems design. There are performance metrics that describe the effectiveness or efficiency for the fin-riser-fluid ensemble to exchange thermal energy, with the primary goal being intentional heat removal from the solar thermal collector to the heat exchange fluid leaving the panel.

Metals are not good absorbers of irradiation, we need to coat them with a selective surface (ideal, low-e opaque) or a black carbon surface to absorb light.

The engineering goal of solar thermal system design is to minimize thermal losses, and to optimize the useful energy yield for the client in their locale, given a prioritized season for solar thermal performance. The net useful optocaloric or thermal response from a solar thermal flat plate systems is determined under steady state conditions, although real systems will observe fluctuations in optical and thermal responses due to variable solar and ambient temperature conditions. The useful energy collected (Q_u) is termed the useful gain, in proportion to the area of the collector (A_c), calculated as the remainder of the absorbed energy ($A_c \cdot S$) considering thermal losses through convection, conduction, and radiation. Given a system for which the optical performance is assumed to be high ($f(\tau,\alpha) \rightarrow 1$) we focus on both the manner in which one can absorb more light and the strategies to decrease the sources of thermal loss.

The simple Eq. (13.19) can be expanded in Eq. (13.22):

$$Q_u = A_c \cdot q_u = Q_{abs} - Q_{loss}, \tag{13.19}$$

$$Q_{abs} = A_c(I_t \cdot f(\tau,\alpha)) = A_c \cdot S, \tag{13.20}$$

$$Q_{loss} = \frac{T_{p,m} - T_a}{R_L}, \tag{13.21}$$

where $T_{p,m}$ signifies the mean plate temperature (a general term describing the absorbing system) of the solar thermal collector. Q_{loss} is represented in terms of the difference between the mean or average temperature of the plate and the surrounding ambient air temperature outside (T_a). The thermal resistance leading to losses is R_L. We can envision that the mean plate temperature would be much hotter than the surrounding air temperature during the day, when sufficient light is being absorbed. The gradient in temperature is a driving force that leads to thermal losses at the perimeter of the collector.

From these equations we arrive at Eq. (13.22):

$$Q_u = A_c \cdot [S - U_L(T_{p,m} - T_a)]. \tag{13.22}$$

The useful energy from the optocaloric system (Q_u) can also be framed by a second equation (Eq. (13.23)), in terms of the heat transfer fluid being mechanically pumped through the collector. The fluid for a solar hot water system is often

$Q_u = A_c \cdot q_{useful}$.

Due to the heat capacity of most fluids thermal systems have longer time constants for remaining in an excited state. Hence we may substitute hourly irradiation on a tilted surface I_t for irradiance G_t in describing system performance.

The **plate** of a solar thermal system is used as a general term for the absorbing system inside the collector.

contained within a closed loop that exchanges thermally with a hot water tank. The heating loop is then filled with a mixture of propylene glycol and water or water and ethanol as an antifreeze medium. The heat transfer fluid will be pumped at a mass flow rate of $\dot{m}$, and will have a known specific heat capacity of c_p (J/(kg K)). We only need to measure the temperature of the fluid entering the system at the *inlet* ($T_{f,i}$) and the temperature of the fluid leaving the system at the *outlet* ($T_{f,o}$).

$$\dot{Q}_u = \dot{m}c_p[T_{f,o} - T_{f,i}]. \tag{13.23}$$

Hence, we have two equations for Q_u. Equation (13.22) that uses inputs of $T_{p,m}$ and T_a, while Eq. (13.23) uses the heat transfer fluid temperatures at the inlet and the outlet of the collector ($T_{f,i}$ and $T_{f,o}$). We shall demonstrate that we can link Eqs. (13.22) and (13.23) with an empirical parameter called the *collector heat removal factor* F_R. This will allow us to only require the measurement of the inlet temperature and ambient air temperature with the irradiance to evaluate solar thermal collector system performance.

Returning to heat loss, the sum of the various heat losses that are affected proportionally by the thermal gradient can be represented as the total or overall heat loss coefficient, U_L with units of W/(m^2 K). The overall heat loss coefficient (*U*-value) times area (*UA*) is the reciprocal of a series of conductive and convective thermal resistances *RL*, seen in Eq. (A.1):

If system efficiency is a ratio of useful energy output relative to irradiance input, then $\eta = \frac{q_u}{G_t} \sim \frac{q_u}{I_t}$.

$$U \cdot A = UA = \frac{1}{R}\ [\mathrm{W/K}]. \tag{13.24}$$

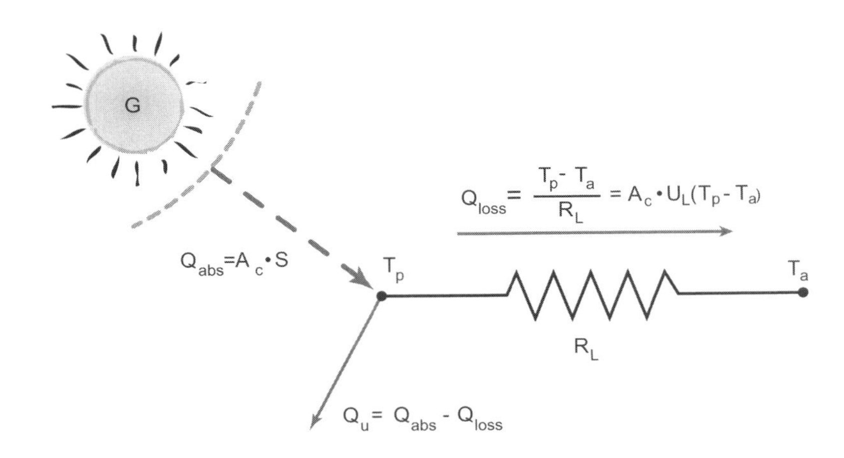

Figure 13.5: Optocaloric diagram of a thermal system indicating Q_u, Q_{abs} or *S*, and Q_{loss}.

We can map Eq. (13.22) to a diagram of the thermal resistance network, seen in Figure 13.6. In turn, by reviewing the schematic cross-section in Figure 13.5, we observe that the sum of Q_{loss} contributions is due to conduction, convection, and radiation energy losses. The total heat loss coefficient is in turn a sum of contributions from the top (U_t), bottom (U_b), and edges (U_e) of the solar thermal collector. This additional assessment can then be diagrammed as a more detailed thermal resistance network displayed in Figure 13.7:

$$U_L = U_t + U_b + U_e. \qquad (13.25)$$

The heat loss coefficient U_L has thermal conductivity units of W/(m² K).

In a flat plate collector system (FPC), the system is designed as a low-profile rectangular box, with a glass cover upon the top and insulation along the sides and the bottom of the box. The U_L values for a FPC will be larger than those for an evacuated tube collector (ETC), and $f(\tau, \alpha)$ metrics will be less important. In contrast, the ETC system has lower thermal losses (as the dominant loss mechanism is from radiative exchange) and $f(\tau, \alpha)$ drive the system performance.

We expect that a cover should be transparent to light, because the *absorber* is the functioning energy conversion device, and covering up the absorber with opaque insulation would negate the SECS function. However, the cover is also the least insulated for thermal performance and has a large exposed area. Thus, U_t, the top loss coefficient is the largest value in a FPC (U_t=6–9 W/(m² K)),

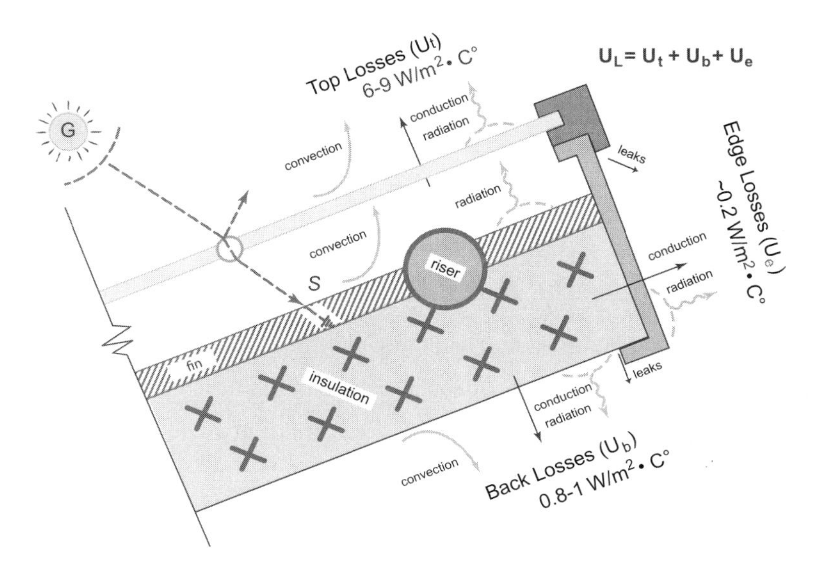

Figure 13.6: Schematic cross-section of a Solar Hot Water module, indicating sources of q_{losses}. (Adapted from Tiwari, 2002.)

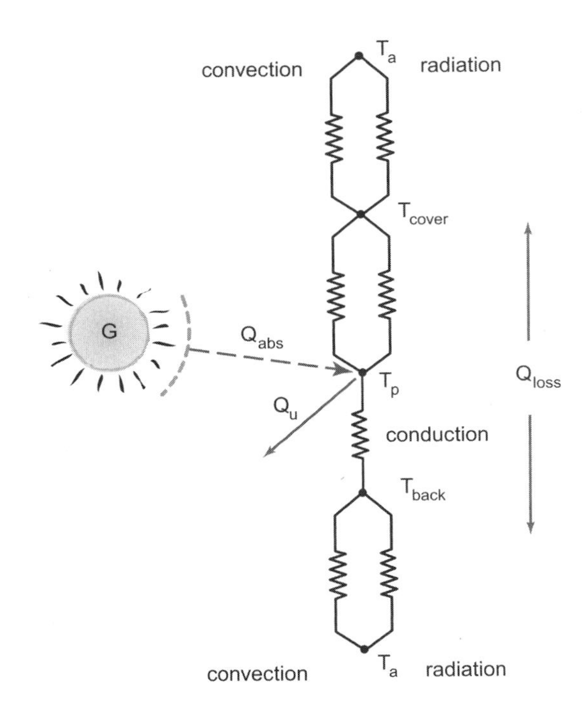

Figure 13.7: Detailed optocaloric diagram of a thermal system indicating Q_u, Q_{abs}, and Q_{loss} for the cover, plate, and surrounding air.

A cover-absorber sub-system can increase the thermal performance relative to a cover-less system, but also slightly decreases optical performance.

followed by bottom and edge losses. Typical back surface heat loss coefficients for FPCs are $U_b = 0.8–1\,W/(m^2\,K)$ and edge coefficients are $U_e \sim 0.2\,W/(m^2\,K)$. The addition of a single or double glass cover allows light to penetrate the system while reducing the major convective losses that occur out of the top without a cover.

We can see from the units of U_L that the rate of energy loss is proportional to the area of the collector as well as the temperature difference between the solar thermal system and the ambient air. This suggests conditions under which very large area collectors will have losses that outstrip the solar gains—something to keep in mind for solar design projects. In essence, the useful energy yield will have limits for a non-concentrating array given the available solar resource, the surrounding air temperature, and the size of the solar thermal array. At very large collector areas, we can envision a scale at which the solar gains match the increase in thermal losses, yielding a net zero (or negative) quantity of Q_u.

COLLECTOR EFFICIENCY FACTOR: F'

There are multiple metrics that describe the performance of a solar thermal system, each of which uses a common letter **F**. The use of F here originates from the word "fin"—as in the fin efficiency to convey thermal energy.[9] The collector efficiency factor F' conveys information about the variation in temperature over a segment when considering the fin-riser connection. We would like the fin, which is connected to the riser tube, *which is connected to the heat transfer fluid* to all have high thermal conductance such that heat exchange is efficient local to the fluid mechanically pumped through the collector. On the small scale, local to a segment of the collector, the riser tube must be in very good thermal contact with the fin, usually through the use of a bonding material with conductance greater than 30 W/(m °C).[10] Full exploration of the thermal exchange among fin-bond-riser assemblies can be found in numerous texts on solar thermal energy conversion:

$$F' = \frac{\text{real energy gain}}{\text{energy gain if } T_{fin} = T_{local}}. \tag{13.26}$$

The efficiency factor of the collector (F') is a representation of how a segment of the solar thermal collector transfers absorbed energy S through the metal fin-riser collector and into the coupled heat exchange fluid locally. As a ratio F' represents the actual or measured useful energy gain (locally) relative to the useful gain that would result under the ideal condition resulting from the absorbing fin surface temperature being the same as the local fluid temperature (T_{local}). This relation can also be framed as a ratio of the thermal resistance between the collector surface and the ambient air (numerator: $\frac{1}{U_L}$) relative to the thermal resistance between the local fluid and the surrounding environment (denominator: $\frac{1}{U_o}$), seen in Eq. (13.27). F' is a design parameter of the actual panel on the market (off-the-shelf), and not something that a design team for SECS project deployment changes. Increased plate thickness and thermal conductivity will increase F', while increased distance between risers will decrease F'[11]:

$$F' = \frac{\frac{1}{U_L}}{\frac{1}{U_o}} = \frac{U_o}{U_L}. \tag{13.27}$$

[9] This F in thermal design is not the same as the *view factor* in geometric relations between graybodies: F_{12}.

Sing! The **fin** bone's connected to the **bond** bone, the **bond** bones's connected to the **riser** bone, the **riser** bone's connected to the **heat transfer fluid**, now shake dem solar thermal bones.

[10] John A. Duffie and William A. Beckman. *Solar Engineering of Thermal Processes*. John Wiley & Sons, Inc., 3rd edition, 2006.

[11] D. Yogi Goswami, Frank Kreith, and Jan F. Kreider. *Principles of Solar Engineering*. Taylor & Francis Group, LLC, 2nd edition, 2000.

The heat transfer fluid enters the collector at a lower temperature than when it exits. So the **local fluid temperature** increases along the path of the fin-riser route. We evaluate the local fluid temperature in segments.

F' then becomes a correction factor reflecting the thermal resistance of the absorber plate relative to the local fluid temperature. Again, the collector efficiency factor is used to illustrate the relationship for a segment of a fin-riser ensemble within the larger collector (like a slice of bread), and defines a ratio of the real thermal losses found in that collector segment relative to the ideal scenario where the collector fin's absorbing surface was the same temperature as the local fluid temperature in the collector. We can evaluate the ratio of the actual useful heat collection rate relative to the useful heat collection rate when evaluating the hypothetical scenario where the temperature of the plate is the same as the local fluid temperature:

$$1/U_o = F' \cdot \left(\frac{1}{U_L}\right). \tag{13.28}$$

$1/U_o$: The resistance thermal loss for a boundary between the *collector fluid* and the ambient air.
$1/U_L$: The resistance thermal loss for the boundary between the collector's *absorbing surface* and the ambient air.

COLLECTOR HEAT REMOVAL FACTOR: F_R

In Eq. (13.22), we note again that the actual useful energy (Q_u) is equal to the optical gains ($A_c \cdot S$) minus the losses due to conduction/radiation and convection, when considering a temperature difference between the ambient air (the environment) and the average temperature of the whole absorber plate—the mean plate temperature ($T_{p,m}$).[12] Measuring $T_{p,m}$ in the field is impractical, given the fin-riser assembly is tucked away inside the insulated cover-absorber flat plate and the temperature will be a range of values across the plate, given the cooler inlet fluid on one side and the hot outlet fluid leaving the plate.

[12] Recall that *radiant exitance* is a unit of energy per area *exiting* from a surface.

$F_R = f(F')$. The collector heat removal factor is a function of the fin efficiency factor.

Researchers developed a method as far back as the 1950s to use a far more convenient measure of the inlet fluid temperature for the module ($T_{f,i}$). The temperature of the inlet fluid is always lower than the mean plate temperature due to the thermal resistance between the absorber plate and the heat transfer fluid (which we just explored for F'), so the system performance requires a correction factor—the collector heat removal factor F_R.

The correction factor F_R can be seen in Eq. (13.32), where the numerator is the useful energy determined as $\dot{Q}_u$ from Eq. (13.23). The denominator is modified from Eq. (13.22), where $T_{f,i}$ replaces $T_{p,m}$[13]:

$$F_R = \frac{\dot{m}c_p(T_{f,o} - T_{f,i})}{A_c[S - U_L(T_{f,i} - T_a)]}, \tag{13.29}$$

$$F_R = f(F') = \frac{\dot{m}c_p}{A_cU_L}\left[1 - \exp\left(\frac{-A_cU_LF'}{\dot{m}c_p}\right)\right], \tag{13.30}$$

$$F_R = \frac{\text{actual useful energy}}{\text{useful gain if } T_{f,i} = T_{p,m}}. \tag{13.31}$$

F_R conveys the ratio of the actual useful energy gain relative to the useful energy gain when the entire collector is at the same temperature as the fluid inlet temperature ($T_{f,i}$). F_R is also a function of F'. Again, $T_{f,i}$ replaces $T_{p,m}$ from Eq. (13.22) in the denominator. The use of F_R in practice allows us to perform a change of variables in our third and final equation for Q_u, Eq. (13.32). Notice how the equation has the same form as Eq. (13.22), but uses measures of the fluid inlet temperature and the air temperature along with S for the input variables:

$$Q_u = A_c \cdot F_R[S - U_L(T_{f,i} - T_a)]. \tag{13.32}$$

Another way to solve the set of problems is in terms of the mean plate temperature:

$$T_{p,m} = T_{f,i} + \left[\frac{\frac{Q_u}{A_c}}{F_R \cdot U_L}\right] \cdot (1 - F_R). \tag{13.33}$$

We shall see that the use of both $f(\tau, \alpha)$ and F_R are essential metrics to convey the performance of various solar thermal collector systems in assessing the long-term performance for systems design.

[13] John A. Duffie and William A. Beckman. *Solar Engineering of Thermal Processes*. John Wiley & Sons, Inc., 3rd edition, 2006; and D. Yogi Goswami, Frank Kreith, and Jan F. Kreider. *Principles of Solar Engineering*. Taylor & Francis Group, LLC, 2nd edition, 2000.

The F_R models were developed in the 1950s by Hottle, Whillier, and Bliss, and updated in 1979 by Phillips. Given most collectors have glass covers, do you think that modern IR cameras would improve the analysis of a mean plate temperature?

Measuring $T_{f,i}$ and T_a are easy, while measuring $T_{p,m}$ would be impractical. We use F_R as a change of variables from $T_{p,m}$ to $T_{f,i}$.

PUTTING IT ALL TOGETHER

We observed the energy balance in Eq. (13.19),and now we can place the elements with their respective contributions in Eqs. (13.34) and (13.35). We're going to assume that we need the collector heat removal factor F_R to make our lives easier in the end (so it pops up in each equation). First, the radiant heat absorbed considering cover-absorber modifier $f(\tau,\alpha)$ for our solar gains (S; using irradiance on a tilted surface as the input G_t). Second, the losses are evaluated given the fluid inlet temperature and the outside ambient air temperature:

$$q_{abs} = F_R \cdot f(\tau,\alpha) \cdot (G_t), \tag{13.34}$$

$$q_{losses} = F_R \cdot U_L \cdot (T_{f,i} - T_a). \tag{13.35}$$

Rewritten: $q_u = F_R \cdot f(\tau,\alpha) \cdot G_t - U_L \cdot (T_{f,i} - T_a)$. The efficiency of energy conversion for any system is the ratio of the useful energy converted relative to the input energy. Hence the efficiency of the collector (η) is simply q_u over G_t:

$$\eta = \frac{Q_u}{A_c \cdot G_t} = \frac{q_u}{G_t}. \tag{13.36}$$

Now we shall expand Eq. (13.36) and cancel out common terms. In this expansion we assume a constant pumping rate of heat transfer fluid, steady state irradiance conditions, and that $f(\tau,\alpha)$ is a cumulative term for the sum of irradiance components ($f(\tau,\alpha)_{avg}$):

$$\eta = \frac{Q_u}{A_c \cdot G_t} = \frac{\cancel{A_c} \cdot F_R \cdot f(\tau,\alpha) \cdot \cancel{G_t}}{\cancel{A_c} \cdot \cancel{G_t}} - \frac{\cancel{A_c} \cdot F_R \cdot U_L \cdot (T_{f,i} - T_a)}{\cancel{A_c} \cdot G_t} \tag{13.37}$$

Through inspection, one observes that the efficiency of a solar thermal system can be written in the linear form of $y = b - m \cdot x$:

The maximum module efficiency occurs at the y-intercept, where $T_{f,i} = T_a$.

$$\eta = F_R \cdot f(\tau,\alpha) - F_R \cdot U_L \left(\frac{(T_{f,i} - T_a)}{G_t} \right). \tag{13.38}$$

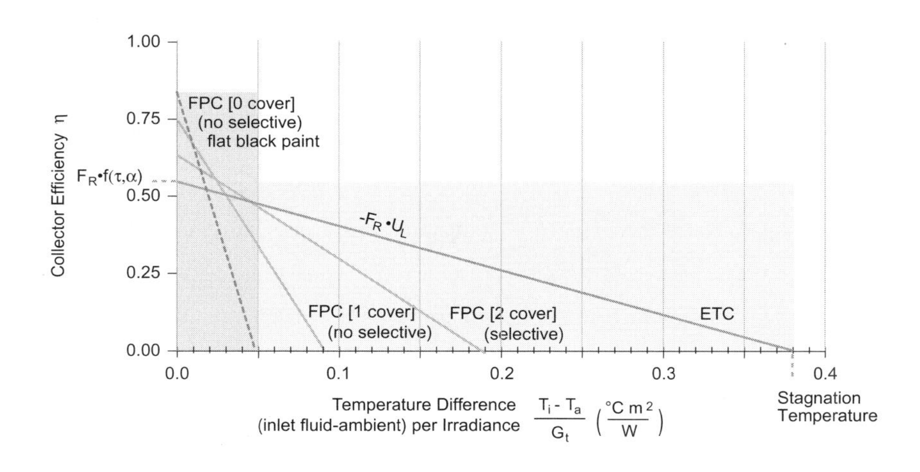

Figure 13.8: Performance curves of solar hot water optocaloric systems evaluated under SRCC testing conditions. Three flat plate collectors are represented, along with one evacuated tube collector system (no concentration). The stagnation condition for the ETC is indicated in this plot.

As seen in Figure 13.8, the intercept of the solar thermal performance curve indicates the maximum module efficiency, at $F_R \cdot f(\tau,\alpha)$, while the negative slope is indicated with $-F_R \cdot U_L$. The x-axis is composed as $(T_{f,i} - T_a)/G_t$, in units of (°C m²)/W. Collectively, the x-axis is a ratio of the temperature gradient between the fluid and the outside air and the irradiance upon the collector. Keep that ratio in mind when assessing collector performance.

The solar thermal collector performance curve is of the form

$$y = b - m \cdot x.$$

y: η

b: $F_R \cdot f(\tau,\alpha)$

$-m$: $-F_R \cdot U_L$

x: $\left(\frac{(T_{f,i}-T_a)}{G_t}\right)$.

At the two extremes of maximum and minimum system efficiency, we observe two limiting conditions. When η is at a maximum $T_{f,i} = T_a$, that is the ambient air temperature is the same temperature of an incoming fluid (say 35–40 °C; quite warm). This indicates that solar thermal collectors do great in warm, sunny weather—but we often demand the highest amount of heat during the Winter, not in the Summer. At the other extreme, the system reaches its maximum operating temperature for a steady state irradiance. The maximum temperature is also called the stagnation temperature, as T_{max} would occur if mechanical pumping is stopped on a clear sunny day, or $Q_u = 0$:

$$T_{max} = \frac{G_t \cdot f(\tau,\alpha)}{U_L} + T_a \text{ [stagnation temp.].} \quad (13.39)$$

So as Winter approaches and the air temperature drops, the system performance will also drop for a given irradiance condition. Of course irradiance varies dynamically over the day, but we have to go back to the balance of absorbed gains and thermal losses in an optocaloric system. Aside from darkness, cold weather conditions will increase $Q_{loss} > Q_{abs}$, resulting in a system that loses as much energy as it gains.

If we think about the balance of gains and losses, we can create a scenario or a special case in Eq. (13.22) where $Q_{abs} = Q_{loss}$, providing a threshold level of irradiance above which the collector provides useful energy. This energetic threshold for a SECS is called the *critical radiation level* ($G_{t,c}$). We set $Q_u = 0$ in Eq. (13.32), separate the solar gain into $S = G_t \cdot f(\tau, \alpha)_{avg}$, and solve for G_t, seen in Eq. (13.40):

$Q_u > 0$ for positive values of $(G_t - G_{t,c})$.

A **closed loop** heating system separates the fluid used for solar heating (e.g., propylene glygol) from the fluid used by the client (e.g., water). A heat exchanger is used inside the storage tank.

$$G_{t,c} = \frac{F_R \cdot U_L \cdot (T_{f,i} - T_a)}{F_R \cdot f(\tau, \alpha)_{avg}}. \tag{13.40}$$

As Q_u can be composed in terms of G_t and $G_{t,c}$, indicating that a net positive quantity of work can be done only when solar gains exceed the threshold of thermal losses, and $G_t > G_{t,c}$:

$$Q_u = A_c \cdot F_R \cdot f(\tau, \alpha)_{avg} \cdot (G_t - G_{t,c}), \tag{13.41}$$

Locale: Fort Collins, CO is home to New Belgium Brewery and Fat Tire Amber Ale—a cherished beer of the 1990s microbrewery explosion. In addition to an eco-aware business strategy, New Belgium Brewery has installed a 200 kW_p photovoltaic array. **Client:** So why not pose a hypothetical solar hot water array here too for the residence of a client that is employed at NBB?

where $Q_u > 0$ for conditions where $G_{t,c} > G_t$. This is a fairly important critical level cross-over in energy transfer, as many solar thermal systems deliver hot air or hot water to a storage tank of some form. Imagine if the collector dropped below the critical irradiance level—the warm fluid entering the collector would be lost to the environment, resulting in a colder fluid leaving the collector than entering it and cold fluid chilling the storage tank instead of heating it. This is why thermal collector systems tend to require an automatic shutoff valve to the pumps when $T_{f,o} < T_{f,i}$.

SHW EXAMPLE

In order to get a sense of modeled system performance over the hours of the day, we develop a solar hot water system example.[14] The hourly useful energy and system efficiency will be calculated for two FPC modules connected in parallel, diagrammed in Figure 13.9. The modules are located in Fort Collins, CO,

[14] John A. Duffie and William A. Beckman. *Solar Engineering of Thermal Processes.* John Wiley & Sons, Inc., 3rd edition, 2006.

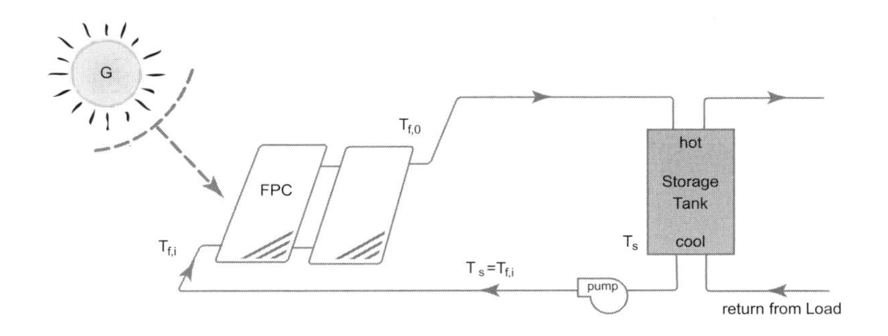

Figure 13.9: A simple diagram of a two-panel FPC for residential solar hot water (SHW). The FPC heating loop is closed (not exposed to the water), using ethylene glycol.

installed on a residential rooftop with a slope of $\beta = 30°$ and collector azimuth of $\gamma = -10°$ (the resident's home was oriented to the southeast—no big deal).

The FPC has one cover consisting of low-iron glass, and the fins are made of thin corrugated copper metal with a copper riser welded on. Each fin-riser assembly has a selective surface of TiNOX.[15] The collector is 2 m long by 1 m wide ($A_c = 2\,m^2$). We assume that our FPC has a very good overall heat loss coefficient U_L of 4.8 W/m^2 °C and 40% propylene glycol mixed with water is the heat exchange fluid (closed loop), mechanically pumped through the collector at a rate of 0.045 kg/s. The collector heat removal factor F_R is assumed to be 0.85 (dimensionless), while the $f(\tau,\alpha)$ is 0.87. The fluid entering the inlet is at 40°, and we assume there is a sensor/controller technology to turn off the pumped flow whenever $q_u < 0$ or if $T_{f,o} < T_{f,i}$.

[15] TiNOX is a selective absorber manufactured by *Almeco Solar*—a composite film of Ti nitride/Ti sub-oxide and an antireflective SiO_2 layer. The optical properties are reported as $\alpha_{sw} \sim 0.90$ (with a small amount of blue light reflected in the visible) and $\epsilon_{lw} \sim 0.04$.

Proposed FPC SHW system:

- A_c: 2 m^2.
- $f(\tau,\alpha)$: 0.87.
- F_R: 0.85.
- U_L: 4.8 W/m^2 °C.
- Fluid: Propylene glycol.
- $T_{f,i}$: 40 °C.

The average loss of energy per area for 1 h (q_{loss}) is calculated in Eq. (13.42) from 9 to 10 in the morning, as 0.25 MJ/m^2. The average solar gain per area in the same hour (S) is 1.77 MJ/m^2. As seen in Eq. (13.44) by calculating the ratio of energy output (q_u) to energy input (I_t; times 100%), we can estimate the average hourly efficiency for the system.

We note in Table 13.1 that the early morning hours before 8 am (solar time) and the evening hours after 5 pm did not yield any useful energy for our hot water system (hence the valve was shut off), although this was also a clear day the entire day. Recall that there is a balance between gains and losses, as well as a critical threshold in gains to be surpass before providing useful solar energy. A total of 8–9 h of solar thermal production is quite good near the equinox time of year, even though we say that there are approximately 12 h of daylight.

Date of evaluation is September 20 (Fall), daylight savings time is "on":

- MDT: UTC—6 h.
- MST: UTC—7 h.
- n: 263 (near equinox).
- δ: 0°.
- β: 30°.
- γ: −10°.

Table 13.1: Table of performance for resident SHW system in Fort Collins on September 20. Energy density values are average integrated values for each hour.

Time (Solar)	T_a (°C)	I_t (MJ/m²)	S (MJ/m²)	$U_L(T_{f,i} - T_a)$ (MJ/m²)	q_u (MJ/m²)	η (%)
6–7a	18	0.32	0.08	0.38	−0.25	–
7–8a	20	1.08	0.33	0.35	−0.02	–
8–9a	22	1.66	1.16	0.31	0.72	43
9–10a	**25**	**2.12**	**1.77**	**0.25**	**1.29**	**61**
10–11a	27	2.62	2.26	0.22	1.73	66
11–12a	28	2.81	2.44	0.21	1.90	68
12–1p	29	2.75	2.39	0.19	1.87	68
1–2p	30	2.56	2.23	0.17	1.75	68
2–3p	30	2.24	1.93	0.17	1.50	67
3–4p	29	1.54	1.29	0.19	0.93	61
4–5p	29	0.86	0.60	0.19	0.35	40
5–6p	27	0.15	0.05	0.22	−0.15	–

The estimated yield over the day can now be summed to 12 MJ/m² on September 20 in Fort Collins for the brewer's home. The day's useful energy must be assessed in proportion to the number of collectors (2 modules) and the surface area of each collector ($A_c = 2\,m^2$):

We assume solar time and clock time are essentially the same here. September 20 is during standard time (no daylight savings), and Fort Collins is just West of the 105° W standard meridian for Mountain Daylight Time (UTC—6 h), so if we were to convert solar time to clock time we would need to subtract 6 min for the wobble of the planet and subtract 60 min to remove daylight stupid time. Love a good location.

$$\begin{aligned} q_{loss} &= U_L(T_{f,i} - T_a) \\ &= 4.8\ \mathrm{W/m^2\ {}^\circ C}(40\text{–}25\ {}^\circ\mathrm{C}) \cdot \\ &\quad 1\ \mathrm{J/(W/s)} \cdot 3600\ \mathrm{s/h} \cdot 1\ \mathrm{MJ}/10^6\ \mathrm{J} \\ &= 0.25\ \mathrm{MJ/m^2}, \end{aligned} \tag{13.42}$$

$$\begin{aligned} q_u &= \frac{Q_u}{A_c} = F_R(S - q_{loss}) \\ &= 0.85(1.77\text{–}0.25\ \mathrm{MJ/m^2}) \\ &= 1.29\ \mathrm{MJ/m^2}, \end{aligned} \tag{13.43}$$

$$\begin{aligned} \eta &= \frac{Q_u}{A_c \cdot I_t} \\ &= \frac{q_u}{I_t} = \frac{1.29\ \mathrm{MJ/m^2}}{2.12\ \mathrm{MJ/m^2}} \times 100\% \\ &= 61\%. \end{aligned} \tag{13.44}$$

For one day of optocaloric conversion, the two-panel system produced 48 MJ of energy, which is equivalent to 45,517 Btus. The average system efficiency was 60% for the day, and provides sufficient heat for the home that day without auxiliary heating from a fuel supplement:

SRCC database: Solar Rating and Certification Corporation certifies solar thermal performance for both liquid and air collectors, including glazed/unglazed FPCs, ETCs, integrated collector systems (ICS), concentrating and thermosyphon systems, along with transpired air collectors.

$$\sum Q_u = 2 \text{ modules} \times \frac{2 \text{ m}^2}{\text{module}} \times 12 \text{ MJ/m}^2 \\ = 48 \text{ MJ}. \tag{13.45}$$

This example was only for solar conversion on a single day, and did not consider the hourly load schedule for hot water demand. When solving this type of problem on an hourly basis for the entire year, including on-site demand for hot water, we use simulation tools such as SAM.[16] The required inputs for collector performance use parametric values of $F_R \cdot f(\tau, \alpha)$ and $-F_R \cdot U_L$, as well as a modifier K_θ (a fraction from 0 to 1) for the changing angle of incidence over the day. An extensive database exists within SAM of solar thermal collector systems that have been tested to identify intercept, slope, and angle of incidence modifiers. The database is developed and maintained by the *Solar Rating and Certification Corporation* (SRCC), a third-party non-profit organization. The SRCC was formed in 1980 in the USA to implement national rating standards and certification programs solely for solar thermal products. The SRCC database allows us to compare systems of solar thermal collectors, including FPC systems with and without covers, with selective surface absorber coatings or non-selective coatings. The SRCC also certifies ETCs, concentrating systems, integrated collector systems (ICS), thermosyphon systems, and transpired air heating systems (solar walls).

[16] P. Gilman and A. Dobos. System Advisor Model, SAM 2011.12.2: General description. NREL Report No. TP-6A20-53437, National Renewable Energy Laboratory, Golden, CO, 2012, 18 pp.

PROBLEMS

1. Sketch a performance plot of efficiency (η) vs. $(T_i - T_a)/G_t$ for the two main solar thermal collector systems described in class (both have selective surface absorbers and glass covers). Label the roles that are played by $U_L, f(\tau, \alpha)$, and F_R (alone or in combination) in describing the performance of one system.

2. Provide the simple equation for useful solar energy given gains and losses, and then separate the gains and losses in separate detailed equations, documenting the contributions of each parameter.

3. Consider three *selective surfaces* having the following properties:

$$\alpha_\lambda = \epsilon_\lambda = \begin{cases} 0.95 \text{ for } 0 < \lambda < \lambda_c \text{ nm} \\ 0.05 \text{ for } \lambda_c < \lambda < \infty \text{ nm} \end{cases} \tag{13.46}$$

where λ_c is separated in three steps from 1800 nm to 3000 nm.

After creating a parametric table of values, calculate the following for each λ_c:

a. We know that the surface of the Sun is at an equivalent temperature of approximately $T_{sun} = 5777$ K. Calculate the absorptance of each surface.

b. The surfaces are heated and maintained at 150 °C each. Calculate the emittance of each surface.

4. Beginning with Eqs. 13.4,13.8, and 13.10, the $f(\tau,\alpha)$ product for two angles in a flat plate optocaloric system (e.g., solar hot water box, solar hot air box): $\theta_1 = 20°$C and $\theta_2 = 60°$C. The flat plate collector has a double-glazed cover and a black absorber plate underneath. The value of the extinction coefficient (k) multiplied by the light path length (d) for each glazing is $kd = 3.7 \times 10^{-2}$. The emittance of the absorber is $\epsilon_p = 0.96$.

5. Consider your results and explain the principle by which the Earth's atmosphere is in fact a selective surface of its own. What is the cutoff wavelength appropriate for the "sky cover," and what is the function of the cover for the absorber/aperture in the system. What factors that you know of will change the performance of the sky cover to decrease the radiative losses of the longwave band into space. Is an increase in efficiency likely from this change?

6. Using Eqs. 13.11,13.12 and 13.13, estimate the radiation absorbed by a single cover optocaloric flat plate system under the following conditions:

- $I = 0.9\,\mathrm{kWh/m^2}$, $I_d = 0.1\,\mathrm{kWh/m^2}$.
- $I_t = 1.6\,\mathrm{kWh/m^2}$.
- $\theta = 21°$, and $\theta_z = 20°$, $\beta = 30°$.
- $\tau = 0.78$, where $kd = 3.7\times10^{-2}$.
- $\alpha = 0.90$
- $\rho_g = 0.2$.

Provide the percent difference of the result os S using τ, α_{avg} vs. using Eq. 13.11

7. A flat plate collector has a single glazing of *low-iron* glass. It is deployed just south of the Canadian border in North Dakota (49° N) with a slope of 35°. For the particular hour of interest here, the angle of incidence of beam radiation is 9° and the zenith angle is 26°. In this hour, we measure $I = 1.44\,\mathrm{MJ/m^2}$ and calculate $I_0 = 1.80\,\mathrm{MJ/m^2}$. The thermal module is stationed over a grass field with $\rho_g = 0.4$, and the glazing has a $kd = 1.3\times10^{-2}$. The absorptance of the plate for total radiation is $\alpha = 0.90$ (assume simple absorptance independent of incidence angle). Given the information for the collector, your knowledge of estimating horizontal components from a clearness index, and the cover-absorber model:

a. What is $f(\tau, \alpha)_b$?

b. What is $f(\tau, \alpha)_d$?

c. What is $f(\tau, \alpha)_g$?

d. Calculate S for the hour using Eq. 13.11.

RECOMMENDED ADDITIONAL READING

- *Solar Rating and Certification Corporation:* Formed in 1980 in the USA, the SRCC is a non-profit organization with a primary purpose to develop and implement national rating standards and certification programs established solely for solar thermal products http://www.solar-rating.org/.

- *Solar Engineering of Thermal Processes*[17]: The third (or fourth) addition to advanced solar thermal engineering. One of the fundamental solar engineering texts for decades.

- *Principles of Solar Engineering*[18]: The second addition to a classic solar text.

- *Fundamentals of Solar Energy Conversion*[19]: A robust complimentary text on solar energy engineering analysis, with useful graphics.

- *Power From The Sun*[20]: A complimentary web resource on solar thermal systems.

- *Solar Energy: Fundamentals, Design, Modelling and Applications*[21]

- *Planning & Installing Solar Thermal Systems: A guide for installers, architects and engineers*[22]: An excellent pragmatic guide for in the field guidance.

- *Solar Energy: The State of the Art*[23]: International Solar Energy Society Position Papers from world leaders in the field.

[17] John A. Duffie and William A. Beckman. *Solar Engineering of Thermal Processes*. John Wiley & Sons, Inc., 3rd edition, 2006.

[18] D. Yogi Goswami, Frank Kreith, and Jan F. Kreider. *Principles of Solar Engineering*. Taylor & Francis Group, LLC, 2nd edition, 2000.

[19] Edward E. Anderson. *Fundamentals of Solar Energy Conversion*. Addison-Wesley Series in Mechanics and Thermodynamics. Addison-Wesley, 1983.

[20] William B. Stine and Michael Geyer. *Power From The Sun*. William B. Stine and Michael Geyer, 2001. Retrieved January 17, 2009, from http://www.powerfromthesun.net/book.htm.

[21] G. N. Tiwari. *Solar Energy: Fundamentals, Design, Modelling and Applications*. Alpha Science International, Ltd., 2002.

[22] German Solar Energy Society (DGS). *Planning & Installing Solar Thermal Systems: A guide for installers, architects and engineers*. Earthscan, London, UK, 2nd edition, 2010.

[23] Jeffrey Gordon, editor. *Solar Energy: The State of the Art*. ISES Position Papers. James & James Ltd., London, UK, 2001.

- *System Advisor Model (SAM)*: NREL software to explore the performance of solar hot water and concentrating solar power systems, with financial modeling (free).
- *RETScreen*: Canadian software to explore the performance of solar hot water and concentrating solar power systems, with financial modeling (free).

SYSTEMS LOGIC OF DEVICES: OPTOELECTRONICS

14

I sing the body electric

I celebrate the me yet to come

I toast to my own reunion

When I become one with the sun.

Michael Gore and Dean Pitchford, Fame (1980)

THERE ARE SEVERAL similarities between optocaloric and optoelectronic systems. We recall that all SECSs are thermodynamically open, non-equilibrium systems. The temperature of the cell, in particular the "temperature" of the charge carriers inside the cell, is far from the ambient temperature of the surroundings. The systems-surroundings diagram for a photovoltaic energy conversion device (ECD) in Figure 14.1 should appear quite similar to the solar thermal ECD displayed earlier in Figure 13.1. In reviewing the figure, we observe that the absorbing collector will accept both longwave and shortwave irradiance, although shortwave energies above a threshold called the *band gap* will be the only light driving charge carrier generation. Other wavelengths that may be absorbed only contribute to an optocaloric response (i.e., warming up module) which is undesirable in the case of photovoltaics. Again the system will lose some energy (and conversion efficiency) to

Remember that for the purpose of this text, optoelectronics implies the interaction of radiation with matter to elicit an electronic response (by tradition, this has also meant the input of electrons to elicit a irradiation response).

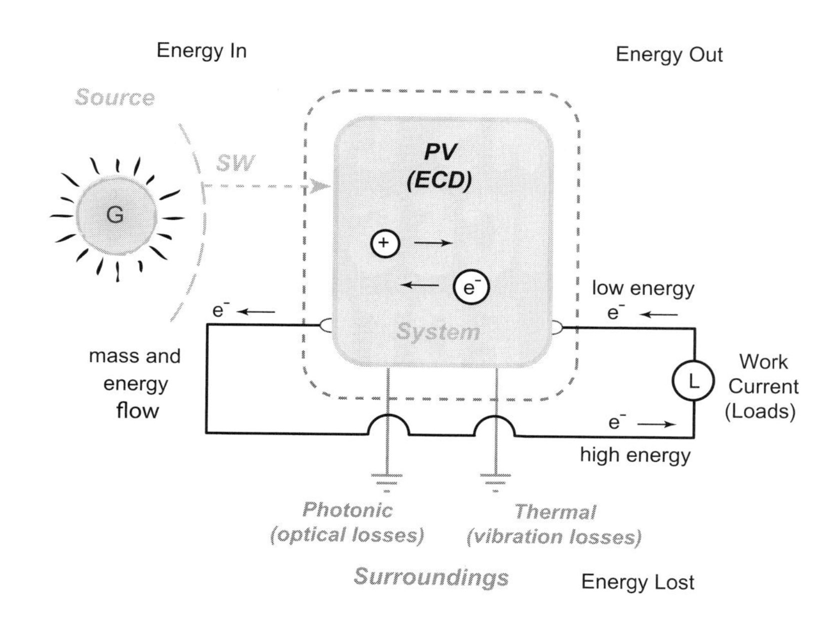

Figure 14.1: Schematic of an optoelectronic device interacting as open energy conversion device. $T_a \neq T_{ECD}$.

both optical and thermal energetic losses. Here, light absorption leads to the photogeneration and separation of positive and negative *charge carriers* (holes h^+ and electrons e^- in a solid state semiconductor). Mass is exchanged across the system/surroundings boundary in the form of electrons, along with the energy imparted from the absorbed photons. Useful energy is delivered initially in the form of direct current electrical power (W_{dc}), which may then be used directly to power a load such as charging a battery, or converted into alternating current power (W_{ac}) to supply power to the grid and/or our buildings.

Consider what can be done with light to elicit a change from a boring material in the dark to an *excited* system that may deliver useful electrons into an circuit. In principle there are two physical mechanisms to generate electric current from light in semiconductors: the *photovoltaic effect* and the *thermoelectric effect*. Note that we do not include the *photoelectric effect* in our simple list of light-harvested optoelectronic energy conversion strategies. This is because the photoelectric effect is not a galvanic process (energy producing), rather it is an electrolytic process (energy demanding). The photovoltaic effect is a galvanic process that generates and separates excited charge carriers with intrinsic potential derived from the photon absorption process. The thermoelectric effect is also an energy producing process that generates and separates charge carriers derived from a strong thermal gradient across a material. We already know that photons can be used as optocaloric currents of work to create a thermal gradient, and we already use thermocouples in our pyranometers to measure irradiance. We are also aware of photodiodes in light sensors that absorb light to generate electric power—photodiodes are photovoltaics (just on a very small scale).

In terms of power generation from current technologies, we will emphasize photovoltaics in this chapter. Photovoltaics are a proven large-scale technology, and PV is the growing technology that is doubling in size ever 1–2 years globally. While we will describe the function of PV devices here, it should be recognized that PV is a very mature technology with over 60 years of industry innovation behind it. The developments of thermoelectric technologies are not insubstantial, and we should expect transformative innovations from that framework in the near future. In society, we already use thermoelectric technologies for sensor development and for smaller-scale energy harvesting.

VIGNETTE: Two words. Very similar etymologies, yet very different meanings: **photoelectric** and **photovoltaic.** The photoelectric effect is a property of materials to emit electrons into a vacuum under bias (an applied potential difference between metal electrodes), now traditionally applied to materials characterization.

The *photovoltaic effect* is **not equivalent** to the *photoelectric effect.* In 1887, then 30-year-old German physicist Heinrich Hertz observed that metal electrodes (with an applied electrostatic bias) separated by a small space of air would "spark" across the gap more easily with the addition of ultraviolet electromagnetic waves (which we now call photons) incident upon the electrode surfaces. The photons had sufficiently high energy (short wavelengths) to cause electrons bound within the metal to be ejected from the confines of the metal into free space, then jumping to the next nearest metal at the opposite electrode. This observation was expanded by a number of well-recognized experimentalists including J. J. Thomson (1899) and Nikola Tesla (1901). Tesla even submitted and received a patent to charge a capacitor using a metal plate (US685957). The Hertz observation was an example of an optoelectronic effect, but it was not the photovoltaic effect.

What if we wanted to absorb a photon of sufficient energy (lower than the energy required to eject it into space) to promote a low energy electron into an open shelf of energy (an orbital or band) while still bound via the attractive force from the atomic nuclei? Such a process, absorbing light and maintaining the excited electron within the material of excitation, describes the first two steps in the *photovoltaic effect*. The final step is to separate excited charge carriers to electrically conductive electrodes (ohmic contacts).

In historical contrast, the photovoltaic effect was observed and documented much earlier, in 1839 by then *19-year-old* French experimental physicist Alexandre-Edmund Becquerel. Becquerel observed a photocurrent (no sparks here, just electronic power) from light-sensitive electrodes immersed in an acidic solution. The electrodes were platinum metal, coated with AgCl- or AgBr-silver salts much like our older photographic film materials. Then in 1877, Adams and Day published their work on selenium metal immersed in an electrolyte bath, and exposed to light. This was followed by C. E. Fritts in 1883, who compacted a selenium photovoltaic cell into a flat plate using gold leaf and brass electrodes.

Perhaps some confusion entered the literature 50 years later, when L. O. Grondahl observed the photovoltaic effect in copper/copper oxide materials, but titled his 1933 research: *"The copper-cuprous-oxide rectifier and photoelectric cell"*. Yes, there is that problem with similar etymologies. Even in their 1954 paper from

Bell Labs research, authors Chapin, Pearson, and Fuller used the more generic title of *photocell*, and would term their device a *photobattery* in everyday speech (*"A new p-n junction photocell for converting solar radiation into electrical power"*). In fact, the term *photobattery* is a perfectly accurate way to describe the devices using the photovoltaic effect. We would even venture that a photocapacitor is a more likely descriptor for Tesla's 1901 device in comparison.

Now where does Einstein come in? Oh yes, in 1905 his paper *"On a Heuristic Viewpoint Concerning the Production and Transformation of Light"* was published. This paper (re-)established the concept of electromagnetic radiation as quanta of photons, giving great weight and validity to the idea that a massless photon (of a known frequency) could exchange energy with an electron to promote the charged species to an excited state. Later, in 1915, Robert A. Milliken demonstrated experimentally how this exchange was possible. Einstein's explanation was the attribution for his 1921 Nobel Prize in Physics.

In fact, the word similarity between "photoelectric" and "photovoltaic" has led to the most commonly confused terms in solar energy, leading to a subsequent stretch of logic that Einstein was somehow responsible for the invention of photovoltaics. *Albert Einstein did not discover photovoltaics, in theory or in practice.* Even major solar companies and texts have made this error, so don't feel bad if you once were susceptible to the overly convenient link between Einstein and PV. A truly great scientist, yes—but Einstein's contributions were much broader than just PV; his work was related to the fundamentals of light as quanta (photons) and light-matter interactions. Einstein was instrumental in enabling other scientists to explore and measure the processes for both the photoelectric and photovoltaic effects.

In summary…

Again, the photoelectric effect **is not** the photovoltaic effect, and Einstein *did not invent PV*. Repeat until it sticks.

The Photoelectric Effect:

- Absorb light (photons)
- Excite electrons sufficiently to eject them from their bound state (about nuclei) to a free, kinetic state in space.

The PE effect describes an energy transformation *from potential energy to kinetic energy*.

The Photovoltaic Effect:

- Absorb light (photons)
- Excite charge carriers sufficiently to promote them from a lower energy bound state to a higher energy bound state.
- Separate charge carriers to ohmic contacts.

The PV effect describes a different energy transformation *from low potential energy to high potential energy*.

CHANGES IN ELECTRONIC STATE

Excited State: describes a high energy potential for an electron or a hole inside of a semiconductor upon light absorption or thermal increase. Excited states are not in a ground state, as they have been [clap hands] "pumped up" (to be declared with an Austrian accent).

Ground State: a lower energy "rest" level for charge carriers at thermal equilibrium.

As the field of solar energy conversion systems is composed of people from diverse backgrounds, let us start by explaining the concept of *state* to those who have not come across it, when connected to electromagnetic and electronic phenomena. Put generally, a *state* here is not a geographic boundary within a nation, but rather another kind of address describing the relative placement and orientation of quantum particles such as the photon and the electron. Remember that things like electrons will only be immobile at absolute zero temperature (a thermodynamic impossibility), so the relative address is more like a pointer in a computer program–even though the real position changes, we know how to describe the object by it's relative address.

We will use simple address descriptors for our electrons, such as being in the *ground state* or the *excited state*. To aid our understanding, we will use the analogy of books on a large bookshelf, where the uppermost shelves are largely empty because a quirky roommate has ordered that books must be arranged from the bottom-up. Now, the ground state describes an electron bumping around among other electrons at a given temperature in the lowest shelves. The excited state describes electrons jumping around at higher energy states, among the upper shelves.

PHOTOVOLTAIC EFFECT

Again, SECSs are Energy Conversion Devices acting as *heat engines*. Heat engines are characterized by being *open systems* (allowing mass and energy transfer across the systems-surroundings boundary) and they are *non-equilibrium systems*. The photovoltaic material responds to the Sun[1] by absorbing photons, pumping up the ensemble of energy levels inside the PV device to excited levels as charge carriers (e.g., electrons and holes). The PV device is also designed to separate the excited charge carriers to their respective metal (ohmic) contacts, so that electrons can leave the system through electrical wiring and ultimately do work before returning back to the PV device—and that is it. No magic, no special physics knowledge required at this level. You may have noticed how there was no explicit mention of a "p-n junction" anywhere. *Photovoltaic action* is just three steps occurring within the PV system, followed by doing work in the surroundings: (1) absorb light, (2) photogenerate charge carriers, and (3) separate the charge carriers to the electrical contacts so that they can (4) do work. Communication of the simple process of PV action is often obscured by popular science literature and general texts on energy due to an emphasis on step 3 alone, which describes the need for a *discriminating mechanism* to drive charge separation. The possible multitude of mechanisms enabling ambipolar charge separation are discussed in great detail in PV device texts (electric field, electrochemical potential, effective fields), and exploring the mechanisms in detail can leave the general practitioner of project development (or the beginning student) "lost in the weeds." Additionally, the full field of PV device research in the laboratory is now much more generalized due to the developments in dye-sensitized PV, quantum dot semiconductors, and organic polymer PV approaches, which do not necessarily require a p-n junction for charge separation.

What is important for the systems designer and the broader integrative design team: process of PV action can be simplified to three steps, with one additional step to do work. We are particularly interested in PV technologies as a *commodity*, where the technology is proven and the performance and unit

[1] Remember, light functions like the current or work for a pump: in this case a *heat pump*. See Appendix A for a more detailed discussion of coupling heat pumps with heat engines.

Three steps of PV Action within the system + work outside of the system:

1. **ABS** (S)
2. **GEN** (V_{oc})
3. **SEP** (J_{sc})
4. WORK (R_{ch})

While the **p-n junction** enables charge separation in many PV technologies, it is only part of the third step of PV action. Keep it simple and learn the three steps of PV action first. Then go take a course on PV device physics if you are still interested!

costs are established—off-the-shelf technology for the most part. The basic functioning of a PV cell and module is described, but mainly in terms of assessing and comparing units to deliver the highest solar utility for the client in their given locale: the goal of solar energy design.

CELL-MODULE-ARRAY

In PV design, the fundamental unit is a *cell*, a singular integrated component that is functional alone to produce the photovoltaic effect, much like the fundamental unit of batteries is a *cell*. By linking cells in a circuit, on a panel, we form an integrated *module*, which is what installers mount on rooftops. Module design will often include cells connected in *series*, such that the cumulative voltage of the singular cells adds up to a net higher voltage unit. The term *panel* is more generic, but can be used synonymously with *module*. Finally, when a network of modules are integrated (installed) as a unified power system, that ensemble is termed an *array*, or assemblies of modules. Arrays can also be connected in parallel or in series, and a strategy is developed to provide high current densities appropriate to the *inverters* for the project (converting DC power to AC power). At a very simplistic level, we typically see cells assembled in series inside of a module, and modules largely deployed in parallel circuits within an array. Hence, the modules provides a net voltage for the array, and

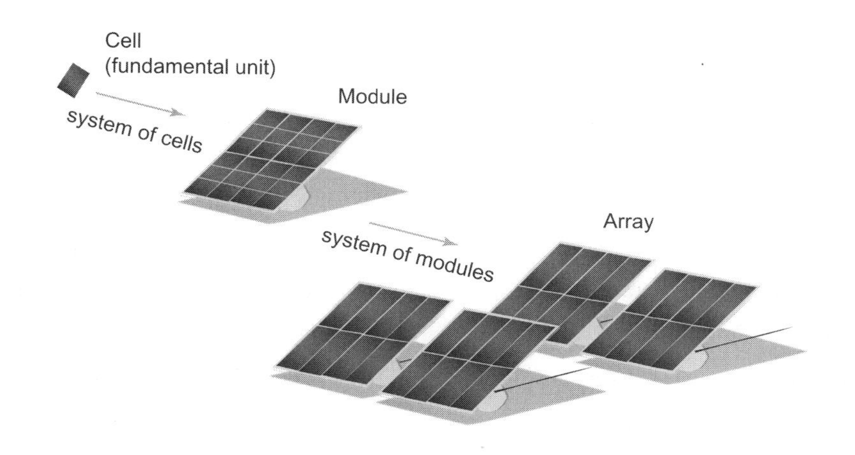

Figure 14.2: Illustration of systems leading from the fundamental unit of the photovoltaic *cell* to an *array* of *modules*.

the assembly of modules provides a net current given a constant solar input (see Figure 14.2).

Again, the PV cell is directly analogous with the battery cell (both are galvanic devices), and original commercialization of the photovoltaic from Bell Laboratories described the technology as the *solar battery*. The power density output from any galvanic cell (photovoltaic, battery, fuel cell; W/m^2) is equal to the current density (J; A/m^2) times the photogenerated cell voltage (V) for a given circuit resistance.

Like batteries, photovoltaic fundamental units are also described as **cells**.

The CEC performance model is from the California Energy Commission, and uses the inputs for the 5-parameter model described by Duffie and Beckman. The original research theses by De Soto (with Klein and Beckman) and by Neises (with Reindl and Klein) are linked within the SAM software from NREL.

$$P_{out} = I \cdot V = [\mathrm{A}] \cdot [\mathrm{V}] = [\mathrm{W}] \qquad (14.1)$$

$$\frac{P_{out}}{\mathrm{area}} = \frac{I}{\mathrm{area}} \cdot V = J \cdot V[\mathrm{A/m^2}] \cdot [\mathrm{V}] = [\mathrm{W/m^2}] \qquad (14.2)$$

When cells are tabbed together (connecting the circuit together by soldering) in series for a module, the voltage is increased in proportion to the number of cells. Hence a reported silicon module voltage will be significantly higher than one-half volt. We can explore the CEC performance model database within the SAM software package from NREL to explore this property.[2] For example, a module with 72 sc-Si cells connected in series has been measured to have a maximum power voltage of ~41 V. Dividing 41/72, we find an average cell voltage of 0.57 V (direct current power only). A module by the same manufacture, using the same sc-Si cell technology, has connected 96 cells in series with a maximum power voltage of ~54 V. Dividing 54/96, we arrive at almost the same value of 0.56 V. Another company produces a third PV module with 72 cells of mc-Si in series for a maximum power voltage of 39.7 V, yielding an average cell voltage at maximum power condition of 0.55 V. All three technologies derive from monolithic silicon material, where the intrinsic property of the Si band gap determines the voltage of the operating cell. In contrast, a manufacturer of CdTe/CdS thin film modules (where CdTe is the analogous absorber to Si) has connected 77 cells in series to produce a module with a maximum power voltage of ~49 V, which delivers and average cell voltage of 0.64 V—significantly higher than those from either sc-Si or mc-Si.

[2] P. Gilman, A. Dobos. System Advisor Model, SAM 2011.12.2: General description. NREL Report No. TP-6A20-53437, National Renewable Energy Laboratory, Golden, CO., 2012. 18 pp; and System Advisor Model Version 2012.5.11 (SAM 2012.5.11). https://sam.nrel.gov/content/downloads. Accessed November 2, 2012.

The **band gap** of crystalline Si is $E_{g(Si)} \sim 1.1$ eV, while the band gap of CdTe is larger at $E_{g(CdTe)} \sim 1.4$ eV.

V_{oc}: intrinsic property of the absorber material (Si, CdTe, CIGS, or dye)

J_{sc}: extrinsic property of the cell, proportional to the area of the cell (A_c) the density and spectrum of the incident photons, the optical properties of the cell, and the collection probability of the cell

η: efficiency of the cell, the ratio of the electricity produced (in W/m^2) to the irradiance absorbed (also in W/m^2)

When photovoltaics are designated for a utility application, then the system can be termed a solar power station, a solar park, or a *solar farm* (or ranch). The scale of a solar farm is rated on the order of MW-peak (MW_p) capacity, or hundreds of MW_p, and are typically composed of many sub-arrays. Areally, a solar park or farm can cover large land areas, and so the pre-assessment and integration of PV in terms of long-term environmental impact and ecosystems services is critical.

PV CELL CHARACTERIZATION

When testing cells or cell assemblies (modules), the system is tested for current and voltage in both *dark* and *light* conditions. Typically, standard bench-top testing is performed under simulated AM1.5 conditions (1 sun, 1000 W/m^2), at 25 °C, and for a short period of analysis (seconds, as opposed to years of operation). For many PV technologies, the diode and photodiode behaviors under dark and light conditions can be represented by a simplified diode equation, called the ideal diode equation.

The value of kT is a link between the macroscopic and the microscopic, yielding the most probable measure of energy for a particle with temperature T. Hence room temperature energy (T = 300 K) is represented by 25.6 meV of energy.

This is analogous to our convention to convert wavelengths to energy using $\frac{h\nu}{\lambda} = 1239.8/\lambda$.

Under dark conditions a PV device will follow the behavior diagrammed symbolically in Figure 14.3 and Eq. (14.3). The use of I is for direct current (in A or mA), while the current density for direct current is represented as J. We convert from J to I by multiplying the area of the collector (A_c).

$$I = I_0 \cdot \left[\exp\left(\frac{qV}{nkT}\right) - 1\right], \tag{14.3}$$

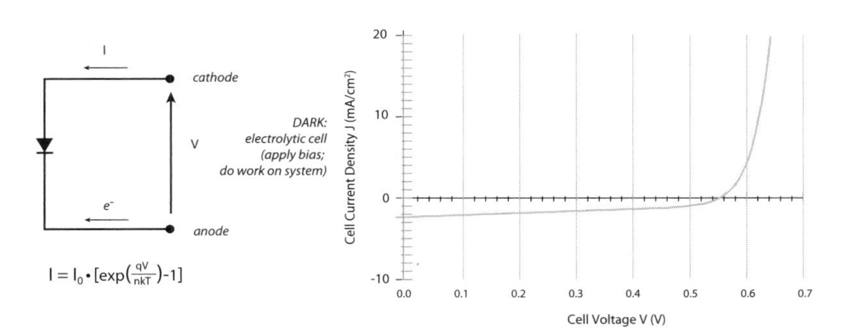

Figure 14.3: Current density vs. potential, or I-V curve for PV cell in the dark.

where I is the measured current given an external applied electrical potential, termed a forward bias. The term I_0 is called the dark saturation current, which is the leakage current for the diode in the dark.

In addition, n is the ideality factor—a modifying value between 1 and 2 that represents the real device's deviation from an ideal diode behavior. k is Boltzmann's constant relating energy with temperature (8.617×10^{-5} eV/K or $k = 1.381 \times 10^{-23}$ J/K) and T is the temperature of the PV cell. The ratio of $\frac{kT}{q} = V_T$ is termed the thermal voltage in semiconductor physics, where q is the electrical charge on a single electron ($q = 1.602 \times 10^{-19}$C). At room temperature of $300\,K$, $V_T = 25.85\,\text{mV}$.[3] From this additional information, we can finally see that the exponential function in Eq. (14.3) acts upon the argument of the ratio of measured voltage V relative to the thermal voltage times the ideality factor $n \cdot V_T$, or $\frac{V}{n \cdot V_T}$.

[3] Christiana Honsberg, Stuart Bowden, Pvcdrom, 2009. http://www.pveducation.org/pvcdrom. Site information collected on Jan. 27, 2009.

The PV device tested in the dark behaves as a type of *electrolytic cell* in generic electrochemical terms, a device that yields significant current and voltage only when we apply work to the system—when we dump energy into the system. In contrast, a PV cell is meant to generate power in the presence of sunlight, and hence behaves as a type of *galvanic cell*, a device that delivers positive current and voltage and hence net positive power to do work in the surroundings (like providing electricity to the grid or powering a home).

In Figure 14.4 we show the behavior of an illuminated PV device diagrammed symbolically, along with the representative I-V curves. Once again, the device can be modeled mathematically using a photodiode equation, seen in Eq. (14.4).

$$I = I_L - I_0 \cdot \left[\exp\left(\frac{qV}{nkT}\right)\right], \tag{14.4}$$

Due to the convention where a power-generating device has a net positive current, the axes are flipped from negative to positive (or from $I = I_0(\ldots) - I_L$ to $I = I_L - I_0(\ldots)$). In the case of an illuminated device, the "-1" term seen in Eq. (14.3) becomes insignificant and is thus dropped from Eq. (14.4).[4]

[4] Christiana Honsberg, Stuart Bowden, Pvcdrom, 2009. http://pvcdrom.pveducation.org/. Site information collected on Jan. 27, 2009.

Figure 14.4 also shows an additional term added to the end of the photodiode equation. These terms represent the real sources of device inefficiency, and are sources for a decreased *fill factor* (*FF*) in assessing device performance.

$$I_{loss} = -\frac{V + IR_s}{R_{sh}}, \tag{14.5}$$

where R_s and R_{sh} represent parasitic losses in current from series resistance in the system and shunt resistance in the cell. Parasitic resistances are effectively thermal losses for the PV system. Parasitic losses exist in contrast to the *characteristic resistance* R_{ch} that occurs from the external *load L* to work.

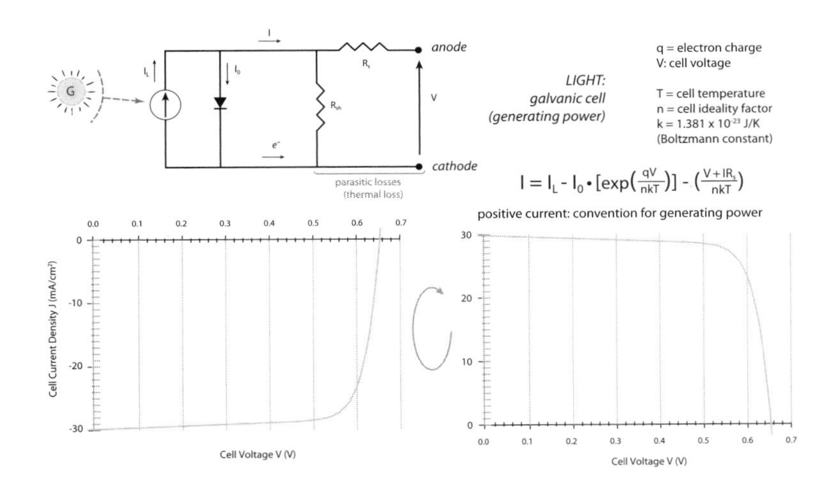

Figure 14.4: Current density vs. potential under illumination of AM1.5 conditions, or I-V curve for PV cell in the light. Under conditions of power generation, the convention is to represent current as positive. We also show that a PV cell has sources of parasitic resistances (R_{sh} and R_s) which are sources of thermal loss in the system.

CHARACTERISTIC MEASURES OF CELLS AND ARRAYS

When we break the three steps of PV action apart, we notice several characteristic measures that may be associated with each. The light absorption step (ABS) is affected by the material's intrinsic absorption coefficient ($\alpha = \frac{4\pi k}{\lambda}$), indicating the depth that light will penetrate into the PV absorber material before being absorbed (for a particular wavelength). The PV absorbing material (like Si or CdTe) will also have a threshold energy level, above which charge carriers will be generated from light, and below which the light will either be transparent or will absorb to raise the thermal temperature of the PV device. Finally, a systems summary value for photovoltaic modules that incorporates the optical losses according to the cover-absorber system ($f(\tau, \alpha)$) is S.

$\alpha = \frac{4\pi k}{\lambda}$ used here for photovoltaics is still distinct from the absorptance use of α in greybody accounting. And yes, it is a repeated symbol with two meanings!

Keep in mind that the photovoltaic system still has a transparent glass cover (often tempered glass) with an opaque absorber. Hence a PV panel also behaves as a flat plate cover-absorber, such that S remains an evaluation of the total energy absorbed from an initial irradiance condition of G_t. Hence, the use of the tau-alpha function can be applied to the PV system. Also, although the *useful* band of light for a photovoltaic module accepts wavelengths smaller than the optoelectronic transition ($\lambda(\text{nm})<\lambda_g(\text{nm})$), meaning eV energies larger than the band gap ($E(\text{eV})>E_g(\text{eV})$), a photovoltaic cell will absorb most of the shortwave and longwave irradiance, increasing the temperature of the system.

Remember, smaller wavelengths (λ(nm)) mean higher energy photons (E(eV)).

E_g is the **band gap energy**, where E(eV) is energy in units of electron volts. Wavelength is proportional to the reciprocal of energy.

$$S = I_b R_b \cdot f(\tau,\alpha)_b + I_d \cdot f(\tau,\alpha)_d \left(\frac{1+\cos\beta}{2}\right) + \rho_g I \cdot f(\tau,\alpha)_g \left(\frac{1-\cos\beta}{2}\right) \tag{14.6}$$

Again, absorption of light in a PV system activates two phenomena, an optoelectronic response and an optocaloric response. As we shall see, only the optoelectronic response will lead to electrical power for work, but almost all solar wavelengths will warm a "dark" opaque material.

PV performance drops when cells become very hot. Thermally cool PV systems that are also exposed to lots of light have happy long-lived cells.

- optoelectronic response: where electrons are excited from the ground state to the conduction band (free)
- optocaloric response: where photogeneration leads to vibrating waves of atoms (phonons).

The photogeneration step (GEN) describes the formation of charge carriers separated by a maximum potential (voltage), without separating the charges. This is effectively associated with the steady state I-V (current-potential) testing condition where the cell is exposed to light with an external resistance ($R \rightarrow \infty$) so large as to forbid current from flowing into an external circuit. Hence, we may associate the *open circuit voltage* (V_{oc}) with step two.

PHOTOVOLTAIC ACTION—Three Steps in the System +WORK:

1. **(ABS)** Light *absorption* process that causes a transition in the material from a ground state (low energy) to an excited state (high energy). This is coupling the PV heat engine to the solar heat pump. [α, E_g, **and** $G \cdot f(\tau, \alpha) = S$]
2. **(GEN)** *Photogeneration* of "free" charge carriers as a negative-positive pair. Traditionally: electrons (e^-) and holes (h^+) (the absence of an electron). [V_{oc}]
3. **(SEP)** *Separation* of charges via an *asymmetry* in the device. The asymmetry serves as a *discriminating transport mechanism* to anode and cathode contacts (also called *ohmic contacts*. [J_{sc}]
4. **WORK** The carriers leave the system as high potential electrons, entering the external leads (wiring). The electrons then lose energy due to the resistance in the circuit and due to the *characteristic resistance* from the Load (L). Ultimately the low energy electrons combine with the positive charge carriers on the opposite contact, returning the charge carrier pair to the ground state, [R_{ch}].

The charge separation step (SEP) describes the discriminating mechanism by which opposite charges are separated to electrical contacts (ohmic contacts). This is effectively associated with the steady state I-V (current-potential) testing condition where the cell is exposed to light with no external resistance ($R \rightarrow 0$), causing a short circuit condition called *short circuit current* (I_{sc}, or J_{sc} for short circuit current density). The process of charge separation incorporates some physical attribute within a PV device that acts much like a selective membrane, allowing electrons to flow only in one direction while holes flow in the opposite direction. This asymmetric movement of opposite charges can be termed ambipolar motility. The motility of charges can be measured in terms of charge *drift* (motility according to a potential in an electric field) and charge *diffusion* (motility according to an effective potential, or chemical potential from a concentration gradient).

All three prior steps occur within the cell as a thermodynamic system, but for work to occur there must be mass and energy exchange with the surroundings. As seen in Figure 14.5, the maximum power (P_{mp}) to be derived from a PV system at standard steady state lighting conditions (standard testing conditions, STC) is found at a characteristic external resistance R_{ch}—so-called to separate intentional resistance to do work from *parasitic resistances* that sap power

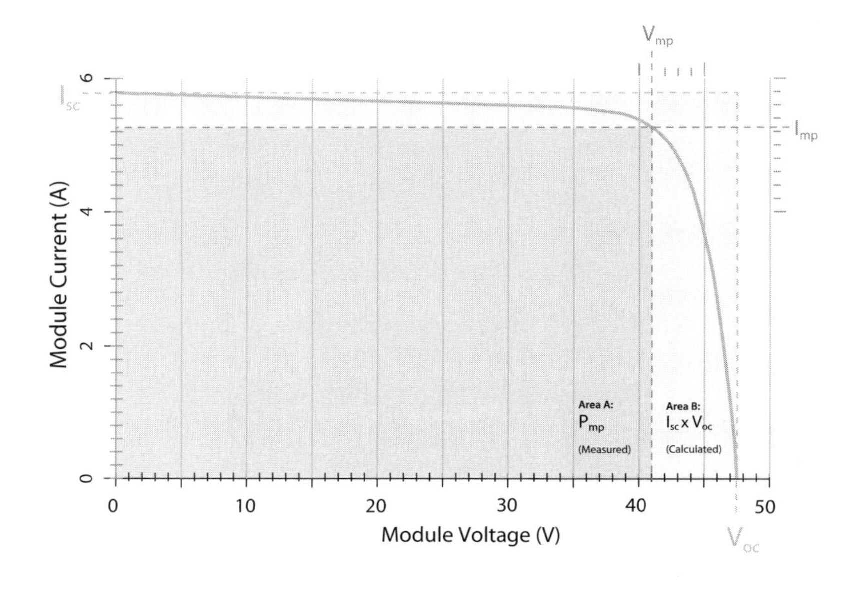

Figure 14.5: I-V curve for a commercially available sc-Si module. Two boxed areas are defined: the measured maximum power conditions ($P_{mp} = I_{mp} \times V_{mp}$), and the calculated product of the measured short circuit current (I_{sc}) and open circuit voltage (J_{oc}). The slope between the origin and coordinates V_{mp}, I_{mp} is defined as $1/R_{ch}$ Data collected from the CEC database in the SAM software (NREL).

A **load** is a demand for work from the client—in the case of PV, loads are in the form of electric power.

from the system due to material flaws and poor electrical contacts in the circuit. Hence we may associate R_{ch} with the delivery of maximum power P_{mp} from a PV device under illuminating conditions and a desired external load.

PV SYSTEMS INTEGRATION AND SIMULATION TOOLS

So far we have discussed photovoltaics as a singular cell. In practice, when designing PV arrays for the client, the team will be called upon to estimate the annual energy yields for installed arrays, accounting for dynamic changes in irradiance, air temperature, and wind effects. Accounting for these factors leads us to a smarter design approach of Systems Integrated Photovoltaics (SIPV).[5] PV systems can be integrated into fields, farms, green roofs, white roofs, parking ramps, and car ports among many options (see Figure 14.6). The integration can

[5] Jeffrey R. S. Brownson *Design and Construction of High-Performance Homes: Building Envelopes, Renewable Energies and Integrated Practice,* chapter 2.2 Systems Integrated Photovoltaics, SIPV. Routledge, 2012.

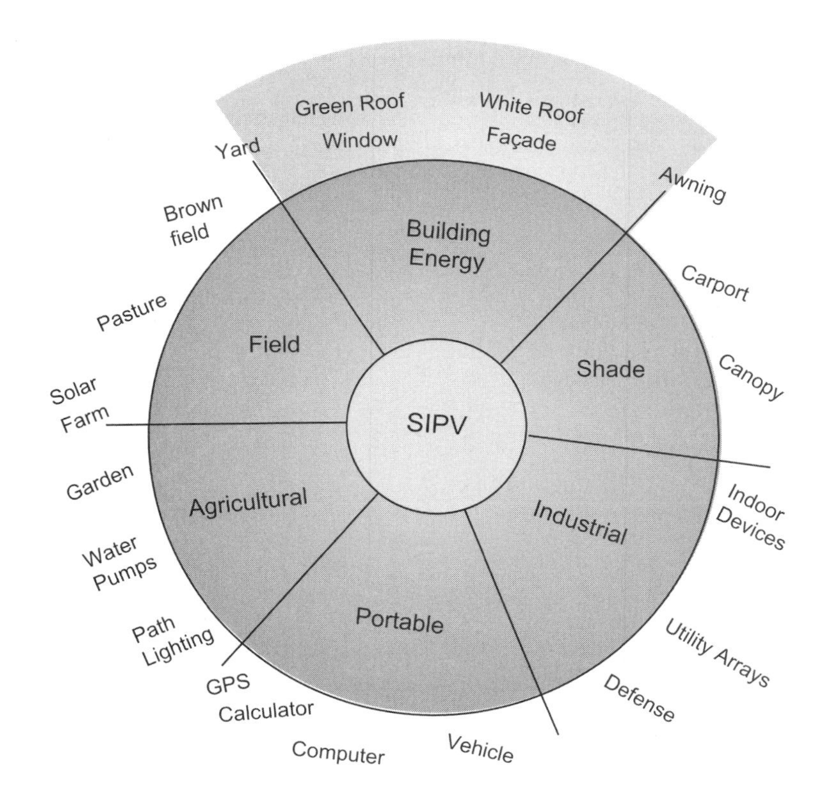

Figure 14.6: Systems integrative portfolio.

take advantage of coupled performance between the system and surroundings. Within each locale there will be microclimate effects from the manner in which the array is integrated into the site. SIPV design addresses the local environment for PV deployment, exploring a systems relationship among the SECS technology and factors such as the physical support structure, the light-reflective properties of surrounding materials, and the thermal properties of the immediate microclimate.

One of the systems that are popular targets for integration are buildings, hence the term Building Integrated PV (BIPV). However, we should note that a building itself is composed of many complex systems such as roofs, windows, awnings, and walls—each system composed of real components that functioned together as a whole. The historical meaning of BIPV originated in the 1980s as a cost reduction strategy, whereby a designer was expected to remove a functional part of a wall system or roof system within the built environment and replace that component with a photovoltaic module. The exchange in cost for the removed component (such as a shingle or a pane of glass) was supposed to defray the net cost of installing the photovoltaic module—in essence, sacrificing a functional piece from the whole to make room for cost. Unfortunately, integration for better building costs does not imply integration with the goals of the system. From this perspective, a window-mounted air conditioning unit would be classified as "building integrated AC," rather than the actual BIAC from a forced air exchange and integrated ducting (see Figure 14.6).

SIPV: Systems Integrated Photovoltaics. Coupling performance between system and surroundings, including microclimate effects of irradiance and temperature. From a sustainability systems perspective, SIPV would include increasing ecosystem resilience.

Time to move past removing parts to make room for PV. Integrate PV into the appropriate **systems** by cost reduction through demand balance, and maintain the fabric of systems within the building. The net present value (NPV) of the system will increase.

From a building's integration perspective, we now have stronger strategies of cost reduction through balance. By decreasing the energy demand across the whole system, we enable a small compliment of on-site PV supply. That is to say, when integrating PV into the functional systems of the built environment such as residential homes, PV should wait until *after* the home has been retrofitted for energy efficiency, and is now a low energy demand system.

> The fundamental first step in any BIPV application is to maximize energy efficiency within the building's energy demand or load. This way, the entire energy system can be optimized.
> –Eiffert & Kiss (2000)[6]

[6] P. Eiffert, G. J. Kiss, Building-integrated photovoltaics for commercial and institutional structures: A sourcebook for architects and engineers. Technical Report NREL/BK-520–25272, U.S. Department of Energy (DOE) Office of Power Technologies: Photovoltaics Division, 2000.

SIPV as Informed Process:

- Integration of energy supply with *Efficient Demand*
- Integration of supply with goal of *Solar Design*
- Integration with the intended *System* (locale and system type)
- Integration with the *Surroundings* (microclimate and services)
- Integration with the functionality and constraints of *Photovoltaic Action*

The challenge for the emerging generation of PV systems designers is to think more specifically regarding the system of interest within the building. The point of integration could be as specific as a green roof (itself a SECS). In a GRIPV (green roof integrated PV), the phase change of water in the plant to water vapor (a latent heat exchange via evapotranspiration) establishes a cool microclimate near the plant. With favorable systems integration, the green roof microclimate may reduce operating temperature of a module on hot summer days, and as a result improve power production and potentially increase the lifetime of both the PV system and the roof structure.[7] Studies have not yet shown this to be a major reduction in module temperature, but it is an example of an increase in the net present value for the entire roof system, while increasing ecosystems services for the area.

[7] Jeffrey R. S. Brownson. *Design and Construction of High-Performance Homes: Building Envelopes, Renewable Energies and Integrated Practice,* chapter 2.2, Systems Integrated Photovoltaics, SIPV. Routledge, 2012.

Imagine **solar farms** that are integrated into the pasture system, or fields. **SIPV** does not stop with buildings.

Ecosystems Services:

- **Supporting:** fundamental services necessary for the production of all other services (e.g., photosynthesis, soil formation, water cycling).
- **Provisioning:** the products obtained from ecosystems, including energy and fresh water.
- **Regulating:** the benefits obtained from regulating ecosystems processes (e.g., air quality, water, and erosion regulation).
- **Cultural:** the non-material benefits that society gains from ecosystems (e.g., cognitive development, reflection, recreation, aesthetic experiences, and ecotourism).

Another system that is pursued for installation is the open field (green or brown field). At present, typical large-scale PV projects do not take full advantage of a systems approach to ecosystems services. Design teams do prepare environmental impact reports for the systems design, and work to minimize impact to water, soil, and biota in the installation. However, systems integrated PV with open spaces could be developed in a sustainable approach to improve ecosystem resilience in the locale as well. Such an approach indeed requires a transdisciplinary team as well as stakeholder buy-in. Envision an integrative design scenario where a properly devised field installation provides electrical power to society and also increases ecosystems services.

We would like to know how those microclimate and meteorological drivers will affect the long-term annual performance of the designed system. To help in the decision-making and planning process, the integrative design team should use available simulation tools to assess energy and finance flows. One accepted method of modeling a system applies a component-based performance model. Component-based models separate parts of the system, such as the inverter and the PV array, for a more detailed approach to control dynamic systems simulations. In contrast, simulation programs like PVWATTS that lump inverter and PV performance are often used for rapid analysis of systems, however details are lost in the process and we recommend a component-based approach when feasible for an integrative design project. The component modeling practice has been spearheaded using TRNSYS software from the University of Wisconsin-Madison, but for many solar projects a team can also explore using core engine elements of TRNSYS inside of the System Advisor Model from the National Renewable Energy Laboratory (NREL).[8]

The 5-Parameter modeling approach was developed by De Soto, Klein, and Beckman to continue work that addresses the manner in which *PV module performance* changes with respect to environmental conditions of ambient air temperature, microclimate affects, wind speed, and irradiance conditions.[9] The 5-Parameter method uses available information that PV module manufactures commonly report in spec sheets as inputs for simulating a PV module in the locale under study. The approach uses information according to NOCT (Nominal Operating Cell Temperature) parameters resulting from a standard testing for entire PV modules, where the mounting conditions are reported according to an irradiance of 800 W/m^2, air temperature of 20 °C, wind speed of 1 m/s, and an open back mounting. Comparison indicates that NOCT conditions for testing modules are quite different from the bench-top testing of PV cells using Standard Testing Conditions (STC) of 1000 W/m^2 and cell temperature of 25 °C. The 5-Parameter model was later adopted by the California Energy Commission, updated by Neises, and is referred as the *CEC model* in the SAM simulation software.[10]

[8] S. A. Klein, W. A. Beckman, J. W. Mitchell, J. A. Duffie, N. A. Duffie, T. L. Freeman, J. C. Mitchell, J. E. Braun, B. L. Evans, J. P. Kummer, R. E. Urban, A. Fiksel, J. W. Thornton, N. J. Blair, P. M. Williams, D. E. Bradley, T. P. McDowell, M. Kummert, and D. A. Arias. TRNSYS 17: A transient system simulation program, 2010. URL http://sel.me.wisc.edu/trnsys.

[9] W. De Soto. Improvement and validation of a model for photovoltaic array performance. Master's thesis, University of Wisconsin, Madison, WI, USA, 2004; and W. De Soto, S. A. Klein, and W. A. Beckman. Improvement and validation of a model for photovoltaic array performance. *Solar Energy*, 80(1):0 78–88, 2006.

[10] Ty W. Neises. Development and validation of a model to predict the temperature of a photovoltaic cell. Master's thesis, University of Wisconsin, Madison, WI, USA, 2011; and System Advisor Model Version 2012.5.11 (SAM 2012.5.11). URL https://sam.nrel.gov/content/downloads. Accessed November 2, 2012.

[11] D. L. King, W. E. Boyson, and J. A. Kratochvill. Photovoltaic array performance model. SANDIA EPORT SAND2004–3535, Sandia National Laboratories, operated for the United States Department of Energy by Sandia Corporation, Albuquerque, NM, USA, 2004.

The **5-Parameter** method was adopted by the CA Energy Commission, and is also named the **CEC method**, with an associated database.

Spec sheet: a short data document provided by the manufacturer, with a summary of device performance and deviations in performance given environmental conditions.

Direct current (dc) flows in one direction, either positive or negative. **Alternating current** (ac) changes between a positive flow and negative flow, and has frequency, magnitude, and waveshape.

In fact two large databases of PV module parameters are available with SAM: the CEC model database and the Sandia model database.[11] The Sandia model developed by King and colleagues to simulate conditions that deviate from NOCT testing scenarios and predates the CEC/5-Parameter model. The Sandia method requires multiple parameters to be characterized specifically for each module as opposed to drawing from current manufacturer data sheets, and so the database list is slightly smaller. However, the Sandia model is also very robust for array simulation in designing a PV system for a client in their desired locale.

Direct current is delivered from the photovoltaic array to an inverter and power conditioning system before integrating the electric power with the utility grid. The amplitude of that current will vary in proportion to irradiance from zero amps (at night) to a maximum current in the circuit on a sunny day, but the flow of power from a PV is in only one direction (hence, direct current). The role of modern inverter systems also includes *power conditioning*, in which the entire *Power Conditioning Unit* delivers inverting along with data-logging, power control and transformation, DC-DC conversion to step a lower voltage up before inverting, and a *maximum power point tracker* (MPPT). An MPPT is now standard in PV design: called for to operate the PV system at the maximum power point (P_{mp}) even under the variable conditions of irradiance and module temperature.[12] Additionally, the progression in microelectronics and solid state devices has led to an expanding class of *microinverters* and *microcontrollers* that integrate into the PV module. Microinverters have the advantage of being able to adapt array configurations with challenging shading constraints. Modern inverter conversion efficiency is now very high, upwards of 95%.

[12] J. P. Dunlop. *Photovoltaic Systems.* American Technical Publishers, Inc, 2nd edition, 2010. The National Joint Apprenticeship and Training Committee for the Electrical Industry.

[13] System Advisor Model Version 2012.5.11 (SAM 2012.5.11). URL https://sam.nrel.gov/content/downloads. Accessed November 2, 2012.

When one wishes to incorporate an inverter in a design simulation, we recommend again using the SAM software from the USA DoE National Renewable Energy Laboratory.[13] SAM uses the *Sandia Performance Model for Grid-Connected Inverters*, to simulate conversion of direct current power (W_{dc}) into alternating current power (W_{ac}) compatible with the eletric power grid. The Sandia performance model for inverters is used with a photovoltaic array performance model in SAM to estimate expected system performance from Typical Meteorological Year inputs.[14]

When sizing the inverter for a design simulation, the selected inverter power rating should have a maximum power output ("Power AC_o") larger than the total W_{ac} power demand from the client or the size of the PV array (with slight oversize recommended for future load additions).[15] Also, the maximum power condition voltage (V_{mp}) for the PV array must be able to operate between the low and high range of the MPPT.

[14] D. L. King, S. Gonzalez, G. M. Galbraith, and W. E. Boyson. Performance model for grid-connected photovoltaic inverters. Tech. Report: SAND2007–5036, Sandia National Laboratories, Albuquerque, NM 87185–1033, 2007.

[15] J. P. Dunlop. *Photovoltaic Systems*. American Technical Publishers, Inc, 2nd edition, 2010. The National Joint Apprenticeship and Training Committee for the Electrical Industry.

INTRINSIC/EXTRINSIC PROPERTIES IN PV

There are both *intrinsic* and *extrinsic* properties to a PV cell. Intrinsic properties are inherent with the materials used to form a photovoltaic device, while extrinsic properties are derived from the surroundings (light and temperature). The *band gap* (E_g, where E(eV) often stands for energy in units of electron volts) of a photovoltaic device is in *intrinsic* material property, and determines the maximum voltage possible from the functional PV cell. The band gap is related to the *density of states* found in the conduction bands and the valence bands. The valence bands are *full* of electrons occupying a density (a spread) of their *ground states* (low energy) The conduction bands are free of electrons, but can be thought of as possible containers for *exited state* electrons, should they be excited by light (photons) or temperature (phonons).[16] In between the valence band and the conduction band is a *gap* (a *band gap*) where electrons cannot "hang out." Again, if the PV absorber material is exited via supra-band gap photons (sufficiently high energy light), or if the PV absorber material receives sufficient thermal energy from its surroundings, electrons will be excited to the conduction band. From the conduction band, electrons can either "fall" back down to release thermal energy (phonons) or light (photons) or they can be separated via a *discriminating mechanism* to electric contacts to then enter a circuit and perform work.

E_g is the symbol for the **band gap**, where *E* stand for energy in eV units.

The *ground state* is a way of saying electrons are in their lowest energy levels.

[16] In a crystal structure, physicists assign the vibrational waves of atoms in the material the title of *phonons*. We think of vibrating atoms as *thermal energy*. So phonons on a collective macroscopic scale are representative of an increase in thermal energy.

Supra-band gap: a fancy way of saying E (eV) > E_g or $\lambda < \lambda_g$, meaning that if absorbed the photon will excite an electron into the conduction band, leaving a hole in the valence band.

We can create an analogy for the conduction and valence bands: think of simple book shelves in a dorm room or flat, formed with wood planks and cinder block (a classic college invention). Your roommate has decided to be creative and leaves a large "gap" between the lowest few shelves and the highest shelves (he claims artistic privilege, and your just happy to have some shelves for books). The books for the shelves are all organized from the bottom-up, and they fill up all the bottom shelves perfectly (you being a fastidious organizer of texts). Now, you can input energy to "pump" a book up to an upper shelf right? The act of lifting up the book to the nearest shelf above the gap is equivalent to the act of absorbing a photon of energy equal to the band gap ($\lambda = E_g$). The act of lifting up the book to a higher shelf above the gap is equivalent to the act of absorbing a photon of greater energy than the band gap ($\lambda > E_g$). In both cases, the book will rest safely on the shelf. However, your roommate does not like books on shelves higher than the lowest upper shelf, and always comes in and moves an upper book immediately down to the lowest upper shelf (this is annoying, but your roommate continues to declare an artistic imperative). You will note that by "pumping" up a book to the upper shelves, you have left a "hole" in your perfectly filled lower book distribution. In this analogy, the act of absorbing a photon and photogenerating an excited electron has also photogenerated the *absence of an electron* in the valence band—we call this absence of an electron a *hole* (actually an excited hole with potential!), and the hole will have a positive charge balancing the negative charge of the electron. Thus, we have photogenerated an *electron-hole pair* with an electrochemical potential between them. The pair are also called *charge carriers*.

Recall that we think of light as a flow of work in a **pump**.

Material absorber examples:

$E_{g(Si)} \sim 1.11\,\text{eV} = 1127\,\text{nm}$
$E_{g(Ge)} \sim 0.66\,\text{eV} = 1878\,\text{nm}$
$E_{g(CdTe)} \sim 1.44\,\text{eV} = 861\,\text{nm}$

Notice that CdTe has over double the band gap energy of Ge.

The act of a higher energy electron falling back down to the lowest energy conduction band level (and releasing phonons, or thermal energy in the process) is normal for a non-quantum optoelectronic transition. Meaning, you will only get a maximum theoretical potential from the band gap potential, even if the photons that were absorbed are of higher energy than the band gap. You can think of this as yet another energy filter in our solar materials. Of course, there are other limitations that will reduce the net voltage acquired from a photovoltaic cell. For instance, the silicon band gap of $E_{g(Si)} \sim 1.1\,\text{eV}$ would seem to provide 1.1 V of

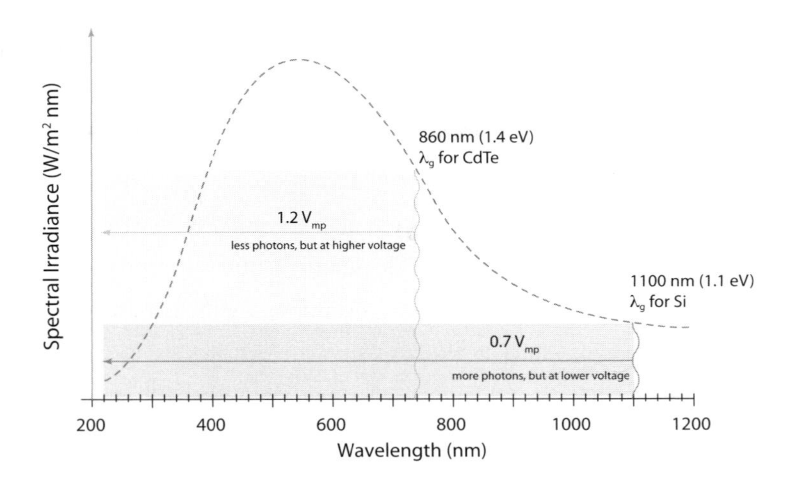

Figure 14.7: Demonstrating the solar spectral distribution for a solar blackbody and the resulting theoretical maximum power voltages for absorber materials Si and CdTe. There is a trade-off between current (via photons accepted) and voltage in choosing photovoltaic materials.

electrical potential in the ideal case of no losses, but the open circuit voltage from the cell will actually be on the order of $\sim 0.6 - 0.7\,\text{V}$ (see Figure 14.7).

Open circuit voltage examples (also see Figure 14.7):

$\text{Si} \sim 1.11\,\text{eV} \rightarrow 0.7\,\text{V}$
$\text{Ge} \sim 0.66\,\text{eV} \rightarrow 0.4\,\text{V}$
$\text{CdTe} \sim 1.44\,\text{eV} \rightarrow 1.2\,\text{V}$

Notice that Ge does not generate a large open circuit voltage per cell.

Let's explore the book shelf analogy one more time. If you were to pick up a book from the lower shelves, and then lift it up into the open gap that your roommate created for aesthetic reasons, then release it—what would happen? Well, the book would come crashing down to fall back to the lower shelf, of course (and perhaps your roommate would ask if you are performing a new art installation on the futility of the printed text in the digital age...your roommate is strange). The act of lifting up the book to a *non-existent shelf below the gap* is equivalent to the act of absorbing a photon of lesser energy than the band gap ($\lambda < E_g$). The electrons have nowhere to go, and cannot be collected from the gap. So, in order for a photovoltaic material to facilitate steps *one* [**(ABS)** Light *absorption* process that causes a transition in the material from a ground state (low energy) to an excited state (high energy).] and *two* [**(GEN)** *Photogeneration* of charge carriers as a negative-positive pair.], the light absorbed needs to be selective to photons with supra-band gap energy/wavelengths, or $\lambda \geq E_g$.

The *absorption coefficient* ($\alpha = \frac{4\pi k}{\lambda}$) is another intrinsic property that provides an indication of the depth (for a particular wavelength) that light will penetrate into the PV absorber material before being absorbed. Materials with low absorption coefficients will not absorb light well, and will need to be significantly thicker to ensure full absorption. Another way to think about it is that materials with low absorption coefficients and thin films will appear semitransparent (higher τ). For PV materials, the suite of thin film systems such as CdTe/CdS and CIGS/CdS (where CdTe and CIGS are absorber materials) have very high absorption coefficients for wavelengths above the band gap (in units of $\sim 10^5\,\text{cm}^{-1}$). The suite of monolithic silicon systems such as multicrystalline silicon (mc-Si) and single crystal silicon (sc-Si) have a lower absorption coefficient than the thin film materials, and are also thicker devices ($\sim 10^4\,\text{cm}^{-1}$).

CdTe/ CIGS: $\alpha \sim 10^5\,\text{cm}^{-1}$ Crystalline Silicon has $\alpha \sim 10^4\,\text{cm}^{-1}$.

A third intrinsic property of a PV absorber material is the *reflectance of the surface* to incoming light. In particular, silicon is a highly reflective material, such that incoming light can be reflected before being absorbed by the material. If we think of silicon wafers from the microelectronics industry, we note that they are shiny and *silver*, not the blue or blue-black of the PV cells that are recognizable in modern silicon modules. This is because the silicon has been coated with a thin film of *anti-reflective* materials. An anti-reflective coating helps photons to transition from the low index of refraction of air ($n = 1$) to the high index of refraction for silicon near the band gap, by introducing a material of graded refraction, thus "guiding" the light into the silicon.

Anti-reflective coatings are also termed *optical impedance matching films.*

It should be noted that many additional light-trapping techniques and photonic manipulation techniques have been developed to increase the ability of a PV device to absorb light. These techniques include texturizing the surface as well as including a rear-side reflector behind the absorbing material. There are also macroscopic light concentration methods that are used to increase the density of photons incident upon the cells, which will be discussed in the following chapter on concentration.

Extrinsic properties are related to the current generation for the cells. The area of a cell, module, or array is an extrinsic parameter that can be increased by the design team to meet a power scale requirement for the client in the locale. The two main parameters occurring outside of the cells are the levels of irradiance on site and the ambient temperature. An increased irradiance (e.g., higher in Dubai vs. Dublin) affects J the total current density by increasing J_L. A decreased ambient temperature (cooling the modules) will affect the cell performance

If you want more power and you are not area-constrained, just increase the size of the array. The array area is an extrinsic parameter.

favorably in compliment with high irradiance conditions. Increased temperature in photovoltaics (where the temperature often occurs due to optocalorics) increases the energy of electrons within a material (an increased density of states), thereby decreasing the band gap (E_g) of the material. A decrease in band gap also leads to a drop in the open circuit voltage, because V_{oc} is derived from I_0, which is temperature dependent.

$$V_{oc} = \frac{kT}{q} \ln\left(\frac{I_{sc}}{I_0}\right). \tag{14.7}$$

The change in open circuit voltage is approximately −2.2 mV per °C in ideal silicon cells.[17] The change in performance tied to an entire photovoltaic module will be larger than this. If we take the example of a silicon absorber material and the terrestrial shortwave spectrum of AM1.5, we find that 81% of the power density from the Sun (UV/ Vis/ IR; $\lambda > 1100$ nm) is acceptable for optoelectronic conversion. However, this is deceiving—much of that absorbed supra-band gap energy will be lost thermally as the excited electrons fall back down to 1.1 eV (adding to the temperature of the cell), while on the other side of the gap the remaining 29% of sub-band irradiance (in the IR) from the shortwave band will still be directly absorbed by the silicon, again increasing the temperature of the cell as well. The bottom line is that on a bright day a PV module will get hot.

[17] Christiana Honsberg and Stuart Bowden. Pvcdrom, 2009. URL http://www.pveducation.org/pvcdrom. Site information collected on Jan. 27, 2009.

In Table 14.1 we put the increased optocaloric response of PV modules at high irradiation conditions into proportion with a relative drop in optoelectronic module efficiency. A photovoltaic cell tends to absorb a broad band of wavelengths, including those wavelengths below the band gap. Those sub-band gap wavelengths do not generate excited electrons, they lead to an expected *optocaloric* response—they make the panel hot! For every +10 °C above the standard testing condition of 25 °C and AM1.5 simulated solar irradiance (1000 W/m^2) the module could lose ~2–5% efficiency. The data in Table 14.1 was extracted from the CEC Performance Module Database within the System Advisor Model software, available from NREL.[18] While these data are reasonable estimates for each material, the actual values can be looked up directly for each commercial product available.

[18] P. Gilman and A. Dobos. System Advisor Model, SAM 2011.12.2: General description. NREL Report No. TP-6A20–53437, National Renewable Energy Laboratory, Golden, CO, 2012. 18 pp; and System Advisor Model Version 2012.5.11 (SAM 2012.5.11). URL https://sam.nrel.gov/content/downloads. Accessed November 2, 2012.

Table 14.1: Optocaloric Response showing drops in optoelectronic performance for typical PV absorber materials. Performance reported as a percent drop from maximum power (W_{mp}) per 10 °C increase above STC (25 °C cell T, AM1.5: 1000 W/m²). Data extracted from CEC Performance Module Database in System Advisor Model from NREL.

W_{mp} loss	PV Absorber Material
−1.7%/ +10 °C	CdTe (non-silicon thin films)
−2.3%/ +10 °C	a-Si (amorphous silicon thin films)
−4.0%/ +10 °C	sc-Si (single crystal silicon)
−4.5%/ +10 °C	mc-Si (multi-crystalline silicon)
−4.7%/ +10 °C	CIGS (non-silicon thin films)

We see that CdTe and amorphous silicon materials have a low susceptibility to thermal increase, while crystalline silicon modules tend to drop in performance by 4–5% from their maximum power conditions per 10 degree increase in temperature (naturally associated with a high irradiation day). Notice that this drop in optoelectronic performance is relative to the increase in the *cell temperature* (or the panel temperature), not the surrounding air temperature, which could be significantly cooler than the modules themselves. On a bright day (800–1000 W/m²) a module can be found to have temperatures 25–32 °C *above the surrounding air temperature!* Should the panel have poor cooling strategies (e.g., integrated into a rooftop without systems consideration) and thus be insulated, the increases could be >40 °C above the surrounding air temperature.[19]

[19] M. W. Davis, A. H. Fanney, and B. P. Dougherty. Prediction of building integrated photovoltaic cell temperatures. *Transactions of the ASME*, 123(2):200–210, August 2001.

In addition to lost performance on hot days, hot modules tend to decrease the lifetime performance of a photovoltaic array. Consider systems strategies to keep the PV system cooler on bright days.

Let us explore an example case of single crystalline (monocrystalline) silicon cells within a 5.38 kW_p PV array (25 modules, rated at 215 W_p), mounted horizontally in sunny Phoenix, Arizona. Drawing from a typical meteorological year, on May 28 at 1:30p (solar time) the irradiance on a horizontal surface is recorded as 1000 W/m², and the air temperature (dry bulb) is found to be 40 °C with a wind speed of about 2.5 m/s. Assuming an approximate NOCT model response, the internal cell temperature could be estimated to be about 75 °C, or +50 °C above standard cell testing conditions. In such an example a single PV module in that 5.38 kW array would operate at 215 W_{dc} under maximum power conditions, STC. However, with the thermal increase from the optocaloric response, the module would experience a 20% drop from maximum power conditions (5 × −4.0%/ +10 °C). Hence a 215 W module would temporarily drop down to 172 W_{dc}. From another perspective, the 25 panel array on the Arizona rooftop with a 20% drop would also produce nearly a kilowatt of power less than peak performance during the afternoon of May 28. However, if a design team were to consider a systems integrative PV approach to incorporate some measures of passive cooling to the PV array, perhaps that optocaloric response could be managed better, providing higher performance and longer lifetime to the system.

In contrast, a similar PV array mounted in Philadelphia on May 22, at 11:00a (solar time). The irradiance is still 1000 W/m^2, but the air temperature (dry bulb) is found to be 20 °C with a wind speed of about 4.5 m/s. The estimated PV cell temperature would then be about 50 °C, or +25 °C above standard cell testing conditions. With the thermal increase from the optocaloric response, the sc-Si module would experience a 10% drop from maximum power conditions (2.5 × −4.0%/ +10 °C), and a 215 W module would temporarily drop down to 194 W_{dc}. The Philadelphia PV array (5.38 kW_p) would only lose half a kilowatt compared to the same scale system in Phoenix. This is not to say that one locale is better than the other. Each has its own trade-offs and local patterns to adapt and integrate with. What is important is that we learn to integrate the PV array with the systems of interest, within the locale and on behalf of the client—for both optoelectronic and optocaloric responses.

ROBOT MONKEY DOES MINORITY CHARGE CARRIERS!

Welcome to topsy-turvy thinking! Where being negative means positive.

Yet another in-depth section in the PV chapter! When assessing the photo-generation of usable charge carriers from non-concentrated lighting conditions

For **p-type semiconductors**: **holes** are the **majority carriers**; **electrons** are the **minority carriers**.

One photon absorbed generates two charge carriers: one electron plus one hole.

For **n-type semiconductors**: **electrons** are the **majority carriers**; **holes** are the **minority carriers** and the drivers of PV action.

(e.g., regular flat plate PV operation), one should note that the *minority carriers* run the PV process, *not* the majority carriers. In a semiconductor like crystalline Si, the material will have a manufactured doping that leads to higher carrier population values (per volume) in thermodynamic equilibrium (no incident light) for either electrons (n-type, for "negative), or holes (p-type, for "positive") by many orders of magnitude. The carrier having higher populatin per volume in equilibrium is called the *majority carrier*. When the semiconductor is pumped up by light, out of thermodynamic equilibrium, additional electrons and holes are generated in pairs, each from a single photon absorption, Yet only the minority carrier powers the PV process. You would really expect the opposite, without prior knowledge of device physics, right? But there it is, *minority carriers drive the PV process*. Minority carriers are effectively the rate-limiting portion of the electrochemical process in PV action (to use a chemistry analogy). Thus, for the p-type (positive) portion of the cell the *electrons* dominate the separation process; while for the n-type (negative) portion of the PV cell the *holes* drive the separation process.

Let's perform a Fermi estimate of the problem to hopefully bring clarity from a few general sources of data.[20] Given regular sunshine on the surface of Earth (considered "low levels of light"), $G \sim 700$–$1400\,\mathrm{W/m^2}$, why is it that the minority charge carriers are the only population that matters in a photovoltaic device? We will perform a quick (back of the envelope) estimation of photon density in a semiconductor. We need to estimate the number of charge carriers that exist in an irradiated PV cell under daylight conditions. Such an estimate will require the number of photons "packed" into a volume of semiconductor.

[20] A Fermi problem is an estimation process given limited information to explore the dimensions and scale in a systems question, and to deliver an approximate answer using justified guesses for portions of a problem that are very dense to compute directly. Named after Nobel Laureate and Physicist Enrico Fermi, who would derive artful approximations for complex problems from very limited or even absent sources of data.

How many photons are there per cubic centimeter for AM1.5 conditions?

- Given the AM 1.5 spectrum, adding up all the photons in the shortwave band, allows us to estimate $\sim 10^{17}$ photons per $\mathrm{cm^2}$ per second impinging upon the semiconductor surface (the photon flux integral).[21]

- The absorption coefficient of Si is known ($\alpha \sim 10^4$).

[21] Rolf Brendel. *Thin-Film Crystalline Silicon Solar Cells: Physics and Technology.* John Wiley & Sons, 2003.

- By multiplying the values for the photon flux integral and the absorption coefficients ($10^{17} \times 10^4 = 10^{21}$), the units combine to yield an approximate order of the additional electron and hole populations (over and above those present in thermal equilibrium) generated per volume (cm^3) per second. A similar number can be arrived at more precisely using the generation rate of carriers per wavelength.

- In the bulk, a large number of both carriers will recombine via radiative recombination, Auger recombination, and Shockley-Read-Hall (SRH) recombination. These collective acts of recombination allow only 10^{14} photogenerated electrons and holes to survive per cm^3 volume.

- Hence, 10^{14} charge carries per cm^3 volume is the net steady state balance, combining charge carrier generation from absorbed photons and charge carrier losses from recombination.

Quick estimation of orders of magnitude:

1. For Silicon, we will observe approximately 10^{14} excess electron-hole pairs generated per cm^3 from AM1.5 conditions.
2. In general, there are on the order of $\sim 10^5\,cm^3$ charge carriers in the minority carrier population.
3. In general, there are on the order of $\sim 10^{15}\,cm^3$ charge carriers in the majority carrier population.

Hence, the presence of 10^{14} photons absorbed in a volume (if most were absorbed) would dramatically change the minority charge carrier population (because $10^{14}\,cm^3 \gg 10^5\,cm^3$). At the same time, the presence of 10^{14} photons absorbed in a volume (even if all were absorbed) would not significantly budge the apparent numbers of the majority charge carrier population (because $10^{14}\,cm^3 \ll 10^{15}\,cm^3$).

Congratulations!
Robot Monkey grants you one glowing banana for getting through the mindstretch

METRICS: PERFORMANCE AND UNIT COSTS

There are no magic bullets or technologies that will abruptly change the commercially available photovoltaics in today's market, and we encourage the design team to work with the materials at hand as a sustainable integrative design approach.

Stop waiting for magic bullets to solve our energy challenge. Technologies are advancing every day in the lab, but the lab is not the marketplace for PV. Commercial photovoltaics are deployed in today's market from off-the-shelf technologies, sold as a commodity, and at very low risk to the buyer. That is not to say that an amazing technology will not emerge, nor that the amazing science developing in labs for PV is without merit. Solutions for society must be built using the *technologies of the present, in the market* rather than waiting for a possible disruptive change in the future. Be that change by using integrative design and sustainable systems thinking! Hence, we only deal with commercially available photovoltaic technologies in this text. Commodities can be thought of as goods without significant product differentiation on the market—meaning there is a diversity of similar technologies available at a similar market price. There are essentially only four technologies available in significant amounts on the market (or near the market): multicrystalline silicon wafer cells (mc-Si), single crystal silicon wafer cells (sc-Si), thin film CdTe/CdS heterojunction cells (CdTe), and a small segment of thin film CIGS/CdS heterojunction cells (CIGS). Other technologies may be available as specially crafted products (discernible goods) for niche markets such as for individual architectural designs, but their percentage of the market is negligible at present.

A **commodity** is a generic good that satisfies a need or desire by a client. For the most part, the client does not differentiate a multi-crystal silicon PV from a CdTe PV, provided they both satisfy the same demand and *price per peak watt*.

With respect to integrative design of solar energy conversion systems for your client, the client will typically only accept a low-risk solution, a systems approach designed to be physically stable for the PV industry-accepted time horizon of 20–30 years (if not longer). The scope of this text relies on the economic availability of a technology with a time horizon of more than 20 years, and de-emphasizes emerging technologies that have not yet made an impact on the market. The market for PV cell technologies is very competitive, and PV systems emerging beyond the core technologies are effectively considered as far away as jet packs in terms of a systems design for a client.[22] Keep in mind that your client *cannot purchase laboratory materials*, and the development of benchtop ideas into commercially available commodities is often decades away. Companies may start up advertising a fringe technology, but they frequently fail shortly afterward due to the intense competition for the commodity. It is unlikely that there will be economically competitive alternative PV technologies (that obey a similar time horizon of 20–30 years) within the next 5 years for dye-sensitized cells, polymeric or organic PV devices, and for PV devices relying solely on quantum dot absorbing materials. Even technologies once available such as amorphous Si (a-Si) and ribbon silicon have been marginalized in recent years due to competitive options from mc-Si, sc-Si, and CdTe modules.

Rumors and news reports abound of "amazing technologies" emerging from national laboratories and universities that *may* change the face of photovoltaics. These scientific discoveries are often overstated or ill defined by popular science reporting, and will not likely have a rapid impact on the PV industry, which is conservative and very slow to change. Consider that it took more than two decades for CdTe and CIGS thin film PV to emerge as a market option in the commercial PV world.

[22] Anyone who has watched The Jetsons cartoon wonders where those jet packs and flying cars of the future are now that we *are* in the future.

Keep in mind that the characteristic measure for economic merit is not simple systems efficiency (η). The performance metric that we use to assess devices in systems design encompasses both the efficiency of the device under standardized conditions and the unit cost of that technology. The particular unit cost used for cell technologies is the *cost per unit power*. In US dollars, we establish the cost per power in terms of dollars per *peak watt* ($\$/W_p$). A peak watt is a term for a cell tested under controlled laboratory conditions, which reflects the cell efficiency. Cells are tested under (Standard Testing Conditions) of simulated AM1.5 spectral lighting conditions, 1000 W/m^2, at 25 °C for a short period of time (e.g., seconds). The cell efficiency is derived from the power density output at the maximum power point (P_{mp}; using a characteristic resistance, R_{ch}) in W/m^2.

$$\eta = \frac{\text{output energy}}{\text{input energy}} = \frac{P_{max}\ [\text{W/m}^2]}{1000\ \text{W/m}^2} \times 100\%. \tag{14.8}$$

CAPACITY FACTORS: USA

The resource for solar electric production has been evaluated for many states in the USA by the Department of Energy. The metric of performance used in this case is the *capacity factor*, which is defined as if the power generated came from a central power plant. The capacity factor is the ratio of the power produced over monthly period relative to the power generated when operating at full capacity over the entire month. Solar systems will already operate at significantly reduced capacity due to diurnal phenomena of nighttime and monthly/seasonal changes in the length of daylight over the course of the year, linked to the latitude of the locale. In addition, meteorological phenomena (synoptic scale, mesoscale, and microscale) will further diminish the capacity factor below 25% in the majority of the continental US and mid-latitudes. We list the state-wide estimates of PV capacity factors in Table 14.2. By analyzing these data for all states in Figure 14.8, we observe that all the states but Alaska have significantly higher capacity factors than Germany ($CF = 0.11$).

State	Capacity Factor	State	Capacity Factor
Alaska	0.105	New Jersey	0.200
West Virginia	0.172	Virginia	0.200
Michigan	0.173	Tennessee	0.201
Ohio	0.173	South Carolina	0.202
Rhode Island	0.176	Georgia	0.203
Vermont	0.176	North Dakota	0.203
Pennsylvania	0.177	North Carolina	0.206
Maryland	0.179	Arkansas	0.207
Wisconsin	0.180	Florida	0.209
Connecticut	0.182	Hawaii	0.210
Massachusetts	0.182	Montana	0.212
Indiana	0.184	South Dakota	0.214
New Hampshire	0.184	Nebraska	0.217
New York	0.184	Texas	0.218
Delaware	0.186	Idaho	0.220
Illinois	0.186	Oklahoma	0.223
Kentucky	0.186	Oregon	0.227
Minnesota	0.189	Wyoming	0.229
Maine	0.191	Kansas	0.238
Missouri	0.193	Utah	0.248
Louisiana	0.196	California	0.252
Mississippi	0.197	Colorado	0.259
Iowa	0.199	Arizona	0.263
Washington	0.199	Nevada	0.263
Alabama	0.200	New Mexico	0.263

Table 14.2: A list of States and Capacity Factors for Utility PV Generation.

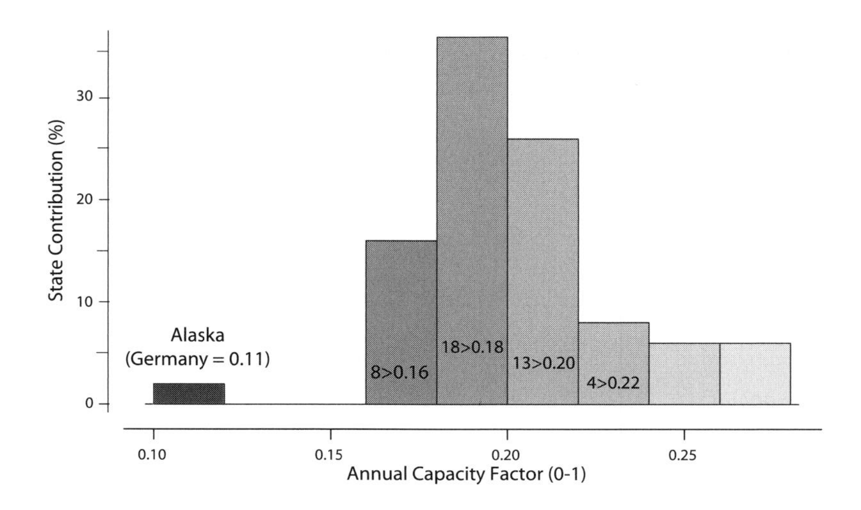

Figure 14.8: Histogram of State Capacity Factors. Notice how most states have capacity factors much higher than Germany.

PROBLEMS

1. List the three steps of photovoltaic action, and the fourth step once the electrons leave the system. Provide the associated characteristic measures for each step.

2. Germanium seems to absorb a broader band of light than silicon or cadmium telluride. Why would I want CdTe as a single junction terrestrial absorber over Ge?

3. Use SAM (Flat Plate PV–Residential component model) to estimate the annual power production for three locations from 10x horizontal PV modules connected in an array with the following specifications:

    - **Location:** San Diego, CA; Madison, WI, and Albany, NY
    - **Module:** sc-Si (mono-c-Si in SAM), 72 cells in series, $W_{mp} \sim 210\,W$ (search "210" in the CEC database)
    - **Inverter:** Use a simplified single-point efficiency inverter set to $2200\,W_{ac}$
    - **Array:** The array will be composed of two parallel strings of five modules

- **Orientation (PV Subarrays):** These modules are horizontal, so tilt is 0° and azimuth is not applicable.

Leave the rest of the model at the default settings. Run the simulations. From the output, select "Tables," then from the output variables select "monthly data": "Net ac output (kWh)." Plot the average monthly kWh of electric power from a typical meteorological year for each of the three cities.

4. Repeat the SAM models using $\beta=\phi$ and $\gamma=180°$ (alternate azimuth convention where North is 0°).

5. Repeat the SAM models using β and γ are set to there optimal orientations.[23]

[23] M. Lave and J. Kleissl. Optimum fixed orientations and benefits of tracking for capturing solar radiation in the continental United States. *Renewable Energy*, 36:1145–1152, 2011.

RECOMMENDED ADDITIONAL READING

While many of the detailed principles of photovoltaic device behavior and device design are beyond the scope of this text, there are several excellent resources recommended to engage in lifelong learning. The text "Solar Cell Device Physics" by Prof. Stephen Fonash was developed as an educational and professional resource for emerging researchers in photovoltaic device science and engineering.[24] The Fonash text details the basic physics of PV function and the engineering principles for implementing discriminating charge separation mechanisms (e.g., homojunctions and heterojunctions) in a cell. The device structures covered include homojunction cells and heterojunction cells, multi-junction solar cell structures, novel structures for organic polymer and dye-sensitized cells, and Schottky barrier devices with optional back layers.[25] Bear in mind that a practical photovoltaic device resource is distinct from a typical microelectronics text, in that a photovoltaic researcher must collectively study the multiple phenomena of optical physics (for light absorption), both dark and light-based electronic conditions in a device (for charge generation and rest states), positive and negative charge carriers transport behavior (for separation of holes and electrons in opposing directions), as well as defect behaviors

[24] Stephen Fonash. *Solar Cell Device Physics*. Academic Press, second edition, 2010.

[25] J Arch, J Hou, W Howland, P McElheny, A Moquin, M Rogosky, F Rubinelli, T Tran, H Zhu, and S J Fonash.*A Manual for AMPS 1-D BETA Version 1.00*. The Pennsylvania State University, University Park, PA, 1997.

AMPS-1D—created by Prof. Stephen Fonash and the following students and visiting scholars: John Arch, Joe Cuiffi, Jingya Hou, William Howland, Peter McElheny, Anthony Moquin, Michael Rogosky, Francisco Rubinelli, Thi Tran, and Hong Zhu. http://www.ampsmodeling.org/.

resulting in parasitic resistances. In complement with the text, Fonash's research team at The Pennsylvania State University (sponsored by the Electric Power Research Institute; EPRI) also developed the open software *AMPS* (Analysis of Electronic and Photonic Structures) for combined simulations of optics and electronic transport physics in solid state devices. The parametric solver uses the first-principles continuity equations and Poisson's equation to analyze the transport behavior of photovoltaic device behaviors. And while other similar softwares are available at cost for industry, AMPS presents a general and versatile resource that is free for device analysis of microelectronic diodes, light-based sensors and photo-diodes, as well as photovoltaic devices (http://www.ampsmodeling.org/).

Researcher professionals now at the Arizona State University, Christiana Honsberg and Stuart Bowden also developed the open digital resource *PVCDROM* (sponsored by the National Science Foundation, and with peers from the University of New South Wales). While the focus on optoelectronics in this text is directed toward project development and finance for PV technologies, the resource at *PVCDROM* should be explored by those who wish to delve deeper into the ecosystem of photovoltaics. PVCDROM has served as a strong and growing international resource for general PV systems concepts, from solar resource concepts to silicon cell design, to module performance and testing.[26] The digital resources found at *PVCDROM* are some of the best made openly available to the public for the general education of the properties and function of photovoltaic materials and devices, focusing mainly on the most common *silicon p-n homojunction cell*, the prevailing technology used in industry today. The resource also includes numerous interactive modules and short videos, including information on the manufacture of various silicon cell technologies.

[26] Christiana Honsberg and Stuart Bowden. Pvcdrom, 2009. URL http://www.pveducation.org/pvcdrom. Site information collected on Jan. 27, 2009.

At one point in time, PVCDROM really was intended to be distributed in hard CDROM disks. It has since transcended that medium while retaining the title, which solar educators know well.

For practitioner information on installation, system sizing, mechanical and electrical integration, and permitting there are several strong texts available. The texts from Solar Energy International and the NJATC (National Joint Apprenticeship and Training Committee for the Electrical Industry) are recommended here.

- Solar Cell Device Physics[27]
- Third Generation Photovoltaics: Advanced Solar Energy Conversion[28]
- Physical Foundations of Solid-State Devices[29]
- Solar Electric Handbook: Photovoltaic Fundamentals and Applications[30]
- Photovoltaic Systems[31]
- Building-Integrated Photovoltaics for Commercial and Institutional Structures: A Sourcebook for Architects and Engineers[32]

[27] Stephen Fonash. *Solar Cell Device Physics*. Academic Press, second edition, 2010.

[28] Martin Green. *Third Generation Photovoltaics: Advanced Solar Energy Conversion*. Springer Verlag, 2003.

[29] E. F. Schubert. *Physical Foundations of Solid-State Devices*. E. F. Schubert, Renasselaer Polytechnic Institute, Troy, NY, 2009.

[30] Solar Energy International. *Solar Electric Handbook: Photovoltaic Fundamentals and Applications*. Solar Energy International, 2012.

[31] J. P. Dunlop. *Photovoltaic Systems*. American Technical Publishers, Inc, 2nd edition, 2010. The National Joint Apprenticeship and Training Committee for the Electrical Industry.

[32] P. Eiffert and G. J. Kiss. Building-integrated photovoltaics for commercial and institutional structures: A sourcebook for architects and engineers. Technical Report NREL/BK-520–25272, U.S. Department of Energy (DOE) Office of Power Technologies: Photovoltaics Division, 2000.

15 CONCENTRATION—THE PATTERN TO MANIPULATE LIGHT

CONCENTRATION of light means effectively manipulating photons collected from a large area aperture into a smaller area receiver. So why do we need to concentrate light? Concentration can be used for all three solar conversion processes: optoelectronic, optocaloric, and photoelectrochemical methods. We use concentration when trying to accumulate (or net) a larger number of photons into a smaller receiving area. When referring to concentration of solar energy to yield an optocaloric gain—thermal energy or power—we use the phrase *concentrating solar power* (CSP). When referring to concentration of solar energy to provide enhanced irradiance for photovoltaic action we use *concentrating photovoltaics* (CPV). Please note the distinction for future reading of solar market updates. CSP is distinct from the field of photovoltaics, and calls for a set of skills that are closely linked to mechanical engineering. CPV is a specialized field within photovoltaics, and calls for a set of skills in materials science, physics, electrical engineering, and mechanical engineering (to address dissipation of high thermal gains). Thermal energy is highly desirable in CSP, and undesirable in CPV.

The number of photons absorbed by any SECS is a key extrinsic parameter.

CSP: abbreviation for concentrating solar power. Within the field CSP only applies to solar-thermal conversion from concentration.

CPV: abbreviation for concentrating photovoltaics—when light is being concentrated to yield a photovoltaic work response.

As a reminder, all SECSs are composed of the same general components: an *aperture*, or opening to accept and direct light; a *receiver* containing the absorber for light conversion; a *distribution mechanism* conveying mass from the system to the surroundings to do work; and a *control mechanism* guiding the performance and shutting the system down when called for. Additionally, many SECSs use *storage* systems–largely depending on the correlation between the times of gains and loads.

> **SECS—Basic components:**
>
> ***Aperture:*** the opening to accept and direct light.
>
> ***Receiver:*** the absorber used for solar conversion to heat or electric power.
>
> ***Storage:*** stores useful energy for periods of intermittent gains that are negatively correlated with client loads (not always present).
>
> ***Distribution Mechanism:*** function for the open energy conversion device to convey mass and energy from the system to the surrounding to do work.
>
> ***Control Mechanism:*** guides system performance such as tracking, pumping, power management, and shuts down a system when called for.

Gain/Load correlation:

Positive: light conversion occurs when demand occurs (e.g., electric air conditioning with PV gains)—less call for storage.

Negative: light conversion occurs opposite of when demand occurs (e.g., night parking lights)—storage is crucial.

Zero: no correlation between gains and demand on the scale of analysis—may call for small storage combined with controls.

In this chapter we list the area of the aperture in equations as A_a, and the area of the receiver as A_r. The receiver contains the absorber material, and so for light concentration $A_a > A_r$. For non-concentrating SECSs such as typical residential FPCs and ETCs for solar hot water, the area of the collector is the same as the area of the receiver and the aperture.

For concentrating systems: $A_a > A_r$.

For non-concentrating systems: $A_a \doteq A_r = A_c$.

CONCENTRATION LIMITS

The most common practical metric for concentration is the *geometric ratio* or area concentration ratio C_g. This is the ratio of the aperture area relative to the receiver area.

$$C_g = \frac{A_c}{A_r}. \tag{15.1}$$

The alternate, more precise ratio for energy balances is the optical concentration ratio or the flux concentration ratio C_{optic}. The optical concentration ratio is the measure of the average irradiance impinging upon the area of the receiver relative to the irradiance collected by the larger aperture. Obviously this is a more difficult metric to establish, and the optical concentration ratio varies in proportion with the geometric ratio, so for comparison purposes we use C_g.

$$C_{optic} = \frac{\frac{1}{A_r}\int G_r dA_r}{G_a}. \tag{15.2}$$

Low concentration levels are on the order of $C_g > 1 - 10x$, while high concentration levels can rise up to $C_g \sim 10{,}000x$ for solar furnace research applications. The higher the level of concentration, the more the system requires only the beam normal component of light ($G_{b,n}$ or DNI), and the more precise the optics and tracking system needs to be. If we refer back to our chapter on measurement and estimation, we recall that pyrheliometers (DNI sensors) typically have acceptance angles of 2.5°, while high concentrating SECS (two-axis tracking systems) for CPV or CSP tolerate even smaller angles of acceptance (on the order of 0.5–1°). Hence, even robust measurements on site will tend to overestimate the DNI available to a high concentrating SECS.

Remember also from the chapter on the Sky, that DNI is strongly influenced by **aerosol** contributions. **Aerosol optical depth (AOD)** measures become very important for CPV and CSP.

So what are the limits to light concentration? In order to answer this question we must return our focus to the Sun.[1] First assume the Sun and the receiver are perfect blackbodies (a reasonable assumption for the Sun, a decent approximation for a receiver surface). Use the Stefan-Boltzmann Law to calculate the two directions of radiant energy exchange from the Sun to the receiver ($Q_{S\rightarrow r}$) and from the receiver to the Sun ($Q_{r\rightarrow S}$), where the radius of the Sun is known (r) as well as the half angle of acceptance for the receiver pointing at the Sun (θ_m = 0.27°). As diagrammed in Figure 15.1, the distance to the Sun from our receiver is R, and the aperture and receiver areas have already been defined (A_a, A_r). Only a fraction of that energy from the Sun reaches the receiver according to the view factor F_{S-r}.

[1] Wow, that was a brilliant pun…and that was another one! Feel free to laugh or cringe.

$$Q_{S\rightarrow r} = (4\pi r^2)\sigma T_{Sun}^4 \cdot F_{S-r}, \tag{15.3}$$

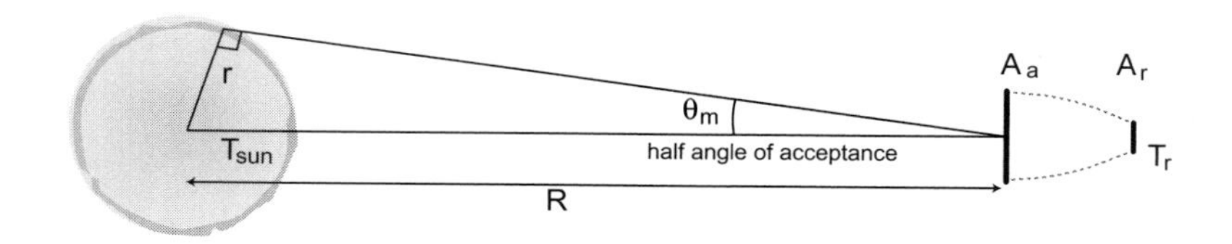

Figure 15.1: Schematic for calculating concentration limit of an aperture-receiver system using the power of the Sun. Assume $T_{Sun} = T_r$ and $F_{r\text{-}S} = 1$. The half angle of acceptance $\theta_m = 0.27°$, about 1/4°.

$$F_{S-r} = \frac{A_a}{4\pi R^2}. \tag{15.4}$$

By combining Eqs. (15.3) and (15.4), we arrive at the radiant energy exchange from the Sun to the receiver in Eq. (15.5).

$$Q_{S\to r} = A_a \frac{r^2}{R^2} \sigma T_{Sun}^4. \tag{15.5}$$

A perfect receiver will be focused to radiate with energy $A_r \sigma T_r^4$ only in the direction of the Sun, therefore we assume a view factor $F_{r-S} \to 1$. Also under ideal conditions of a blackbody and thermodynamic exchange according to Kirchoff's Law, $T_r = T_{Sun}$ and $Q_{S\to r} = Q_{r\to S}$.

$$Q_{r\to S} = A_r \sigma T_{Sun}^4 \cdot F_{r-S}. \tag{15.6}$$

By taking the ratio of Eqs. (15.5) and (15.6) (assuming $F_{r-S} = 1$), we arrive at a familiar geometric ratio formulation equal to the ratio of squares in Eq. (15.7). Also, through a solid understanding of trigonometry developed earlier in the text, we observe that the sine of the half angle of acceptance for the receiver pointing at the Sun (θ_m) is the ratio of the Sun's radius relative to the radial distance from the Earth to the Sun.

$$C_g = \frac{A_a}{A_r} = \frac{R^2}{r^2}, \tag{15.7}$$

$$\sin\theta_m = \frac{r}{R}. \tag{15.8}$$

Hence, the maximum theoretical concentration achievable about a focal point (a 3-D concentration) in vacuum or air is the inverse of the sine function squared.

If we relax our concentration to a focal *axis* or line (2-D concentration) then the concentration limit is the inverse of the sine function.

$$C_{g,max,3-D} = \frac{n}{\sin^2 \theta_m}. \quad (15.9)$$

For air ($n=1$) the theoretical maximum of solar concentration on a point is ~46,000×, but if we add a lens of glass ($n=1.55$) the theoretical maximum increases to ~103,500×.

$$C_{g,max,2-D} = \frac{n}{\sin \theta_m}. \quad (15.10)$$

In this case, for air the theoretical maximum of solar concentration on a line is ~216×, while for glass the theoretical maximum increases to ~324×. We caution the reader that C_g is not a linear figure of merit, and by looking ahead to Figure 15.6 we see that system thermal gains have a non-linear response.

METHODS OF CONCENTRATION

There are two main physical methodologies to concentration: imaging and non-imaging (or anidolic) concentration. While imaging optics like parabolic mirrors/lenses are in use, many modern thermal systems for CSP use non-imaging methodologies. Welford and Winston, as well as O'Gallagher and others have published extensive texts with analysis of the theory and practice of non-imaging optics, which we encourage the enthusiastic solar professional to pursue. Probably the most recognizable architecture is that of the Compound Parabolic Concentrator (CPC), developed independently by Winston in 1965 and Baranov in 1966.[2] The CPC is constructed of two parabolic segments, each of which works to create a wide angle of acceptance for light, and can achieve the ideal concentration levels described as theoretical limits $C_{g,max}$ described for imaging optics.

In either development (imaging/non-imaging), the higher one wishes to concentrate photons, the more precise the optics and the tracking of the Sun become. Additionally, the higher one wishes to concentrate photons, the more important clear sky conditions become, and the more the system calls for normal

Pay attention to artificial lighting design: whatever is used to spread light out from a point source to a broad area can be used in reverse! Which is why we also see CPCs (with diffuse white coatings) in the top covers of our commercial fluorescent lighting fixtures in renovated schools, restaurants, and hospitals.

[2] Josesph J. O'Gallagher. *Nonimaging Optics in Solar Energy (Synthesis Lectures on Energy and the Environment: Technology, Science, and Society)*. Morgan & Claypool Publishers, 2008.

Beam irradiance normal to the Sun: DNI or $G_{b,n}$. High concentrating SECS call for clear skies and good DNI.

beam irradiance conditions ($G_{b,n}$). High concentrating solar power is more limited under diffuse sky irradiance conditions. So DNI is increasingly important as a project design team considers escalating systems design from fixed axes, to systems that track the Sun, to systems that concentrate sunlight and are required to track the Sun.

Declination (δ) changes rapidly about the Equinoxes, and the slowest over the Solstices.

However, one does not necessarily need to track the Sun every day, moment by moment. Some low-concentrating systems do not call for frequent tracking, and may only require adjustment two to four times a year (recall the rate of change in declination that occurs annually).

MIRRORS AND LENSES

We manipulate and guide light to gather irradiance from a large aperture area into a smaller receiver area. Two macroscopic material properties can be employed for concentration: refraction (high transmittance, τ) and reflectance (ρ). In the case of light passing through an optically transparent medium (like glass, fused silica, or PMMA), we have seen how the index of refraction n of the material will indicate the degree to which light is refracted by an angle different from the angle of incidence (θ). Upon leaving a transparent medium back into air or vacuum, the light will be refracted again with an exiting angle similar to the angle of incidence. In the case of light reflecting off of a shiny surface, light is reflected at the surface and the reflection can be either *diffuse* (like white paint, chalk, or plants) or *specular* (like a shiny mirror). For optical concentration methods in CSP specular surfaces are called for, but we also recommend considering the importance of diffuse reflectors for low concentration of light and minor performance boosts in SECS design.

High efficiency CPV can achieve $\eta > 25\%$ system efficiencies, compared to 15–16% for non-concentrating sc-Si cells.

Unlike the absorber materials in a receiver, *lenses* do not get very hot from optical interactions because they are largely *transparent* to the light—they are designed to refract light, not absorb light. Large-scale lenses are physically and economically impractical, and so are not used in CSP applications like parabolic trough collectors, where mirrors are employed. However for CPV applications, where the device uses multi-junction or tandem PV cells made of multiple III–V semiconductor materials. A typical PV device has only one junction, which can be thought of as *one engine*, while a multi-junction PV will be composed of three different photovoltaic absorber materials, each absorbing successively longer wavelengths (lower energy) in the stack. Hence, a multi-junction cell is analogous to linking multiple engines in series and collecting the net efficiency of a high energy engine, a medium energy engine, and a low energy engine.

CPV devices use **III–V** (three–five) semiconductor materials in a **stack** of tandem PV junctions.

In Figure 15.2 we show cross-sections and ray tracing for both a simple biconvex lens, and for a compact and practical Fresnel lens.[3] Fresnel lenses

[3] William B. Stine and Michael Geyer. *Power From The Sun.* William B. Stine and Michael Geyer, 2001. Retrieved January 17, 2009, from http://www.powerfromthesun.net/book.htm.

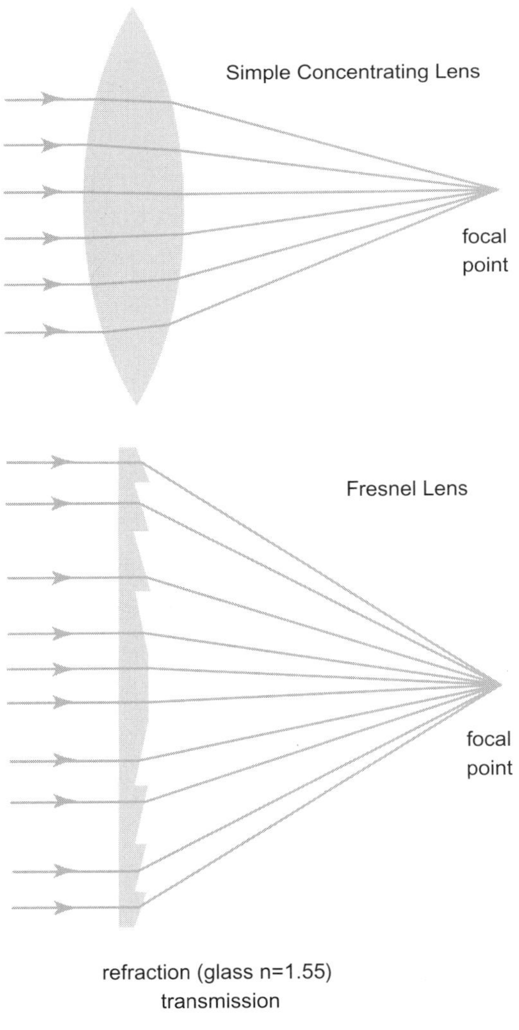

Simple Concentrating Lens
focal
point
Fresnel Lens
focal
point
refraction (glass n=1.55)
transmission

Figure 15.2: Showing both a simple lens and a Fresnel lens. In this case the lens materials are assumed to be glass ($n = 1.55$).

Remember: there is no "frez-nell" pronunciation that is acceptable. The name is French origin and pronounced "frehn-*el*." Don't get this one wrong and embarrass yourself in a crowd.

are widely used due to the reduction in material costs, easy processing of polymers into complex molds for flat lens design, and short focal distance possible. Common materials are polymer plastics or fused silica (SiO_2). When concentrating light, we can use a lens that focuses light into a point (3-D concentration) or we can develop a lens that focuses light onto a line or cylindrical axis (2-D concentration). The limit of concentration increases with 3-D methods, and in the case of CPV, a 3-D Fresnel lens of polymer is used to focus light into a few square millimeters. With high concentration levels of light, PV cells will luminescence or "glow" as a light-emitting diode (an LED) due to recombination phenomena.

Next, we also understand that *mirrors* may also direct light to a focal point or line (Figure 15.3). From our understanding of material reflectance properties, metals like silver and aluminum have been used to provide optically favorable reflector technologies. Mirrors are often formed by coating a metallic thin film on the rear surface of glass or a thick polymer sheet (curved or flat), in order to protect the reflective surface from corrosion and abrasion. Mirrors can also be constructed on metal substrates.

CSP focuses light for applications in electrical power generation, industrial process heating, driving mechanical steam pumps, as well as solar cooking. Both convex mirrors or arrays of flat mirror arranges in compact form according to Fresnel concentration are used to increase the density of absorbed photons from A_a to A_r. Again, the major advantage of CSP for thermal applications is that the total heat loss (U_L) can be reduced due to the significantly smaller surface area of the receiver portion of the collector A_r. The thermal mass of the heat transfer fluid is smaller, and transient thermal effects within the system can be reduced.[4] Also, a high delivery temperature for the acting fluid implies a better match of energy to the demanded application at hand by the client/stakeholders, particularly in using solar thermal for industrial heat applications (e.g., chemical and food processing uses hot water and steam extensively, all powered by fuels). Finally, concentration has the potential to reduce costs of a total system by replacing highly expensive receiver/absorber materials (the engines) with less expensive reflecting surfaces.

[4] William B. Stine and Michael Geyer. *Power From The Sun.* William B. Stine and Michael Geyer, 2001. Retrieved January 17, 2009, from http://www.powerfromthesun.net/book.htm.

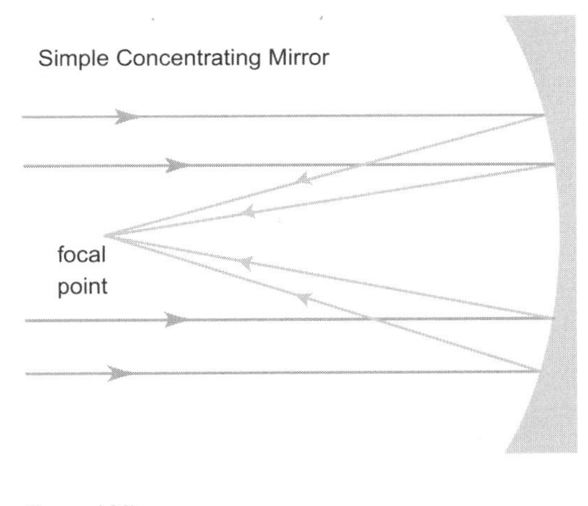

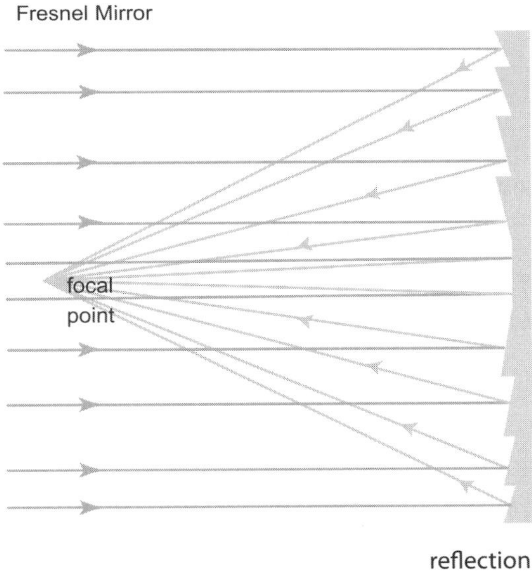

Figure 15.3: Showing both a simple concave mirror and a Fresnel mirror. The mirror materials are assumed to be coated on the back side of a protective transparent layer with a thin film of high reflectance in the shortwave (ρ_{sw}=0.98).

THE CASE FOR STEAM

One of the most useful fluids in energy conversion is *steam* derived from boiling water. Practically speaking however, a non-concentrating solar thermal system will not exceed the boiling point of water at atmospheric pressures without significantly decreased performance. Given a solar thermal heating system, if one adds more modules to the array there is more surface area (A_c) to increase solar gains, but there is also more surface area on all sides of the collectors to increase thermal losses. At some point the temperature of the collector (which is also the *receiver*) increases until the convective and radiative losses from the collector are equal to the absorbed energy S. The increase in losses with ΔT is the driving factor to reduce the efficiency of the system seen in the performance plot in Figure 15.4.

$$Q_u = A_c F_R [S - U_L \Delta T], \tag{15.11}$$

$$T_{max} = \frac{G_t \cdot f(\tau, \alpha)}{U_L} + T_a. \tag{15.12}$$

The maximum thermal temperature of the system can occur with zero mass flow (no mechanical pumping), the stagnation temperature, which results in zero efficiency. Even with pumping, the drop in performance with increasing ΔT is due to the large thermal losses to the exterior. Recall from Eq. (13.34) that losses increase in proportion to the temperature difference between the mean plate temperature and the ambient air.

There is a good thought experiment to explore the limits of connecting solar hot water panels in series for non-concentrated solar thermal performance. The concept

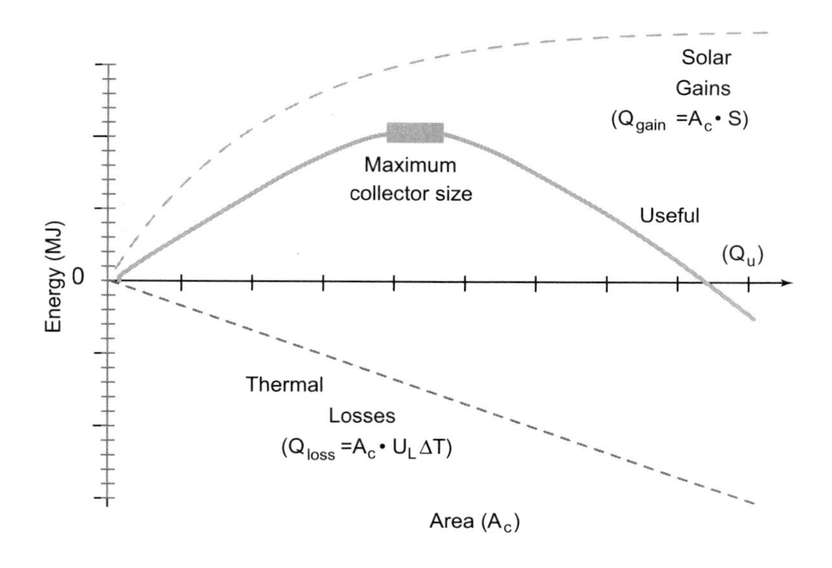

Figure 15.4: Relative gains and losses in a non-concentrating optocaloric collector, according to increasing area A_c. Figure adapted from Stine and Geyer from collector parameters: $G_b = 1000\,W/m^2$; $T_a = 298\,K$; $U_L = 60\,W/m^2K$; $\eta_{opt} = 0.9$; $\varepsilon = 0$.

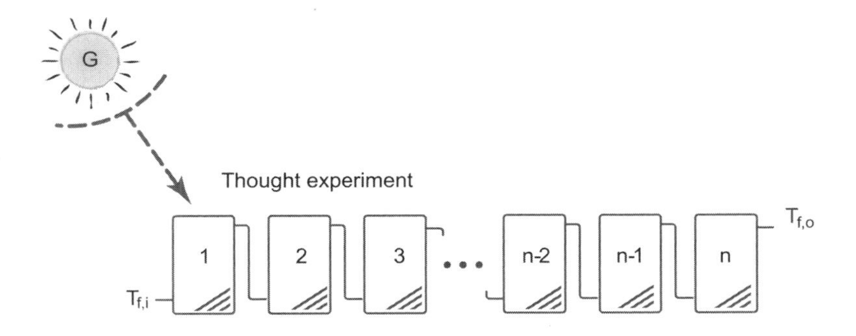

Figure 15.5: Schematic of flat plate modules connected in an absurdly long series as a thought experiment. At some point in the chain $T_{f,i} = T_{max}$ and the efficiency is very low.

of an absurdly long chain of solar hot water panels is presented in Figure 15.5. We temporarily suspend our disbelief about the pressure drop that would occur over such a long system. Given a sunny day and a moderate flow of heat transfer fluid, the fluid enters the system at $T_{f,i[1]}$ conditions, and leaves module #1 with $T_{f,0[1]}$. The next module (#2) will accept the fluid directly from module #1, where the outlet and subsequent inlet fluid temperatures are effectively the same ($T_{f,0[1]}=T_{f,i[2]}$). This repeats for module #3, where $T_{f,0[2]}=T_{f,i[3]}$. The process will proceed through n modules, so what is happening? With each additional module the temperatures $T_{f,i}$ and $T_{p,m}$ are increasing, while the ambient air temperature and the irradiation conditions remain the same. Eventually, q_{loss} for the nth module will exceed q_{abs}.

Increased operating temperature is brought about by concentrating light, which reduces the area of the receiver relative to the aperture.

Hence, the maximum temperature of the optocaloric SECS cannot be exceeded by simply connecting more and more panels in series. For solar thermal performance, an increase T_{max} requires a reduction in the surface area of the system, while increasing the density of light upon the receiving surface. In other words, light concentration is called for. In terms of our linearized relation for solar thermal performance, we are decreasing the slope of the performance plot by a factor of the geometric concentration ratio C_g.

CSP PERFORMANCE

For a concentrating system in CSP, the efficiency measures can be separated into optical and thermal attributes, η_{opt} and η_c.[5]

$$\eta_{opt} = \frac{S}{G_b}, \tag{15.13}$$

[5] William B. Stine and Michael Geyer. *Power From The Sun.* William B. Stine and Michael Geyer, 2001. Retrieved January 17, 2009, from http://www.powerfromthesun.net/book.htm.

where S includes factors related to shape of the receiver, reflection and transmission losses, accuracy of the tracking system, shading effects, the tau-alpha function of the receiver, and the angle of incidence from the beam component of light (θ from G_b). Deeper insight into system efficiency analysis is beyond the scope for this text, but there are numerous resources available and emerging for concentration assessment and design.

$$\eta_c = \frac{q_u}{G_b}. \tag{15.14}$$

The application for heating a fluid like steam to run a thermodynamic power cycle is very common in CSP. This is also called using solar for "spinning power," because the steam is used to bring about a power cycle of spin to a turbine that in "turn" operates a generator of alternating current power. The Rankine cycle the most common used for base load electricity in large power plants. The Rankine power cycle is predominantly utilized with steam, but may also be employed at lower temperatures with the alternative Organic Rankine Cycle, where lower temperature fluid is accepted to also yield lower power density. The Stirling cycle is not very common, but is an attractive CSP where the working fluid is air or helium. Rankine cycle temperatures can be achieved with both 2-D parabolic concentration and 3-D concentration (solar power towers). The Brayton cycle is typically used in gas turbines to serve "peaking power" on the grid. It requires very high temperature/pressure conditions. 3-D concentration is required for both the Stirling and Brayton cycles.

In Figure 15.6 the concentration levels are shown as C_g, required to achieved these ranges of operating temperatures for power cycles given some nominal concentration parameters.[6] An important thing to notice is that C_g is not a linear figure of merit with respect to thermal yield, and should not be treated as such. Also noted in the figure is the trade-off between the power cycle of the engine (which performs at higher efficiency with higher ΔT) and the thermal efficiency of the collector/receiver, which decreases with increasing ΔT due to increased thermal losses. This leads to a combined system efficiency at which a CSP power system can perform optimally.

[6] William B. Stine and Michael Geyer. *Power From The Sun.* William B. Stine and Michael Geyer, 2001. Retrieved January 17, 2009, from http://www.powerfromthesun.net/book.htm.

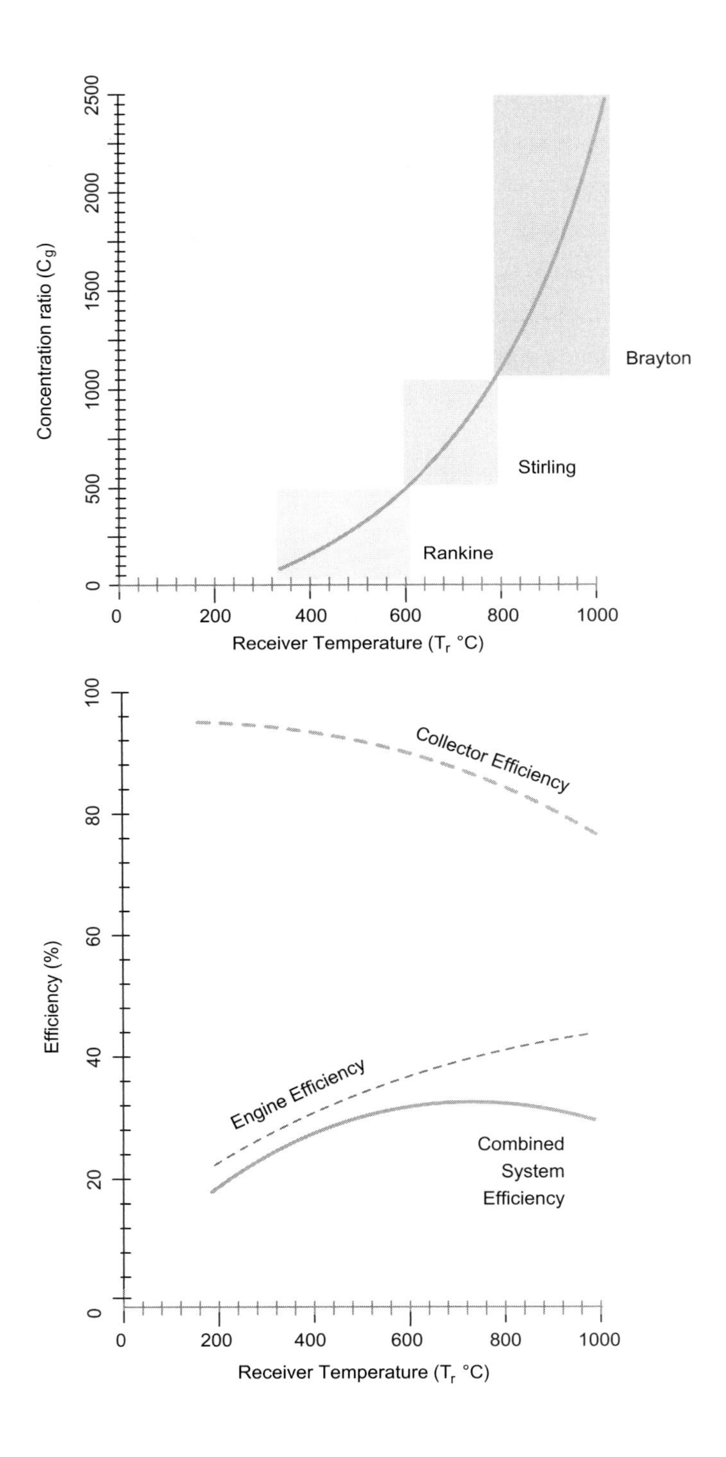

Figure 15.6: CSP system performance according to concentration ratio and the balance between thermal cycle performance in an engine and the collector performance.

The System Advisor Model software from NREL has several CSP power simulation options available to explore.7 While the CSP systems designs are more complicated than addressing flat plate systems, the emerging professional will observe that many of the principles developed for non-concentrating systems can be extended to investigate concentration applications. Finally, CPV performance and finance systems are also available from SAM for future learning and design opportunities. We recommend diving in and trying both, as the common user interface allows one to learn by doing, and experimentation is largely low effort and fun.

[7] System Advisor Model Version 2012.5.11 (SAM 2012.5.11). URL https://sam.nrel.gov/content/downloads. Accessed November 2, 2012.

SCADA

A final note on managing large-scale systems that track the Sun and generate large-scale power and heat. These systems must be managed using modern computing strategies. SCADA (pronounced as a word: *skay*-da) is an acronym for an industrial scale controls and management system: Supervisory Control and Data Acquisition. Although this is a generic term, SCADA systems are used in large-scale solar field, where tracking and mechanical pumping are required to adjust to the solar conditions over the day. When SECSs become large, when they attempt to hybridize systems of power from fuels and from CSP, the control and distribution mechanisms, as well as storage strategies, are all monitored and managed from a SCADA platform.

PROBLEMS

1. What is the role of concentration in SECS design?

2. Explain the manner in which the reflectance of the ground (albedo) can act as a low concentration system.

3. Explain how the inverse squared law of light can limit the range of solar concentration.

RECOMMENDED ADDITIONAL READING

- Optics.[8]
- Nonimaging Optics in Solar Energy (Synthesis Lectures on Energy and the Environment: Technology, Science, and Society).[9]
- Principles of Solar Engineering.[10]
- Solar Engineering of Thermal Processes.[11]
- Power From The Sun.[12]
- Thermal Energy Storage: Systems and Applications.[13]
- Planning & Installing Solar Thermal Systems: A guide for installers, architects and engineers.[14]
- Photovoltaic Conversion of Concentrated Sunlight.[15]

[8] Eugene Hecht. *Optics.* Addison-Wesley, 4th edition, 2001.

[9] Josesph J. O'Gallagher. *Nonimaging Optics in Solar Energy (Synthesis Lectures on Energy and the Environment: Technology, Science, and Society).* Morgan & Claypool Publishers, 2008.

[10] D. Yogi Goswami, Frank Kreith, and Jan F. Kreider. *Principles of Solar Engineering.* Taylor & Francis Group, LLC, 2nd edition, 2000.

[11] John A. Duffie and William A. Beckman. *Solar Engineering of Thermal Processes.* John Wiley & Sons, Inc., 3rd edition, 2006.

[12] William B. Stine and Michael Geyer. *Power From The Sun.* William B. Stine and Michael Geyer, 2001. Retrieved January 17, 2009, from http://www.powerfromthesun.net/book.htm.

[13] Mark A. Rosen Ibrahim Dinçer. *Thermal Energy Storage: Systems and Applications.* John Wiley & Sons Ltd, 2002. Contributions from A. Bejan, A. J. Ghajar, K. A. R. Ismail, M. Lacroix, and Y. H. Zurigat.

[14] German Solar Energy Society (DGS). *Planning & Installing Solar Thermal Systems: A guide for installers, architects and engineers.* Earthscan, London, UK, 2nd edition, 2010.

[15] V. M. Andreev, V. A. Grilikhes, and V. D. Rumyantsev. *Photovoltaic Conversion of Concentrated Sunlight.* John Wiley & Sons Ltd, Ioffe Physico-Technical Institute, Russian Academy of Sciences, St. Petersburg, Russia, 1997.

PROJECT DESIGN

16

THE IMPORTANCE OF INTEGRATIVE DESIGN

The only principle that does not inhibit progress is: anything goes.

Paul Feyerabend

IN THE DESIGN of solar energy conversion systems, we are working with a common theory that DESIGN IS PATTERN WITH A PURPOSE. Think about that. Whereas *art* and *science* provide mechanisms to ultimately open windows into apparent patterns about us, *design* and *engineering* are purposeful approaches to establish *systems that fit the revealed pattern.*

The fundamental underpinning of **design** is the purposeful approach fitting the system or context.

NO WORK SHOULD OCCUR IN ISOLATION! In analogous language to the challenges in designing a solar energy conversion system, Alexander et al. emphasize the fundamental underpinning of design as a systems process in which one takes a purposeful approach to fitting solutions into the context of the problems and patterns in nature:

> "When you build a thing, you cannot merely build that thing in isolation, but must also repair the world around it, and within it, so that the large world at the one place becomes more coherent, and more whole; and the thing which you make takes its place in the web of nature, as you make it"
>
> –Christopher Alexander (1977)[1]

[1] C. Alexander, S. Ishikawa, and M. Silverstein. *A Pattern Language: Towns, Buildings, Construction.* Oxford University Press, 1977.

The future of solar energy conversion systems will be tied to an innovation enterprise, bridging the spectrum of responsibilities and goals in Research, Development, Demonstration, and Deployment (RDD&D), via the iterative process of *integrative design*. The phrase *integrative design process* is already closely associated with the design and deployment of green buildings.[2] Indeed, this chapter draws strongly from design tied to the built environment as residential and commercial buildings and the encompassing environments. Even so, integrative design can be applied as an effective template in the field of solar energy conversion as well. The building focus should not limit the solar design team to think only of rooftops and façades. There are great expanses of field and pasture that we should also bear in mind—new locales with diverse ecosystems, multi-stakeholder strategies, and complex financing methods that will need to be assessed for the design and deployment of an ethically sound, sustainable, systems integrative solar farm. How can we ensure robust ecosystems services are maintained or enhanced with these new projects?

[2] J. Boecker, S. Horst, T. Keiter, A. Lau, M. Sheffer, B. Toevs, and B. Reid. *The Integrative Design Guide to Green Building: Redefining the Practice of Sustainability.* John Wiley & Sons Ltd, 2009.

The process of *integrative design* encompasses and includes:

- Shared understanding of a concept or systems goal, resulting from iterative and collaborative working sessions,
- Expectation that ALL collaborative members or stakeholders will extend beyond their "comfort zones" in the early discovery phase of the design, offering supportive input even beyond immediate areas of expertise,
- Clear expectations of members' work in developing targeted systems performance,
- Attention to maps of the process, and iteration through the process continually improving the map,
- Interdependent tasks (stakeholders co-solve problems with designers): RRD&D does not occur in isolation,

- Encouraging a greater sense of ownership for the whole enterprise, rather than individual thrusts or components, and

- Innovative solutions that challenge our established "rules-of-thumb."

DESIGN

In the design fields, professional projects have the potential to avoid an integrative process, and devolve into arcs of *accelerated aggregation* of technologies. Accelerated aggregation is typical when independent components or unit assemblies are pieced together without due respect for the context and concept for deployment. The potential exists for elements or components to be optimized rather than the whole system, which then tends to cause the system to work less effectively, or become *pessimized*. Sounds bad right? At best, the design and development team will arrive at an incomplete or incorrect answer (a poor use of resources); at worst it could hold back innovation—particularly important to the re-emergent field of solar energy conversion technology. In an integrative effort, the *Discovery Phase* encourages members to develop ownership of the entire innovation enterprise (through iterations), being encouraged to participate and improve upon the initial concept and performance goals of the enterprise, with respect to evolving social, technical, and environmental constraints.

A *charrette* can be thought of as an intensive brainstorming workshop for the discovery phase. Charettes are common to the design process, and are especially useful to establish early transdisciplinary communication and common concepts deliverables for a project. The charrette event is also systems-based, where all parties (client, design team, financial officer, builder, even community representatives) are brought to the round table to form the guiding goals and scope of the project.

charrette: pronounced "shah-*rett*"

POLICY

Solar project developments are strongly affected by local, regional, and federal regulations, access rights, and laws. Integrative systems approaches that involve communication across skill sets will be required to modify or create policies that are ethically sound, helpful to the commons and industry, and speed up transition of spaces to solar ready projects. If policy is not implemented, design can be stunted or delayed by years, often due to a lack of familiarity with a topic rather than a true technological or market barrier.

ECONOMICS

Solar project drivers abound in each locale across the planet, so long as the project development team seeks out the true needs of the client and the constraints of local finance, policy, meteorology, and infrastructure. An integrative systems approach will seek out the highest solar utility to the client given the constraints, and will construct project finance approaches to mesh with the design process.

THE DISCOVERY PHASE: PREDESIGN

"The discovery phase is the foundation of an integrative process"
—7GROUP

Key to the initial predesign process is the *iterative alignment* of stakeholders to a unified *concept* or *conceptual map*. Recall that in *design*, we are looking to create a *a purpose to the pattern*. For integrative design, particularly in the early *Discovery Phase*, we first establish coordinated working sessions among the multiple stakeholders necessary to identify the underlying pattern and an aligned goal or purpose fitting that pattern. Coordinated work sessions can be exhibited as regular events of directed brainstorming (also called *charettes*) and can be documented by constructing and revisiting the conceptual map of the full design and development process. In the discovery phase, the design team holds stakeholders to the concept of the *Four Es*: **E**verybody **E**ngaging **E**verything **E**arly.

J. Boecker, S. Horst, T. Keiter, A. Lau, M. Sheffer, B. Toevs, and B. Reid. *The Integrative Design Guide to Green Building: Redefining the Practice of Sustainability.* John Wiley & Sons Ltd, 2009.

A well-designed automobile or a modern jet aircraft are both interesting systems analogies that use discrete parts, involve many stakeholders that insist upon multiple criteria/constraints influencing the design process. The resulting technology tends to function as a rewarding system for the majority of the stakeholders. One may note that automotive industry is also well known for "concept" cars, as well as the transformation of emerging concepts into the main stream: electric vehicles, hybridized energy storage. We are aware of the next generation of energy efficient jets using carbon nanofiber composites instead of aluminum, and running on biofuels.

At present, the building industry is not typified of a systems design process, in that each step from design to deployment and operation involves a separate stakeholder (the architect, the electrical/mechanical engineer, the builder, the policy author for building codes), isolated from almost all of the remaining stakeholders. The resulting products tend to be highly energy inefficient, have poor indoor air quality, and are designed largely independent of the surrounding solar conditions and shading influences. Each stage of design and deployment is subject to isolated *value engineering*, a set of procedures that seek out optimal economic costs relative to the function of the product. Unfortunately, we again encounter the trouble with optimizing parts or components that are otherwise strongly linked with each other (coupled): the performance of the system as a whole is *pessimized*.

The systems process linked with the built environment should be important to the solar design professional, as the building itself is a solar technology (even unintentionally), and many solar technologies are linked to built structures in the residential and commercial sector of energy demand. Indeed, the broad field of solar energy is characterized by multiple stakeholders. The solar energy field holds an implicit normative concept of the Sun as *commons*, and a strong coupling among the technology, the solar resource, the supporting environment, and society. Hence, we encourage students and professionals to investigate and adapt to the *integrative design process* for future solar energy conversion enterprises.

The predesign phase tends to include the following:[3]

- Preparation: Research and Discovery
- Evaluation of Concept
- Conceptual Design

[3] J. Boecker, S. Horst, T. Keiter, A. Lau, M. Sheffer, B. Toevs, and B. Reid. *The Integrative Design Guide to Green Building: Redefining the Practice of Sustainability.* John Wiley & Sons Ltd, 2009.

THE DESIGN PHASE

Once a conceptual map has been established by the multiple stakeholders, the design team can compare alternatives for solar energy conversion systems and their hybridization with current fuel technologies. Solar energy conversion is a *systems* field, and knowledge of the locale, the quantity and character of the solar resource, and the economic setting is demanded to devise technologies relevant to each region. For example, the solar resource of the mid-atlantic states is different from that of the American southwest (in quantity and character), but it may be more useful than the remote southwest regions due to pre-existing transmission lines in the mid-atlantic region, which plays a significant role in the *balance of systems* costs for solar technology.

The design phase tends to include the following steps:

- Schematic Design
- Design Development
- Construction Documents
- Bidding and Construction, followed by
- Occupancy, Operations, and Performance Feedback

SOLVING FOR PATTERN

The concept of "Solving for Pattern," coined by Berry in his essay of the same title, is the process of finding solutions that solve multiple problems, while minimizing the creation of new problems. The essay was originally published in the Rodale Press periodical *The New Farm*. Though Mr. Berry's use of the phrase was in direct reference to agriculture, it has since come to enjoy broader use throughout the design community.

"A bad solution is bad, then, because it acts destructively upon the larger patterns in which it is contained. It acts destructively upon those patterns, most likely, because it is formed in ignorance or disregard of them. A bad solution solves for a single purpose or goal, such as increased production. And it is typical of such solutions that they achieve stupendous increases in production at exorbitant biological and social costs.

A good solution is good because it is in harmony with those larger patterns—and this harmony will, I think, be found to have a nature of analogy. A bad solution acts within the larger pattern the way a disease or addiction acts within the body. A good solution acts within the larger pattern the way a healthy organ acts within the body. But it must at once be understood that a healthy organ does not—as the mechanistic or industrial mind would like to say—"give" health to the body, is not exploited for the body's health, but is a part of its health. The health of organ and organism is the same, just as the health of organism and ecosystem is the same. And these structures of organ, organism, and ecosystem—as John Todd has so ably understood—belong to a series of analogical integrities that begins with the organelle and ends with the biosphere."[4]

[4] Wendell Berry. *Home Economics*. North Point Press, 1987.

SOLAR UTILITY FOR THE CLIENT AND LOCALE

Consider, the Sun is our nearest nuclear fusion reactor,[5] and about 90% of its energy is found in its center. However, observers have also found that the surface of the Sun (the photosphere) is an opaque (*yes, opaque!*) collection of ionized gases[6] existing at approximately 5777 K, and able to absorb and emit a continuous spectrum of radiation, as seen in Figure 3.8.

[5] A mere 93 million miles away...

[6] Strongly ionized gases are also called PLASMAS.

The great thermal energy present at the exterior surface of the Sun allows it to be an efficient emitter of radiant energy. In a somewhat similar fashion, non-solar lighting sources can be emitters of radiant energy, although the mechanism can be thermally derived or due to quantum emission of photons from the relaxation or recombination of excited electronic states. When we measure light for the tight range (called a *spectrum*) of sensitivity that the human eye detects, we call the measurements *photometry*. When we measure the much broader spectrum generated by the Sun or by cooler thermal bodies like the atmosphere, the Earth, our homes, or our bodies, we call the measurements *radiometry*. For the purposes of most Solar Energy Conversion Systems that this text focuses on, we use the concepts and language derived from radiometry.

However, lighting for human perception and comfort is a well developing field already, and it is worth taking time to point out the similarities among lighting concepts and solar power concepts. Lighting design is an example of applying a purposeful pattern to achieve a high degree of utility (usefulness) relative to a client in a specific location. An example would be designing lighting for a museum hall or art display that satisfies elements of comfort and vision, while simultaneously minimizing exposure to UV bands of light that may damage. The design can be accomplished in numerous ways with natural and artificial light sources, complementary wall reflectances, and interior surfaces (see Figure 16.1, and review Figure 3.8 from earlier).

Figure 16.1: A *radiometric* image of the Sun within the ultraviolet spectrum. Photo from NASA.

Similarly, in SECS, we seek to maximize the solar utility of the resource for a client in a given locale. However, we are able to consider a much *broader spectrum* of energy to harvest for use to society. In particular, we define high energy light (within which visible light is contained) as *shortwave spectra* (for short wavelengths), and low energy light (typically emitted by terrestrial or atmospheric objects) as *longwave spectra* (for low energy, long wavelengths). The sub-band of light applied to vision and photometry is the Visible, within the larger shortwave band.

REVISITING THE GOAL OF SOLAR DESIGN

Design is pattern with a purpose, and solar design is a whole-systems approach. By emphasizing the integrated three systems of *solar utility*, *client and stakeholder*, and *locale* for the goal of solar design, we have actually opened up many possibilities for pattern solutions. Each of the three cannot be addressed without engaging the frameworks of the other two. The concept of solar utility should drive one to compare solar energy conversion systems and contrast their degree of acceptance or feasibility, return on investment, and ecological impact with respect to the client in the locale at hand. And the client lives or works within a specified locale, not distributed across an entire continent and not with a uniform demand 525,600 minutes a year. The client and stakeholders will define and constrain the bounds for solar utility by describing their needs or wants to be supplied in part by a solar energy conversion system. For example, we can envision stakeholders as a small family desiring electrical power in a residential home in contrast to a client as a commercial company in need of thermal energy for industrial food processing in the same locale. The two scenarios place different bounds on what is "useful" from a solar energy technology, and will call for different solar fractions to supply a solar load in combination with a load provided by conventional fuel-based resources.

In actuality, for complex systems thinking and wicked problems in particular a *maximum* solar utility is more of a figurative concept than a reality. However, aspiring for a much greater integration of SECS in society, in a sustainable manner, is valid within the goal of solar design.

How do you measure, measure a year? In daylights, in sunsets, in midnights, in cups of coffee …

Finally, the locale has been identified repeatedly as a space or an address in *time and place* within which the client occupies and demands energy resources. The solar resource is determined by the locale, as the climate regime affects

A **locale** is an address in both time and place.

the seasonal (synoptic scale weather) and daily irradiation patterns (meso- and microscale weather) and frequencies of intermittency. The character or quality of the solar resource will in turn constrain the design team's options for technological solutions that compete with conventional fuel-based technologies. We tend to avoid high concentration solar solutions in locales with a high degree of diffuse sky conditions, right? The locale is also the address for the local and regional code and policies for adoption of solar technology solutions, and the address for incentives and comparative costs of fuel-based heat and power. The locale is both the place and the placement of a SECS, within the greater environment of the microclimate, neighborhood, urban jungle, field, or watershed. Consider each locale as an adaptive space for integration of a solar energy conversion system, and your integrative design should follow the constraints of the time horizon of the client as well as the greater time horizon for adaptive sustainable energy solutions contributing to the broader society and the biome that support us.

Good luck in your design futures!

As a design team, you are each agents of change for a sustainable future. Be active and engaged participants in a transition to a new energy approach in society, which uses solar energy extensively to increase well-being and ecosystems services. Develop and practice your agency for change by working closely with all stakeholders in the design project (early, and often), and by committing to a lifetime of continued learning. Explore the goal of solar design and engineering and create your own patterns for our enriched future.

PROBLEMS

1. Why does high fuel availability seem to change our social perceptions of solar energy conversion as being available and sufficient?

2. Where were the first solar hot water heating systems commercially sold?

3. Where was the first industrial use of concentrating solar power for pumping employed?

4. If our *system* is open to the flow of energy and matter (like fluids or electrons), how would you define the permeable boundaries of that system to help you in the design of solar energy conversion systems?

5. What does a team do in the discovery phase of the integrative design process?

6. Why is the field of smaller rocks in a Zen rock garden greyish-white?

7. What is the reason for designing a home with a patio (portico) in a mediterranean climate?

RECOMMENDED ADDITIONAL READING

- The Integrative Design Guide to Green Building: Redefining the Practice of Sustainability[7]
- Sustainability Ethics and Sustainability Research[8]
- Systems, Messes, and Interactive Planning[9]
- The Visual Display of Quantitative Information[10]
- Sustainable Energy: Choosing Among Options[11]

[7] J. Boecker, S. Horst, T. Keiter, A. Lau, M. Sheffer, B. Toevs, and B. Reid. *The Integrative Design Guide to Green Building: Redefining the Practice of Sustainability.* John Wiley & Sons Ltd, 2009.

[8] Chrstian U. Becker. *Sustainability Ethics and Sustainability Research.* Dordrecht: Springer, 2012.

[9] Russell Ackoff. *The Social Engagement of Social Science*, volume 3: The Socio-Ecological Perspective, chapter Systems, Messes, and Interactive Planning. University of Pennsylvania Press, 1997.

[10] Edward R. Tufte. *The Visual Display of Quantitative Information.* Graphics Press, Cheshire, Connecticut, 2001. ISBN 0961392142.

[11] Jeffreson W. Tester, Elisabeth M. Drake, Michael J. Driscoll, Michael W. Golay, and William A. Peters. *Sustainable Energy: Choosing Among Options.* MIT Press, 2005.

APPENDIX A: ENERGY CONVERSION SYSTEMS

Energy Conversion Systems (ECS) are our bread and butter. You can read much more about ECS in the US Department of Energy's Energy Information Administration site online: Energy Explained. Much of what we have used for common terminology (such as non-equilibrium conditions, steady state conditions), will involve language from *Modern Thermodynamics* developed by Ilya Prigogine and Lars Onsager, expanded by Ji-Tao Wang.[1]

ROBOT MONKEY'S LAST STAND: HEAT PUMPS, HEAT ENGINES, ENTROPY, AND MODERN THERMODYNAMICS

Modern Thermodynamics is concerned with both spontaneous processes, and the direction opposite of the spontaneous process (what is not forbidden is allowable). Energy systems and materials get very interesting, and permit innovation when you begin to *couple* heat engines with heat pumps.[2] In the end, we are essentially talking about coupling of the solar pump to numerous heat engines that will provide work for use in the environment, to the additional benefit of society.

Transformation: The *First Law of Thermodynamics* delineates a conservation of Energy. Systems can only undergo *transformations* from an initial state (Form$_1$) to a final state (Form$_2$). Given transformations of states of *Heat* and states of *Work*, the summed total energy (here, labeled U) is said to be independent

[1] Dilip Kondepudi and Ilya Prigogine. *Modern Thermodynamics: From Heat Engines to Dissipative Structures*. John Wiley & Sons Ltd, 1998; and Ji-Tao Wang. *Non-equilibrium Non-dissipative Thermodynamics: with Application to Low-Pressure Diamond Synthesis*, volume 68 of *Springer Series in Chemical Physics*. Springer, 2002.

[2] *All impossibilities other than* (1) the perpetual motion machine of the first kind (forbidden), (2) the perpetual motion machine of the second kind (forbidden), and (3) the absolute zero of temperature (forbidden) *may be valid under special conditions*. It is our job as creative teams to identify the special conditions and design materials that make those conditions possible.

of the path (meaning the types of transformation). In a hypothetical cyclic process, the integral of the energy change is zero (Eq. (A.1)).

$$\oint dU = 0. \tag{A.1}$$

Form: A *state* of energy derived from *Sources* like photons coming as an electromagnetic form of energy from the Solar resource of our nearest star (93 million miles away).

Source$_1$: A phenomenological manifestation (observable) of a state of energy. This use of the term includes a force combined either with mass (matter) or electromagnetism (light).

Heat: *Not a term of thermal energy or energy content.* Refers to the transfer of energy from a source to a sink due to a difference in temperature.

Temperature: Conjugate of heat. Internal energy or *Enthalpy* of a state. Temperature is determined by latent heat and heat capacity in matter, and has a connotation and conversion for the latent energy in a photon as well.

System: Term used in thermodynamics for confining an investigation to have boundaries for a part/section/segment or whole world being evaluated.

Reservoir: Description for the surroundings in a *closed system* or *open system*. Often used in connection with a description of a *heat engine* or *heat pump*.

Surroundings: Term for whatever is not contained in the *system*; the connecting parts of the whole world.

Source$_2$: This is a use of the term to be related with a *Sink* in a *heat engine* or *heat pump*. The Source/Sink relation is particularly useful when referring to a thermal temperature differences across a device.

A *heat engine* is an *energy conversion device* that facilitates transformation of states from a high energy *system source* to a low energy *reservoir sink* while collecting some work (see Figure A.1). A flux of *entropy* is associated with this conversion, such that the entropy flux leaving the device is always positive (generating entropy). This exchange does not constrain the flux of localized entropy into the system (entropy input). Efficiency (η) is the value of merit in a heat engine (Eq. (A.2)). At the extreme of ideality, a perfect heat engine

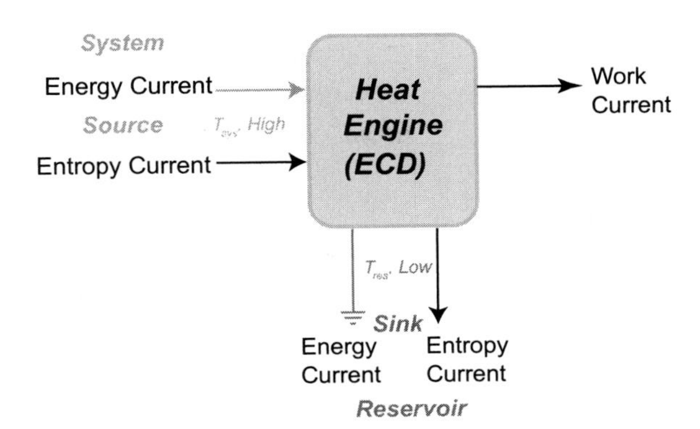

Figure A.1: Schematic of the energy conversion device: heat engine. Arrows indicate currents, or fluxes of energy or entropy.

involves no change in entropy and will provide a maximum efficiency related to the transfer of all energy from temperatures of the high energy *Source* to the low energy *Sink*. This special case is termed the *Carnot efficiency*[3]

$$\eta = \frac{W}{Q_H}, \tag{A.2}$$

$$\eta_{Carnot} = \frac{T_H - T_L}{T_H}, \tag{A.3}$$

where the temperatures T_H and T_L are given in degrees Kelvin (K, not °K).

We shall see in the next Appendix that the Carnot limit based on temperature is a special case of efficiency, historically bound by our tradition of an internal combustion engine, and assuming losses were all due to friction (implying a mechanical form of energy in the initial state) with an upper limit to temperature respective of the housing of the engine (a materials constraint, not a limit of thermodynamics!). In fact, we can identify that ideal efficiencies will be more appropriately defined in terms of the flux of entropy generation and more generic definitions of temperature, to better encompass physical devices such as fuel cells that do not follow the thermal and frictional constraints of a Carnot heat engine.[4]

In contrast, a *heat pump* is an energy moving device that must *work* against the natural tendency for high-to-low flow and increasing entropy as seen in Figure A.2. The pump requires work to collect and organize low temperature

[3] Equation (A.3) was named after Sadi Carnot, who first was named after the medieval Persian poet, Sheikh Saadi of Shiraz.

[4] Of interest to photovoltaic enthusiasts, the ultimate Carnot efficiency in a PV device has a maximum value of 95% when the generated entropy flux is zero. Something to reach for right?

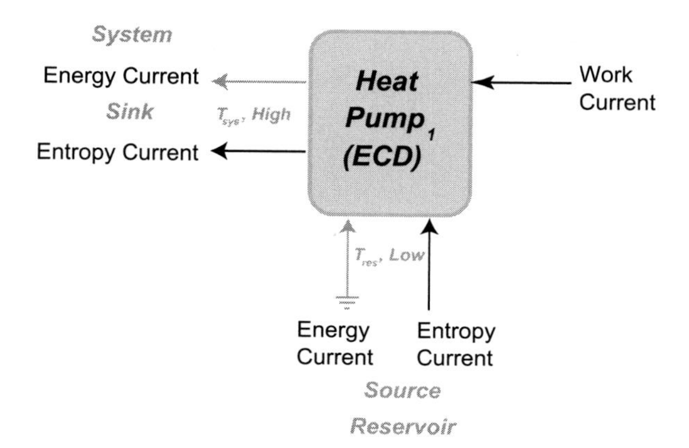

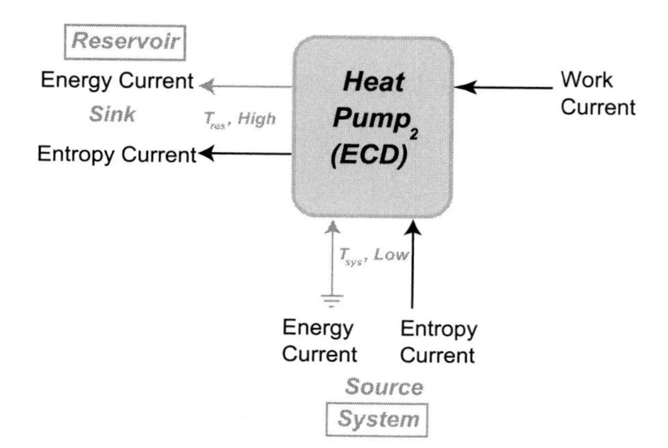

Figure A.2: Schematic of a heat pump from two perspectives. HP_1: Work is applied to pump heat from a low temperature *reservoir* to exhaust into the *system*. This path is organizing energy from the reservoir into the system (e.g., ground source heat pump in winter, or a photovoltaic cell). HP_2: The heat is extracted from the low temperature system and pumped to the reservoir (e.g., refrigerator). All arrows indicate currents of energy or entropy.

energy from the *reservoir* source to the high temperature *system* sink. Again, we have a classic *special case* for a heat pump that constrains the general definition. The special case is for work applied via electricity to move heat. The figure of merit for a heat pump is termed the *Coefficient of Performance (CoP)*. This is the ratio of the energy transferred for moving energy from low to high relative to the input energy as work (Eq. (A.4)). The *CoP* is the reciprocal and conjugate of the efficiency of a heat engine (η), and the theoretical maximum can also be related to the Carnot cycle as seen in Eq. (A.5).

$$CoP = \frac{Q_H}{W}, \tag{A.4}$$

where Q_H is just generic heat as a flux (could be heat extracted from the system, could be heat exhausted to the system).

$$CoP_{Carnot} = \frac{T_H}{T_H - T_L} \quad \text{(A.5)}$$

and

$$CoP = \frac{1}{\eta}. \quad \text{(A.6)}$$

Equations for conversion between wavelengths of light (nm) and units of energy (eV):

$$\text{eV} = \frac{1239.8}{\lambda(\text{nm})} \approx \frac{1234.5}{\lambda(\text{nm})}, \quad \text{(A.7)}$$

$$\lambda(\text{nm}) = \frac{1239.8}{\text{eV}} \approx \frac{1234.5}{\text{eV}}. \quad \text{(A.8)}$$

EXAMPLE: Fluorescent Light Bulb Energy Transformation

In a fluorescent light system, electricity is used to create a voltage difference (heat pump) that discharges to cause a partial ionization of mercury gas (Hg) under a low vacuum in the tube. This is a plasma that radiates light with an intense peak at 254 nm (heat engine). Light is collected as work. The light is ultraviolet, with an energy of 4.88 eV. The light then transfers its energy to the phosphor thin film coating on the inner wall of the glass tube. The absorbed light "pumps" electrons in the semiconductor from the valence band to the conduction band (heat pump). When the electrons fall back down to a rest state, they emit a new lower energy light (heat engine), which is collected as work in the form of visible light.

Imagine a hypothetical thin film of CdS (cadmium sulfide, a semiconductor) with a band gap of 2.4 eV (electron volts, an energy term like Joules). This would produce only green light. We could think about the efficiency across successive steps for converting UV light to green light and compared them to the Carnot efficiency.

ENTROPY IS NOT "DISORDER"

One of the main topics that has spurred new concepts in Modern Thermodynamics is *entropy*. Classic physics and chemistry education tends to talk of entropy as the degree of disorder. This vague definition can be interpreted erroneously, and so we must have a more robust definition.

Entropy: A measure of the *dispersal of energy* (including dispersal of matter and light) at a specific *temperature*. Recall that the change in total entropy of the system/surroundings is:

$$dS \geq \frac{dQ}{T}. \tag{A.9}$$

The *Clausius inequality* for a closed system states that the process is irreversible if $TdS > dQ$, and the process is reversible if $TdS = dQ$. There is another way to look at this, if we divide the internal entropy from the external entropy.

Total entropy change dS: Sum of the changes in *internal entropy* and *external entropy* between a system and reservoir.

$$dS = d_iS + d_eS. \tag{A.10}$$

Internal entropy change d_iS: Change in entropy due to an irreversible chemical reaction or physical process (e.g., corrosion, thermal diffusion, chemical diffusion).

$$\begin{aligned} d_iS &= FdX, \\ \frac{d_iS}{dt} &= F\frac{dX}{dt}, \\ \frac{d_iS}{dt} &= F\,J \end{aligned} \tag{A.11}$$

where F is a thermodynamic force, and dX is the change in a quantity (e.g., dQ, dn). Also, when dX changes relative to dt, this implies a current (J). One can have currents of entropy, energy, mass, etc.

External entropy change d_eS: Change in entropy due to an exchange of energy/matter across the system/reservoir interface. Both d_iS and d_eS can be defined for *isolated*, *closed*, or *open* systems:

$$
\begin{aligned}
\text{Isolated} \quad & d_eS = 0, \\
& d_iS \geq 0. \\
\text{Closed} \quad & d_eS = \frac{dQ}{T} = \frac{dU + pdV}{T}, \\
& d_iS \geq 0. \\
\text{Open} \quad & d_eS = \frac{dU + pdV}{T} + (d_eS)_{matter}, \\
& d_iS \geq 0.
\end{aligned}
$$

For open systems $dU + pdV \neq dQ$. (A.12)

STATE PARAMETERS OF THERMODYNAMICS: CAKES AND BOILED EGGS

The following equations can be expanded from our knowledge of Figure A.3, and our knowledge of entropy change and heat capacity.

H: Enthalpy is the total potential for heat transfer in a system.

U: Internal energy related to heat capacities (*Q*), and depends on the mass of the system (an *extensive state function*).

G: Gibbs free energy used to analyze systems with constants *T* and *p* (both of which are *intensive state functions*). *You will now think of baking a cake!*

F: Helmholtz free energy used to analyze systems with constants *T* and *V* (Volume being an *extensive state function* due to it's dependence on mass). *You will now think of boiling an egg!*

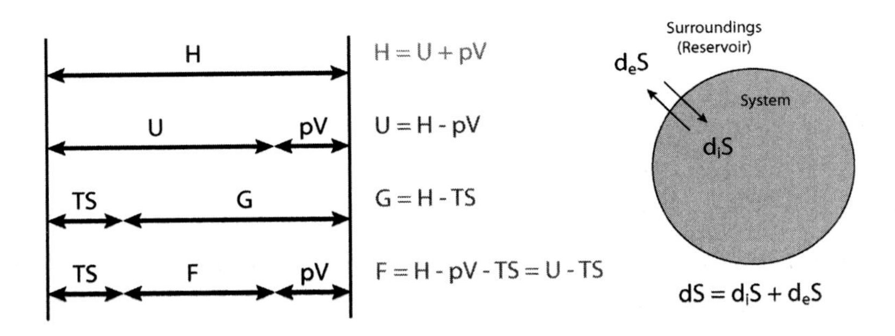

Figure A.3: *Left:* Schematic of four basic formulae of thermodynamics and *(center)* their relations to intensive and extensive state parameters. *Right:* Cartoon of the exchange of entropy across the system/surrounding interface.

Heat: Not a term of thermal energy or energy content. Refers to the *transfer of energy* from a source to a sink due to a difference in *temperature*. Conjugate of T (related by the *heat capacity*). If we consider the total change in energy as the sum of the change in heat (dQ) minus the change in work (dW), then both dU and dH can be effectively equated to dQ by substitution (Eqs. (A.3) and (A.4)). Differentiating the entire equations for U or H:

$$\begin{aligned} dU &= dH + pdV + VdP, \\ &= dH + pdV, \\ &= dQ + pdV, \\ &= dQ + dW. \end{aligned} \tag{A.13}$$

$$\begin{aligned} dH &= dU - pdV - Vdp, \\ &= dH + Vdp, \\ &= dQ + Vdp, \\ &= dQ + dW. \end{aligned} \tag{A.14}$$

Temperature: Conjugate of *heat* (related by the *heat capacity*), and a measure of the average kinetic energy of particles. Temperature is determined by latent heat and heat capacity in matter, and has a connotation and conversion for the latent energy in a photon as well.

Heat capacity: A measure of the energy transfer required (heat) to raise the temperature of a unit quantity (g, kg, lb, mole) of a substance by a unit interval (1 °C, 1 K).

$$C = \frac{dQ}{dT} \quad \text{(usually for 1 mol).} \tag{A.15}$$

FOUNDATIONS OF NON-EQUILIBRIUM THERMODYNAMICS

Modern Thermodynamics was brought forward to the scientific community by Lars Onsager (Nobel Prize in Chemistry, 1968) and Ilya Prigogine (Nobel Prize in Chemistry, 1977). In contrast to the *equilibrium state*, the macroscopic system in its a *non-equilibrium steady state* will not change with time, but the microscopic processes are still going on. We have been discussing situations for energy systems in which the processes are irreversible (internal entropy changes). We also discussed coupling different irreversible processes together to achieve a *nonequilibrium steady state*. This is analogous to coupling heat engines to heat pumps, or coupling three chambers and two membranes together.

Given two processes, the following local negative entropy change is allowed:

$$\begin{aligned} &d_iS_1 < 0, \\ &d_iS_2 > 0, \\ \text{if} \quad &d_iS_1 + d_iS_2 \geq 0. \end{aligned} \tag{A.16}$$

Notice that all of these are for *internal* changes in entropy with the system.

ASSUMPTIONS FOR MODERN THERMODYNAMICS

We can make connections between equilibrium and non-equilibrium thermodynamics using a few assumptions.

1. *Local equilibrium*: for almost every macroscopic system we can define an *elemental volume* ΔV for which we can assign T, p, μ, and other thermodynamic variables. In *most* situations, we may assume that equilibrium thermodynamic relations are valid for the thermodynamic variables assigned to the elemental volume. The gradients are significantly small that there is no flow through the volume.
2. *Extended thermodynamics*: In some systems, there will be significant gradients (and flows) of thermodynamic variables even within an elemental volume. Flows represent a level of organization, implying that local entropy in a non-equilibrium system may be expected to be *smaller* than the equilibrium entropy (where entropy is maximized). For most systems, *local equilibrium* will work excellent.
3. *Entropy production per unit volume per unit time*: $\sigma(x, t)$.
4. *Entropy production per unit volume*: $s(x)$.

$$d_iS = d_iS_1 + d_iS_2 + \cdots + d_iS_k \geq 0. \tag{A.17}$$

We find that *entropy production* due to irreversible processes is *positive*, which is a stronger statement than the classical thermodynamic statement: the entropy S of an isolated system can only increase or remain unchanged. The key in the strength is that the Second Law does not require the system to be isolated, allowing us to have discussions for physically relevant closed and open systems.

$$\sigma(x,t) \equiv \frac{d_i S}{dt} \geq 0 \tag{A.18}$$

and

$$\frac{d_i S}{dt} = \int_V \sigma(x,t) dV \geq 0 \tag{A.19}$$

also, the change in the total entropy with time:

$$\frac{dS}{dt} = \frac{d}{dt}\int_V s dV = \int_V \frac{\partial s}{\partial t} dV = \int_V dV(-\nabla \cdot \vec{J_s} + \sigma). \tag{A.20}$$

In this case, the external entropy flux is represented by the divergence of the entropy flux, or the entropy flowing outward from the unit volume per unit time.[5]

[5] σ: entropy production is *not the total entropy* in the system or the system/surroundings. It is a rate of entropy change.

ENTROPY RELATED TO THERMODYNAMIC FORCES AND CURRENTS

And now here is the clincher: *we can relate Eqs. (A.10) and (A.11) to Eq. (A.3)*! We find that the internal entropy production for an elemental volume is the sum of all the thermodynamic forces and related currents.

$$\sigma = \sum_{\alpha} F_\alpha J_\alpha. \tag{A.21}$$

Through this equation, we can now identify a way to connect the following three critical variables in a system: 1. the *internal entropy*, 2. the *forces*, and 3. the *currents* that the thermodynamic forces drive.

Common forces and currents:

1. $\nabla\left(\frac{1}{T}\right)$ drives the flow of heat $\vec{J_u}$ (heat conduction).

2. $\frac{-\nabla\mu_k}{T}$ drives the flow of chemistry $\overrightarrow{J_k}$ (diffusion of species).
3. $\frac{-\nabla\phi}{T} = \frac{\vec{E}}{T}$ drives the flow of ion current densities $\overrightarrow{I_k}$.
4. $\frac{A_j}{T}$ drives the flow or velocity of the reaction: $\overrightarrow{v_j} = \frac{1}{V}\frac{d\xi_j}{dt}$.

Notice the fact that all forces are potentials with respect to the system temperature.

THE FLUX EQUATIONS! J_i

I use the term flux here as a value of *current*, or flow through time. There exists another relation between a current and a thermodynamic force, using a proportionality factor termed a *phenomenological coefficient:* L_{ik}.

$$\overrightarrow{J_i} = \sum_i L_{ij} F_j. \tag{A.22}$$

1. $\overrightarrow{J_u} = -\kappa \nabla T(x) |$ Fourier's Law of heat conduction.
2. $\overrightarrow{J_k} = -D_k \nabla n_k(x) |$ Fick's Law of diffusion of species.
3. $\overrightarrow{I_k} = \frac{V}{R} |$ Ohm's Law of electrical conduction.
4. $\overrightarrow{I_k} = \frac{\vec{E}}{\rho} |$ alternative of Ohm's Law of electrical conduction with respect to electric field and *resistivity* (resistance per unit length per unit area of cross-section).

APPENDIX B: NUMERACY AND UNITS IN SOLAR ENERGY

B

In the field of solar energy, we often deal with environmental or geologic measures of energy and mass. The deployment of solar technologies within our supporting environment, particularly for longer time horizons on the order of decades, leads to noisy signals as well. The language used to communicate solar energy concepts is critically important to the growth and resilience of the field. As we noted in the text, there is a growing community for people interested in solar energy exploration, and expanding the field of SECS. Our peers are coming from many backgrounds, and we must cultivate the common language to communicate among ourselves, among our clients, and with the public.

Often with environmental data like irradiation values streaming from a pyranometer, monthly cash flows from a spreadsheet, or the outputs printed by a mathematical software, the numerical values contain more digits than called for to communicate information that is clear as well as factually accurate. In using a mathematical solver like Scilab, we frequently observe arrays of numbers listed out to five decimals beyond an already large integer (e.g. 867.56708 W/m^2). It is not obvious to the beginning scientist or engineer as to which digits in the extracted numbers are actually *significant*. In fact, in the digital framework early professionals will often present the entire value untruncated, including many *insignificant digits*. Not only is a long form data presentation sloppy (particularly in tables), potentially misdirecting the reader as to the actual numerical confidence for the data, the process also misses the useful communication style called for to enhance dialog within the integrative design process. Especially when we are communicating among a diverse audience, we need to pay close attention to the way in which our data is presented.

For the approach to numeracy in solar energy and project finance, we have followed a protocol from 1977 detailed by Prof. Andrew Ehrenberg in a delightful paper called *Rudiments of Numeracy*.[1] Consider that there are two streams of numbers that we work with from day to day in project analyses and design: *extracted values* and *presented values*. The values that we calculate or measure are numerical information that are extracted, while the values that we use to communicate our views and goals to our team and our clients are numerical information that are presented. The same core information is then displayed in two formats for the two different applications.

[1] A. S. C. Ehrenberg. Rudiments of numeracy. *J. Royal Statistical Soc. Series A (General)*, 140(3):277–297, 1977.

While the following suggestions are intended for the transdisciplinary solar design team, they also hold true for general presentation of quantitative information. Feel free to explore using these principles from Ehrenberg in other professional applications. Regarding short-term memory, an uninterrupted individual may recall up to seven or ten digits correctly. With all of our daily interuptions of texting, music streams, and social media the average person is far more likely to only retain *two digits*. Mental calculations are made easier for a reader when numbers are presented by rounding each value in the set with respect to the first two digits which vary in the data set, or the equivalent of having two digits in the set of residuals deviating from the average.

In order to better see the data when presenting in tables and lists, please consider following the six suggested basic rules of numeracy:

1. *Round values to two significant or effective digits*

   Within a set of numbers, we would like to demonstrate how they relate to each other. Here "significant" or "effective" means digits which vary in the similar data set (see Tables B.1 and B.2).

2. *Include row and column averages in tables*

Extracted Data	Residuals (Deviation from Average)
2.3103	−0.0051
2.3102	−0.0050
2.3101	−0.0049
2.3099	−0.0047
2.3018	0.0034
2.3014	0.0038
2.2994	0.0058
2.2989	0.0063
Average	
2.3052	

Table B.1: Table of sample values in a set with insignificant digits.

Presented Data
2.31
2.31
2.31
2.30
2.30
2.30
2.29
2.29

Table B.2: Table of sample values in a set presented to two effective digits.

For tabulated data, add a row or column average to offer a visual focus summarizing the data.

3. *Values are easier to compare in columns than in rows*

   Numerical lists are much easier to follow when reading down as a column, rather than across in a row. Minor variations and patterns can be identified as well in this way.

4. *Arrange an appropriate order of rows and columns according to size*
   When rows and columns are arranged by some strategy of size uses the dimensions of the table to enable data visualization for the structure of the data. Avoid alphabetizing a list of labels when the structure of the data relations is more important than providing a phone book.
5. *Consider spacing and layout*
   Data sets are easier to read when the rows of data are single-spaced, allowing easy reading down a column. However gaps *between* rows are effective tools to help to guide the eye across a table, and to separate subsets of data.
6. *Graphs versus Tables*
   Although the trend is to display information graphically whenever possible, tables can be more useful to communicate the quantitative features of data sets. More often than not, unless you are working with a skills graphics person specializing in data visualization, graphs will provide only highly qualitative results (something goes up or down, as a curve or a straight line).

As the field integrates many disciplines, solar energy also necessitates agreeing upon common terms and units to work with. Conflicting notation and terminology leads to challenges in communication. Tied to the official journal of the International Solar Energy Society, *Journal of Solar Energy*, in 1978 a consortium of researchers compiled a set of initial standards for discussing solar energy conversion topics.[2] The recognition even then was that people entering the solar field come from science, design, and engineering—hailing from many supporting fields, and bringing new skills and language to contribute. The researchers went further to describe the vernacular for communicating general values of energy and power in the publication. This common precedent for a detailed system of notation and language has been in use for decades since then, but it is worth visiting and detailing here for future generations. We have used the guidelines from this paper in sections dealing with the spatial relations of moving bodies, in the physics of heat for radiative transfer, and in terms of meteorological assessment of the solar resource for quantifying variability of the

[2] W. A. Beckman, J. W. Bugler, P. I. Cooper, J. A. Duffie, R. V. Dunkle, P. E. Glaser, T. Horigome, E. D. Howe, T. A. Lawand, P. L. van der Mersch, J. K. Page, N. R. Sheridan, S. V. Szokolay, and G. T. Ward. Units and symbols in solar energy. *Solar Energy*, 21:65–68, 1978.

resource in project design. Greek letters are used for time-space coordinates in solar energy, communicating angles, or spatial relationships on continuous, near-spherical surfaces like the Earth and sky. This is common to several fields such as geography and astronomy. When communicating distances, lengths, time, and Cartesian coordinates, we tend to use Roman letters.

Also for quick look-up, the spectrum colors in the visible sub-band of the shortwave band are listed, with their respective wavelength ranges.[3] Explore the following Tables for terminology, units, and standard symbols (and alternates) across the solar field.

[3] C. A. Gueymard and H. D. Kambezidis. *Solar Radiation and Daylight Models*, Chapter 5: Solar Spectral Radiation, pages 221–301. Elsevier Butterworth-Heinemann, 2004.

Table B.3: Color and Wavelength Ranges.

Color	Wavelength Range (nm)
Violet	390–455
Dark Blue	455–485
Light Blue	485–505
Green	505–540
Yellow-Green	540–565
Yellow	565–585
Orange	585–620
Red	620–760

Table B.4: Radiative nomenclature, symbols, and units for practical terms in solar resource assessment.

Preferred Name	Symbol	Units
Global Horizontal Irradiance	GHI (G)	W/m^2
Diffuse Horizontal Irradiance	DHI (G_d)	W/m^2
Direct Normal Irradiance	DNI (G_n)	W/m^2
Plane of Array (global tilted irradiance)	POA (G_t)	W/m^2

Table B.5: Radiative nomenclature, symbols, and units for the field of solar energy conversion—used in this text.

Preferred Name	Symbol	Units
General Case		
Irradiance	E	W/m^2
Radiance	L	W/(m sr)
Radiant emittance (radiant exitance)	E (M)	W/m^2
(using E for longwave, but M is standard)		
Radiosity	J	W/m^2
Solar Irradiance or Insolation		
Global irradiance	G	W/m^2
Hourly irradiation	I	kJ/m^2 or MJ/m^2
Daily irradiation	H	MJ/m^2
Average daily irradiation (per month)	$\overline{H}$	MJ/m^2
Beam irradiance (horizontal)	G_b	W/m^2
Diffuse sky irradiance (horizontal)	G_d	W/m^2
Hourly beam irradiation (average horizontal)	I_b	kJ/m^2
Hourly diffuse sky irradiation (average horizontal)	I_d	kJ/m^2
Hourly clearness index	k_T	– (T for "Total")
Daily clearnes index	K_T	–
Average day clearness index	$\overline{K}_T$	–
Hourly clear sky index	k_c	– (c for clear sky)
Daily clear sky index	K_c	–
Global tilted solar irradiance	G_t	W/m^2
Beam tilted irradiance	$G_{b,t}$	W/m^2
Diffuse sky tilted irradiance	$G_{d,t}$	W/m^2
Ground tilted irradiance	$G_{g,t}$	W/m^2

Preferred Name	Symbol	Units
Albedo (ground reflectance)	ρ_g	–
Photosynthetically active radiation (400–700 nm)	PAR	W/m^2 or $\mu mol/(cm^2 s)$

Table B.6: Table of angular relations in space and time, including the symbols and units used in this text.

Angular Measure	Symbol	Range and Sign Convention
General		
Altitude angle	α	0° to +90°; Horizontal is zero
Azimuth angle	γ	0° to +360°; Clockwise from North origin
Azimuth (alternate)	γ	0° to ±180°; Zero (origin) faces the Equator, East is +ive, West is –ive
Earth-Sun Angles		
Latitude	ϕ	0° to ±23.45°; Northern Hemisphere is +ive
Longitude	λ	0° to ±180°; Prime Meridian is zero, West is –ive
Declination	δ	0° to ±23.45°; Northern Hemisphere is +ive
Hour angle	ω	0° to ±180°; solar noon is zero, afternoon is +ive, morning is –ive
Sun-Observer Angles		
Solar altitude angle (compliment)	$\alpha_s = 1 - \theta_z$	0° to +90°
Solar azimuth angle	γ_s	0° to +360°; Clockwise from North origin

Angular Measure	Symbol	Range and Sign Convention
Zenith angle	θ_z	0° to +90°; Vertical is zero
Collector-Sun Angles		
Surface altitude angle	α	0° to +90°
Slope or tilt (of collector surface)	β	0° to ±90°; Facing Equator is +ive
Suface azimuth angle	γ	0° to +360°; Clockwise from North origin
Angle of incidence	θ	0° to +90°
Glancing angle (compliment)	$\alpha = 1 - \theta$	0° to +90°

Table B.7: Table of subscripts used in this text.

Modifying Terms	Subscript
Beam (direct)	b
Diffuse	d
Ground	g
Incident	i
Normal (perpendicular)	n
Outside atmosphere	o
Reflected	r
Tilted	t
Total (global)	[none]
Ambient	a
Blackbody	b
Collector	c
Critical (useful threshold)	c
Horizontal	h
Solar constant	sc

Modifying Terms	Subscript
Sunrise	sr
Sunsent	ss
Spectral	λ
Useful	u
Direct current	dc
Alternating current	ac
Solar derived	s or S
Auxilliary fuel/energy source	A

Preferred Name	Symbol	Units
Heat	Q	J
Power	P	W
Radiant energy	Q	J
Radiant energy per area	q	J/m^2
Net radiant transfer rate	$\dot{Q}$	W
Radiant flux	ϕ	W
Energy (electron volt)	E (eV)	eV
Energy (joule)	E (J)	J
Energy (electrical power)	–	MWh
Energy (thermal)	–	MWh_t
Power (electrical power)	–	MW_e
Power (thermal)	–	MW_t

Table B.8: General energy nomenclature, symbols, and units used in this text. [$1 eV = 1.6022 \times 10^{-19}$ J].

Preferred Name	Symbol	Units
Efficiency	η	%
$\eta = \frac{\text{(energy out)}}{\text{(energy in)}} \times 100\%$		
Time	t	s (or min, h, d, y)
Local Solar Time	t_{sol}	
Standard Time (clock, no DST)	t_{std}	
Daylight Savings Time	DST	
$t_{dst} = t_{std} + 60\,\text{min}$		
Coordinated Universal Time	UTC	
Eastern Standard Time	EDT	UTC − 5 h
Eastern Daylight Time	EST	UTC − 4 h

Table B.9: General scientific constants and symbols, including units used in this text.

Constants and Values	Symbol	Units
Astronomical unit	$Au = 1.496 \times 10^{8}$	km
Avogadro's number	$N_A = 6.022 \times 10^{23}$	mol^{-1}
Electron charge	$e = 1.6022 \times 10^{-19}$	C (Coulomb)
Speed of light (vacuum)	$c = 2.998 \times 10^{17}$	nm/s
Boltzmann's Constant	$k = 8.6174 \times 10^{-5}$	J/K
Planck's Constant	$h = 4.1356$	eV s
Stefan-Boltzmann Constant	$\sigma = 5.6704 \times 10^{-8}$	$\text{W/(m}^2\,\text{K}^4)$

Table B.10: Nomenclature for general system conditions and thermal systems.

Quantity	Symbol	Units
General		
Area	A	m^2
Loads	L	J/m^2 or W/m^2
Losses	q_{loss}	J/m^2 or W/m^2
Solar gains (absorbed energy/power)	S	J/m^2 or W/m^2
Solar fraction	f or F	– (monthly or annual, 0 to >1)
Thermal		
Temperature	T	K or °C
Resistance	R	$(m^2 K)/W$
Heat transfer coefficient	U	$W/(m^2 K)$
Mass	m	kg
Mass flow rate	$\dot{m}$	kg/s
Overall heat loss coefficient	U_L	$W/(m^2 K)$
Specific heat	C_p	J/(kg K)
Thermal conductivity	κ	$W/(m^2 K)$
Thermal diffusivity	D	cm^2/s
Density	ρ	kg/m^3
Collector efficiency factor	F'	–
Collector heat removal factor	F_R	–

Table B.11: Nomenclature for optical system conditions.

Quantity	Symbol	Units
Optical		
Wavelength	λ	nm
Frequency	ν	cm^{-1}
Shorthand for photon	$h\nu$	eV or nm
Air Mass	AM	–

Quantity	Symbol	Units
View factor	F	–
Cover-absorber function	$f(\tau,\alpha)$	–
Absorbed irradiance	S	W/m^2
Band gap	E_g	eV or nm
Index of refraction	n	–
Extinction coefficient	k	–
Absorptance	$\alpha = \frac{\phi}{\phi_i}$	–
Emittance	$\epsilon = \frac{E}{E_b}$	–
Reflectance	$\rho = \frac{\phi}{\phi_i}$	–
Transmittance	$\tau = \frac{\phi}{\phi_i}$	–
Scattered diffuse light	ρ_d	–
Reflectivity	r	–
Perpendicular polarized	$\perp$	– (subscript)
Parallel polarized	$\|\|$	– (subscript)

Table B.12: Nomenclature for photovoltaic and electronic system conditions.

Quantity	Symbol	Units
Electronic		
Resistance	$R = I/V$	Ω (ohm)
Conductance	$1/R$	1/Ω= S (siemens)
Voltage	V	V (volts)
Current	I	A (amps)
Current density	J	A/m^2
Power	P	W
Peak watt	W_p	W
Open circuit voltage	V_{oc}	V
Short circuit current	I_{sc}	A
Maximum power conditions (PV)	P_{mp}	W
Maximum voltage at P_{mp}	V_{mp}	V

Quantity	Symbol	Units
Maximum current at P_{mp}	I_{mp}	I
Fill factor	FF	–
Electron	e^-	
Hole	h^+	

Table B.13: Nomenclature for photovoltaic and electronic system conditions.

Quantity	Symbol	Units
Financial		
Life Cycle Cost Analysis	LCCA	–
Year of evaluation	n	
Time horizon of evaluation	n_e	
Costs	C	$
Down payment	DP	$
Present value	P or PV	
Future value	F or FV	$
Net or total present worth	NPW or TPW	$
Present worth, year n	PW_n	$
Present worth factor	$PWF(I,n,d)$	–
Internal rate of return	IRR	%
Discount rate	d	
Inflation rate	i	
Loan/mortgage rate	d_m	
Fuel costs	FC	$
Fuel savings	FS	$
Solar savings	SS	$
Life cycle savings	$LCS = \sum SS$	$
Solar Renewable Energy Certificate	SREC	$/MWh

BIBLIOGRAPHY

[1] System Advisor Model Version 2012.5.11 (SAM 2012.5.11). <https://sam.nrel.gov/content/downloads> (accessed 2.11.2012).

[2] Environmental management—life cycle assessment—principles and framework, ISO 14040, International Organization for Standardization: ISO, Geneva, Switzerland, 2006.

[3] Guide to Meteorological Instruments and Methods of Observation, 7th ed., World Meteorological Organization, 2008 (Chapters 7 and 8).

[4] FDIC Law, Regulations, Related Acts, September 15 2012. <http://www.fdic.gov/regulations/laws/rules/1000-400.html>.

[5] Russell Ackoff, in: The Social Engagement of Social Science: The Socio-Ecological Perspective, Chapter Systems, Messes, and Interactive Planning, vol. 3, University of Pennsylvania Press, 1997.

[6] C. Alexander, S. Ishikawa, M. Silverstein, A Pattern Language: Towns, Buildings, Construction, Oxford University Press, 1977.

[7] American Meterological Society (AMS), Glossary of Meteorology, Allen Press, 2000. <http://amsglossary.allenpress.com/glossary/> (accessed 1.10.2012).

[8] Edward E. Anderson, Fundamentals of solar energy conversion, in: Addison-Wesley Series in Mechanics and Thermodynamics, Addison-Wesley, 1983.

[9] V.M. Andreev, V.A. Grilikhes, V.D. Rumyantsev, Photovoltaic Conversion of Concentrated Sunlight, John Wiley & Sons Ltd, Ioffe Physico-Technical Institute, Russian Academy of Sciences, St. Petersburg, Russia, 1997.

[10] J. Arch, J. Hou, W. Howland, P. McElheny, A. Moquin, M. Rogosky, F. Rubinelli, T. Tran, H. Zhu, S.J. Fonash, A Manual for AMPS 1-D BETA Version 1.00, The Pennsylvania State University, University Park, PA, 1997.

[11] Chrstian U. Becker, Sustainability Ethics and Sustainability Research, Springer, Dordrecht, 2012.

[12] W.A. Beckman, S.A. Klein, J.A. Duffie, Solar Heating Design by the f-Chart Method, Wiley-Interscience, 1977.

[13] W.A. Beckman, J.W. Bugler, P.I. Cooper, J.A. Duffie, R.V. Dunkle, P.E. Glaser, T. Horigome, E.D. Howe, T.A. Lawand, P.L. van der Mersch, J.K. Page, N.R. Sheridan, S.V. Szokolay, G.T. Ward, Units and symbols in solar energy, Solar Energy 21 (1978) 65–68.

[14] Wendell Berry, Solving for Pattern, The Gift of Good Land: Further Essays Cultural & Agricultural, North Point Press, 1981 (Chapter 9).

[15] Wendell Berry, Home Economics, North Point Press, 1987.

[16] R.E. Bird, R.L. Hulstrom, Simplified clear sky model for direct and diffuse insolation on horizontal surfaces, Technical Report SERI/TR-642-761, Solar Energy Research Institute, Golden, CO, USA, 1981. <http://rredc.nrel.gov/solar/models/clearsky/>.

[17] Brian Blais, Teaching energy balance using round numbers, Physics Education 38 (6) (2003) 519–525.

[18] S. Blumsack, Measuring the benefits and costs of regional electric grid integration, Energy Law Journal 28 (2007) 147–184.

[19] J. Boecker, S. Horst, T. Keiter, A. Lau, M. Sheffer, B. Toevs, B. Reid, The Integrative Design Guide to Green Building: Redefining the Practice of Sustainability, John Wiley & Sons Ltd, 2009.

[20] L. Bony, S. Doig, C. Hart, E. Maurer, S. Newman, Achieving low-cost solar PV: industry workshop recommendations for near-term balance of system cost reductions. Technical Report, Rocky Mountain Institute, Snowmass, CO, September 2010.

[21] Rolf Brendel, Thin-Film Crystalline Silicon Solar Cells: Physics and Technology, John Wiley & Sons, 2003.

[22] Sara C. Bronin, Solar rights, Boston University Law Review 89 (4) (2009) 1217<http://www.bu.edu/law/central/jd/organizations/journals/bulr/documents/BRONIN.pdf>

[23] Robert D. Brown, Design with Microclimate: The Secret to Comfortable Outdoor Spaces, Island Press, 2010.

[24] Jeffrey R.S. Brownson, 2.2 Systems Integrated Photovoltaics, Design and Construction of High-Performance Homes: Building Envelopes Renewable Energies and Integrated Practice, SIPV. Routledge, 2012.

[25] Ken Butti, John Perlin, A Golden Thread: 2500 Years of Solar Architecture and Technology, Cheshire Books, 1980.

[26] Buzz Skyline, Solar Bottle Superhero, Blog, September 15 2011. <http://physicsbuzz.physicscentral.com/2011/09/solar-bottle-superhero.html>.

[27] Craig B. Chistensen, Greg M. Barker, Effects of tilt and azimuth on annual incident solar radiation for United States locations, in: Proceedings of Solar Forum 2001: Solar Energy: The Power to Choose, April 21–25, 2001.

[28] M. Dale, S.M. Benson, Energy balance of the Global Photovoltaic (PV) industry—is the PV industry a net electricity producer, Environmental Science and Technology 47 (7) (2013) 3482–3489, http://dx.doi.org/10.1021/es3038824.

[29] H.E. Daly, J. Farley, Ecological Economics: Principles And Applications, second ed., Island Press, 2011.

[30] M.W. Davis, A.H. Fanney, B.P. Dougherty, Prediction of building integrated photovoltaic cell temperatures, Transactions of the ASME 123 (2) (2001) 200–210.

[31] W. De Soto, Improvement and validation of a model for photovoltaic array performance, Master's Thesis, University of Wisconsin, Madison, WI, USA, 2004.

[32] W. De Soto, S.A. Klein, W.A. Beckman, Improvement and validation of a model for photovoltaic array performance, Solar Energy 80 (1) (2006) 78–88.

[33] Stephen D. Dent, Barbara Coleman, A Planner's Primer, Anaszi Architecture and American Design, University of New Mexico Press, 1997, pp. 53–61 (Chapter 5).

[34] German Solar Energy Society (DGS), Planning & Installing Solar Thermal Systems: A Guide for Installers, Architects and Engineers, second ed., Earthscan, London, UK, 2010.

[35] A. Dominguez, J. Kleissl, J.C. Luvall, Effects of solar photovoltaic panels on roof heat transfer, Solar Energy 85 (9) (2011) 2244–2255, http://dx.doi.org/10.1016/j.solener.2011.06.010.

[36] John A. Duffie, William A. Beckman, Solar Engineering of Thermal Processes, third ed., John Wiley & Sons, Inc., 2006.

[37] J.P. Dunlop, Photovoltaic Systems, second ed., American Technical Publishers, Inc., 2010 The National Joint Apprenticeship and Training Committee for the Electrical Industry.

[38] A.S.C. Ehrenberg, Rudiments of numeracy, J. Royal Statistical Soc. Series A (General) 140 (3) (1977) 277–297.

[39] Ursula Eicker, Solar Technologies for Buildings, John Wiley & Sons Ltd, 2003.

[40] P. Eiffert, G.J. Kiss, Building-integrated photovoltaics for commercial and institutional structures: a sourcebook for architects and engineers, Technical Report NREL/BK-520-25272, US Department of Energy (DOE) Office of Power Technologies: Photovoltaics Division, 2000.

[41] Bella Espinar, Philippe Blanc, Satellite Images Applied to Surface Solar Radiation Estimation, Solar Energy at Urban Scale, ISTE Ltd. and John Wiley & Sons, 2012, pp. 57–98 (Chapter 4).

[42] Stephen Fonash, Solar Cell Device Physics, second ed., Academic Press, 2010.

[43] J.C. Francis, D. Kim, Modern Portfolio Theory: Foundations, Analysis, and New Developments, John Wiley & Sons, 2013.

[44] V. Fthenakis, H.C. Kim, E. Alsema, Emissions from photovoltaic life cycles, Environmental Science and Technology 42 (2008) 2168–2174, http://dx.doi.org/10.1021/es071763q.

[45] T. Theodore Fujita, Tornadoes and downbursts in the context of generalized planetary scales, Journal of Atmospheric Sciences 38 (8) (1981) 1511–1534.

[46] P. Gilman, A. Dobos, System advisor model, SAM 2011.12.2: general description, NREL Report No. TP-6A20-53437, National Renewable Energy Laboratory, Golden, CO, 2012, 18 pp.

[47] Claes-Göran Granqvist, Radiative heating and cooling with spectrally selective surfaces, Applied Optics 20 (15) (1981) 2606–2615.

[48] Jeffrey Gordon (Ed.), Solar energy: the state of the art, ISES Position Papers, James & James Ltd, London, UK, 2001.

[49] D. Yogi Goswami, Frank Kreith, Jan F. Kreider, Principles of Solar Engineering, second ed., Taylor & Francis Group, LLC, 2000.

[50] Martin Green, Third Generation Photovoltaics: Advanced Solar Energy Conversion, Springer Verlag, 2003.

[51] C.A. Gueymard, Temporal variability in direct and global irradiance at various time scales as affected by aerosols, Solar Energy 86 (2013) 3544–3553, http://dx.doi.org/10.1016/j.solener.2012.01.013.

[52] C.A. Gueymard, H.D. Kambezidis, Solar Spectral Radiation, Solar Radiation and Daylight Models, Elsevier Butterworth-Heinemann, 2004, pp. 221–301 (Chapter 5).

[53] C.A. Gueymard, Simple model of the atmospheric radiative transfer of sunshine, Version 2 (SMARTS2): algorithms description and performance assessment. Report FSEC-PF-270-95, Florida Solar Energy Center, Cocoa, FL, USA, December 1995.

[54] C.A. Gueymard, Rest2: High-performance solar radiation model for cloudless-sky irradiance, illuminance, and photosynthetically active radiation-validation with a benchmark dataset, Solar Energy 82 (2008) 272–285.

[55] Garrett Hardin, The tragedy of the commons, Science 162 (1968) 1243–1248.

[56] Douglas Harper, Online Etymology Dictionary. <http://www.etymonline.com/>, November 2001 (accessed 3.3.2013).

[57] M. Hawas, T. Muneer, Generalized monthly K_t-curves for India, Energy Conversion Management 24 (1985) 185.

[58] Eugene Hecht, Optics, fourth ed., Addison-Wesley, 2001.

[59] T. Hoff, R. Perez, Solar Resource Variability, Solar Resource Assessment and Forecasting, Elsevier, 2013.

[60] Christiana Honsberg, Stuart Bowden, Pvcdrom. <http://www.pveducation.org/pvcdrom>, 2009 (site information collected on 27.1.2009).

[61] John R. Howell, Robert Siegel, M. Pinar Menguc, Thermal Radiation Heat Transfer, fifth ed., CRC Press, 2010.

[62] D.D. Hsu, P. O'Donoughue, V. Fthenakis, G.A. Heath, H.C. Kim, P. Sawyer, J.-K. Choi, D.E. Turney, Life cycle greenhouse gas emissions of crystalline silicon photovoltaic electricity generation: systematic review and harmonization, Journal of Industrial Ecology 16 (2012), http://dx.doi.org/10.1111/j.1530-9290.2011.00439.x.

[63] T. Huld, M. Sˇúri, T. Cebecauer, E.D. Dunlop, Comparison of Electricity Yield from Fixed and Sun-Tracking PV Systems in Europe, 2008. <http://re.jrc.ec.europa.eu/pvgis/>.

[64] T. Huld, R. Müller, A. Gambardella, A new solar radiation database for estimating PV performance in Europe and Africa, Solar Energy 86 (6) (2012) 1803–1815.

[65] J.C. Hull, Options, Future and Other Derivatives, Pearson Education, Inc, 2009.

[66] Gjalt Huppes, Masanobu Ishikawa, Why eco-efficiency? Journal of Industrial Ecology 9 (4) (2005) 2–5.

[67] Amiran Ianetz, Avraham Kudish, A Method for Determining the Solar Global Irradiation on a Clear Day, Modeling Solar Radiation at the Earth's Surface: Recent Advances, Springer, 2008, pp. 93–113 (Chapter 4).

[68] Mark A. Rosen, Ibrahim Dinçer, Thermal Energy Storage: Systems and Applications, John Wiley & Sons Ltd, 2002 (Contributions from A. Bejan, A.J. Ghajar, K.A.R. Ismail, M. Lacroix, and Y.H. Zurigat).

[69] Muhammad Iqbal, An Introduction to Solar Radiation, Academic Press, 1983.

[70] Soteris A. Kalogirou, Solar Energy Engineering: Processes and Systems, Academic Press, 2011.

[71] Kryss Katsiavriades, Talaat Qureshi. The Krysstal Website: Spherical Trigonometry, 2009. <http://www.krysstal.com/sphertrig.html>.

[72] D.L. King, W.E. Boyson, J.A. Kratochvill, Photovoltaic array performance model, SANDIA REPORT SAND2004-3535, Sandia National Laboratories, operated for the United States Department of Energy by Sandia Corporation, Albuquerque, NM, USA, 2004.

[73] D.L. King, S. Gonzalez, G.M. Galbraith, W.E. Boyson, Performance model for grid-connected photovoltaic inverters, Technical Report: SAND2007-5036, Sandia National Laboratories, Albuquerque, NM 87185–1033, 2007.

[74] S.A. Klein, Calculation of monthly average insolation on tilted surfaces, Solar Energy 19 (1977) 325–329.

[75] S.A. Klein, W.A. Beckman, J.W. Mitchell, J.A. Duffie, N.A. Duffie, T.L. Freeman, J.C. Mitchell, J.E. Braun, B.L. Evans, J.P. Kummer, R.E. Urban, A. Fiksel, J.W. Thornton, N.J. Blair, P.M. Williams, D.E. Bradley, T.P. McDowell, M. Kummert, D.A. Arias, TRNSYS 17: A Transient System Simulation Program, 2010. <http://sel.me.wisc.edu/trnsys>.

[76] Dilip Kondepudi, Ilya Prigogine, Modern Thermodynamics: From Heat Engines to Dissipative Structures, John Wiley & Sons Ltd, 1998.

[77] Greg Kopp, Judith L. Lean, A new lower value of total solar irradiance: evidence and climate significance, Geophysics Research Letters 38 (1) (2011) L01706, http://dx.doi.org/10.1029/2010GL045777.

[78] Frank T. Kryza, The Power of Light: The Epic Story of Man's Quest to Harness the Sun, McGraw-Hill, 2003.

[79] M. Lave, J. Kleissl, Optimum fixed orientations and benefits of tracking for capturing solar radiation in the continental United States, Renewable Energy 36 (2011) 1145–1152.

[80] M. Lave, J. Stein, J. Kleissl, Quantifying and Simulating Solar Power Plant Variability Using Irradiance Data, Solar Resource Assessment and Forecasting, Elsevier, 2013.

[81] M. Lave, J. Kleissl, Solar variability of four site across the state of Colorado, Renewable Energy 35 (2010) 2867–2873.

[82] Annie Leonard, The story of stuff, Story of Stuff, Retrieved October 28, 2012, from the website: The Story of Stuff Project, 2008. <http://www.storyofstuff.org/movies-all/story-of-stuff/>.

[83] Annie Leonard, The Story of Stuff: How Our Obsession with Stuff is Trashing the Planet, Our Communities, and our Health—and a Vision for Change, Simon & Schuster, 2010.

[84] N.S. Lewis, G. Crabtree, A.J. Nozick, M.R. Wasielewski, P. Alivasatos, H. Kung, J. Tsao, E. Chandler, W. Walukiewicz, M. Spitler, R. Ellingson, R. Overend, J. Mazer, M. Gress, J. Horwitz, Research needs for solar energy utilization: report on the basic energy sciences workshop on solar energy utilization, Technical Report, US Department of Energy, April 18–21, 2005. <http://science.energy.gov/~/media/bes/pdf/reports/files/seu_rpt.pdf>.

[85] P.B. Lloyd, A study of some empirical relations described by Liu and Jordan, Report 333, Solar Energy Unit, University College, Cardiff, July 1982.

[86] D. Mahler, J. Barker, L. Belsand, O. Schulz, Green Winners: the performance of sustainability-focused companies during the financial crisis, Technical Report, A.T. Kearny. <http://www.atkearney.com/paper/-/assetpublisher/dVxv4Hz2h8bS/content/green-winners/10192>, 2009 (accessed 2.3.2013).

[87] N. Gregory Mankiw, Principles of Economics, third ed., Thomson South-Western, 2004.

[88] Harry Markowitz, Portfolio selection, Journal of Finance 7 (1) (1952) 77–91.

[89] A. McMahan, C. Grover, F. Vignola, Evaluation of resource risk in the financing of project, Solar Resource Assessment and Forecasting, Elsevier, 2013.

[90] Donella H. Meadows, Thinking in Systems: A Primer, Chelsea Green Publishing, 2008.

[91] Robin Mitchell, Joe Huang, Dariush Arasteh, Charlie Huizenga, Steve Glendenning, RESFEN5: Program Description (LBNL-40682 Rev. BS-371). Windows and Daylighting Group, Building Technologies Department, Environmental Energy Technologies Division, Lawrence Berkeley National Laboratory, Berkeley National Laboratory Berkeley, CA, USA, May 2005. <http://windows.lbl.gov/software/resfen/50/RESFEN50UserManual.pdf>, A PC Program for Calculating the Heating and Cooling Energy Use of Windows in Residential Buildings.

[92] Graham L. Morrison, Solar energy: the state of the art, ISES Position Papers 4: Solar Collectors, James & James Ltd, London, UK, 2001, pp. 145–222.

[93] Baker H. Morrow, V.B. Price (Eds.), Anasazi Architecture and Modern Design, University of New Mexico Press, 1997.

[94] T. Muneer, Solar Radiation and Daylight Models, second ed., Elsevier Butterworth-Heinemann, Jordan Hill, Oxford, 2004.

[95] National Research Council, Review and assessment of the health and productivity benefits of green schools: an interim report, Technical Report, National Academies Press, Washington, DC, USA, 2006, Board on Infrastructure and the Constructed Environment.

[96] Canada Natural Resources (Ed.), Clean Energy Project Analysis: RETScreen Engineering & Cases, Minister of Natural Resources, 2005.

[97] Ty W. Neises, Development and validation of a model to predict the temperature of a photovoltaic cell, Master's Thesis, University of Wisconsin, Madison, WI, USA, 2011.

[98] Gregory Nellis, Sanford Klein, Heat Transfer, Cambridge University Press, 2009.

[99] North Carolina State Unviersity Database of State Incentives for Renewables and Efficiency, DSIRE solar portal. <http://www.dsireusa.org/solar/solarpolicyguide/?id=19>. NREL Subcontract No. XEU-0-99515-01.

[100] Josesph J. O'Gallagher, Nonimaging Optics in Solar Energy (Synthesis Lectures on Energy and the Environment: Technology, Science, and Society), Morgan & Claypool Publishers, 2008.

[101] Elinor Ostrom, Governing the Commons: The Evolution of Institutions for Collective Action, Cambridge University Press, 1990.

[102] R. Perez, R. Stewart, R. Arbogast, R. Seals, J. Scott, An anisotropic hourly diffuse radiation model for sloping surfaces: description, performance validation, site dependency evaluation, Solar Energy 36 (6) (1986) 481–497.

[103] Perez Ineichen, Seals, Modeling daylight availability and irrradiance components from direct and global irradiance [17], Solar Energy J. 44 (5) (1990) 271–289.

[104] John Perlin, Let it Shine: The 6000-Year Story of Solar Energy, New World Library, 2013.

[105] Grant W. Petty, A First Course in Atmospheric Radiation, second ed., Sundog Publishing, 2006.

[106] Nancy Pfund, Ben Healey, What would Jefferson do? the historical role of federal subsidies in shaping America's energy future, Technical Report, DBL Investors, 2011.

[107] Nancy Pfund, Michael Lazar, Red, white & green: the true colors of America's clean tech jobs, Technical Report, DBL Investors, 2012.

[108] Théo Pirard, The odyssey of remote sensing from space: half a century of satellites for Earth observations, Solar Energy at Urban Scale, ISTE Ltd. and John Wiley & Sons, 2012, pp. 1–12 (Chapter 1).

[109] Jesùs Polo, Luis F. Zarzalejo, Lourdes Ramírez, Solar Radiation Derived from Satellite Images, Modeling Solar Radiation at the Earth's Surface: Recent Advances, Springer, 2008, pp. 449–461 (Chapter 18).

[110] Noah Porter, (Ed.), Webster's Revised Unabridged Dictionary, G & C. Merriam Co., 1913. <http://dictionary.reference.com/browse/collusion>, provided by Patrick Cassidy of MICRA, Inc., Plainfield, NJ, USA.

[111] Ari Rabl, Active Solar Collectors and Their Applications, Oxford University Press, 1985.

[112] J. Rayl, G.S. Young, J.R.S. Brownson, Irradiance co-spectrum analysis: tools for decision support and technological planning, Solar Energy (2013), http://dx.doi.org/10.1016/j.solener.2013.02.029.

[113] W.V. Reid, H.A. Mooney, A. Cropper, D. Capistrano, S.R. Carpenter, K. Chopra, P. Dasgupta, T. Dietz, A. Kumar Duraiappah, R. Hassan, R. Kasperson, R. Leemans, R.M. May, T. McMichael, P. Pinagali, C. Samper, R. Scholes, R.T. Watson, A.H. Zakri, Z. Shidong, N.J. Ash, E. Bennett, P. Kumar, M.J. Lee, C. Raudsepp-Hearne, H. Simons, J. Thonell, M.B. Zurek, Ecosystems and human well-being: synthesis, Technical Report, Millennium Ecosystem Assessment (MEA), Island Press, Washington, DC., 2005.

[114] Matthew J. Reno, Clifford W. Hansen, Joshua S. Stein, Global horizontal irradiance clear sky models: implementation and analysis, Technical Report SAND2012-2389, Sandia National Laboratories, Albuquerque, New Mexico 87185 and Livermore, California 94550, March 2012. <http://energy.sandia.gov/wp/wp-content/gallery/uploads/SAND2012-2389ClearSky-final.pdf>.

[115] Tina Rosenberg, Innovations in Light, Online Op-Ed, February 2 2012. <http://opinionator.blogs.nytimes.com/2012/02/02/innovations-in-light/>.

[116] E.F. Schubert, Physical Foundations of Solid-State Devices, Renasselaer Polytechnic Institute, Troy, NY, 2009.

[117] Molly F. Sherlock, Energy tax policy: historical perspectives on and current status of energy tax expenditures, Technical Report R41227, Congressional Research Service, May 2 2011. <www.crs.gov>.

[118] Geoffrey B. Smith, Claes-Goran S. Granqvist, Green Nanotechnology: Energy for Tomorrow's World, CRC Press, 2010.

[119] Anna Sofaer, The Primary Architecture of the Chacoan Culture—A Cosmological Expression, Anasazi Architecture and Modern Design, University of New Mexico Press, 1997, pp. 88–132 (Chapter 8).

[120] Solar Energy International, Solar Electric Handbook: Photovoltaic Fundamentals and Applications, Pearson Education, 2012.

[121] B.K. Sovacool, Valuing the greenhouse gas emissions from nuclear power: a critical survey, Energy Policy 36 (2008) 2940–2953, http://dx.doi.org/10.1016/j.enpol.2008.04.017.

[122] William B. Stine, Michael Geyer, Power From The Sun, 2001. Retrieved January 17, 2009, from <http://www.powerfromthesun.net/book.htm>.

[123] Roland B. Stull, Meteorology for Scientists and Engineers, second ed., Brooks Cole, 1999.

[124] G.I. Taylor, The spectrum of turbulence, Proceedings of the Royal Society London 164 (1938) 476–490.

[125] Jeffreson W. Tester, Elisabeth M. Drake, Michael J. Driscoll, Michael W. Golay, William A. Peters, Sustainable Energy: Choosing Among Options, MIT Press, 2005.

[126] G.N. Tiwari, Solar Energy: Fundamentals, Design, Modelling and Applications, Alpha Science International, Ltd, 2002.

[127] Joaquin Tovar-Pescador, Modelling the Statistical Properties of Solar Radiation and Proposal of a Technique Based on Boltzmann Statistics, Modeling Solar Radiation at the Earth's Surface: Recent Advances, Springer, 2008, pp. 55–91 (Chapter 3).

[128] Edward R. Tufte, The Visual Display of Quantitative Information, Graphics Press, Cheshire, Connecticut, 2001 (ISBN: 0961392142).

[129] US Geological Survey, Mineral commodity summaries 2012, Technical Report, US Geological Survey, January 2012. <http://minerals.usgs.gov/minerals/pubs/mcs/index.html>. 198 p.

[130] G. Van Brummelen, Heavenly Mathematics: The Forgotten Art of Spherical Trigonometry, Princeton University Press, 2013.

[131] F. Vignola, A. McMahan, C. Grover, Statistical Analysis of a Solar Radiation Dataset–Characteristics and Requirements for a P50, P90, and P99 Evaluation, Solar Resource Assessment and Forecasting, Elsevier, 2013.

[132] Alexandra von Meier, Integration of renewable generation in California: coordination challenges in time and space, in: 11th International Conference on Electric Power Quality and Utilization, Lisbon, Portugal, IEEE: Industry Applications Society and Industrial Electronics Society, 2011<http://uc-ciee.org/electric-grid/4/557/102/nested>

[133] W.H. Wagner, S.C. Lau, The effect of diversification on risk, Financial Analysts Journal 27 (6) (1971) 48–53.

[134] Paul Waide, Satoshi Tanishima, Light's Labour's Lost: Policies for Energy-efficient Lighting in support of the G8 Plan of Action, International Energy Agency, Paris, France, 2006 Organization for Economic Co-Operation and Development & the International Energy Agency.

[135] Ji-Tao Wang, Nonequilibrium Nondissipative Thermodynamics: with Application to Low-Pressure Diamond Synthesis, in: Springer Series in Chemical Physics, vol. 68, Springer, 2002.

[136] Susan J. White, Bubble pump design and performance, Master's Thesis, Georgia Institute of Technology, 2001.

[137] Stephen Wilcox, National solar radiation database 1991–2010 update: user's manual. Technical Report NREL/TP-5500-54824, National Renewble Energy Laboratory, Golden, CO, USA, August 2012 (Contract No. DE-AC36-08GO28308).

[138] Samuel J. Williamson, Herman Z. Cummins, Light and Color in Nature and Art, John Wiley & Sons, 1983.

[139] Eva Wissenz, Solar fire.org. <http://www.solarfire.org/article/history-map>.

INDEX

D

E

F

G

Q

R

S

T

U

W